ELEMENTARY
EXPLORATION
SEISMOLOGY

ELEMENTARY EXPLORATION SEISMOLOGY

Clarence S. Clay

University of Wisconsin—Madison

Prentice Hall

Englewood Cliffs, New Jersey 07632

Library of Congress Cataloging-in-Publication Data

CLAY, CLARENCE S. (Clarence Samuel), (date)
    Elementary exploration seismology / Clarence S. Clay.
        p. cm.

    Bibliography
    Includes index.
    1. Seismic prospecting.   I. Title.
TN269.C57 1989   551.2′2—dc19   88–38555
ISBN 0–13–256611–7

Editorial/production supervision and
    interior design: Debra Wechsler
Cover design: Wanda Lubelska
Manufacturing buyer: Paula Massenaro

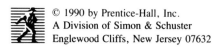 © 1990 by Prentice-Hall, Inc.
A Division of Simon & Schuster
Englewood Cliffs, New Jersey 07632

Printed in the United States of America

10   9   8   7   6   5   4   3   2   1

ISBN 0-13-256611-7

PRENTICE-HALL INTERNATIONAL (UK) LIMITED, *London*
PRENTICE-HALL OF AUSTRALIA PTY. LIMITED, *Sydney*
PRENTICE-HALL CANADA INC., *Toronto*
PRENTICE-HALL HISPANOAMERICANA, S. A., *Mexico*
PRENTICE-HALL OF INDIA PRIVATE LIMITED, *New Delhi*
PRENTICE-HALL OF JAPAN, INC., *Tokyo*
SIMON & SCHUSTER ASIA PTE. LTD., *Singapore*
EDITORA PRENTICE-HALL DO BRASIL, LTDA., *Rio de Janeiro*

to
Andre Jane Clay
and
Arnold, Jo, David, and Michael

Contents

Preface

Seismic geophysics has roots in the physics of wave propagation, signal theory, statistics, and optics. My book combines the roots at the level of an intermediate physics course, beginning with seismic disturbances and travel times, and ending with seismic images of the earth. Along the way, I introduce the art of making seismic computations.

Earthquake seismologists began the development of the seismic root. They established their seismic observatories and collected data directly from earthquakes. In the 1920s, exploration seismologists began using explosions as sources and making observations with portable instruments. Seismic theory, equipment, and field techniques were essentially developed by 1960. The introduction of large computers in the 1960s and powerful desktop computers in the 1980s has completely changed geophysics. Unit I gives the basic theories of wave propagation, seismic measurements, and seismic computations.

The second root, signal or filter theory and filter operations on signals, is in Unit II. In the 1960s, geophysicists started recording seismic signals with digital systems and then processing the signals in digital computers. Digital processing enabled geophysicists to use some of the powerful filter techniques from signal theory. To show how we can process or filter digital signals, I follow the lead of Robert Rice and Enders Robinson and use Laplace's generating functions or polynomials to represent the digitally sampled signals and filters. (A version of the generating functions are known as z-transforms.) Digital filter operations on signals become the algebraic operations of multiplication and division of polynomials. By working directly on digital signals, we can bypass the need to use Fourier transformations in an elementary book. The algorithms are easy to write and run on personal microcomputers. The combination of seismic theory from Unit I, the generating function methods, and machine computations gives realistic synthetic seismograms that include reverberation.

Unit III provides ways of enhancing seismic signals. Since concepts from random signal theory are important, the unit begins with explorations of the properties of a series of random numbers. These explorations lead to signal-to-noise improvement by the stacking of repeated transmissions and matched filters for coded signals. The rest of Unit III presents the third root and shows how to combine the signals from many channels to enhance reflection signals. A chapter on the diffraction and scattering at rough interfaces is included because I think the topic is important. The last chapters show the interconnection of concepts from holography, migration constructions, and seismic theory to convert seismic data to images of the subsurface of the earth. Seismic imaging uses all the roots.

My challenge has been to write an elementary book that develops and combines the roots of seismic geophysics. The prerequisites are a year of college physics, a first course in calculus (three semesters at the University of Wisconsin), and an introductory course in computer science. Additional mathematical techniques are given as part of the development and are expanded in the first appendix. Numerical algorithms and examples of programs for microcomputers are included to facilitate the reader's numerical explorations in geophysics.

I thank the people and organizations who helped me write this book. The Weeks Bequest of the University of Wisconsin granted me a semester of leave when I started writing the class notes that became this book. My colleagues in the Geology and Geophysics Department were supportive of my efforts. Alison Mares prepared these notes for my geophysical classes. I am grateful to M. Reza Daneshvar, Barbara A. Eckstein, Saimu Li, and Martha Kane Savage for their comments and suggestions. The students in my classes are due special thanks because I learned a lot from them. The last version was carefully critiqued by Franklyn K. Levin, John Karl of the University of Wisconsin—Oshkosh, Enders Robinson of the University of Tulsa, and I. J. Won of North Carolina State University. Holly Hodder and Debra Wechsler contributed their professional publishing skills. Lastly, I thank numerous scientists whose research, papers, and books provided me with ideas.

C. S. Clay
University of Wisconsin—Madison
May 1989

Unit I

FUNDAMENTAL SEISMIC MEASUREMENTS

GEOPHYSICS AND PUZZLES

Geophysicists use essentially the same techniques that the fictional Sherlock Holmes used in solving mysteries. The earth, its subsurface structure, and the way it works are mysteries. Like the detective, geophysicists identify a problem or puzzle worthy of attention, collect information, and then deduce an explanation. Some of the information is helpful, and we call this information a *signal*, but much of the information is noise and interferences.

There are three steps in the solution of the puzzle: The first is to separate the signals from the interferences and noise, the second is to solve the puzzle, and the last step is to explain the solution to other people (the jury) and to convince them that the solution is right. These things are much easier to say than do.

The scenario of a geophysical expedition follows that of a mystery novel. There is a large hole in our knowledge of some spot on the earth, and the investigator suspects an interesting hidden structure. The investigator has a strong ego and believes he or she is the person to solve the problem. The kind of problem and purpose usually depends on who is putting up the money. The project proceeds, and the investigators take their instruments to the area and make numerous measurements. The cartoons in Figure 1 show examples of seismic measurements, both fantastic and real. The top cartoon illustrates the saying "putting one's ear to the ground," as in listening for a distant thundering herd or galloping army. The middle cartoon shows a primitive seismograph or detector of earth motions. Sketches of an early geophysical field party are in the lower cartoon.

Seismic signals are recorded and displayed as a function of time. Examples of typical signals are shown in Figure 2. Seismic records from many channels can be very complicated, and the amount of information gathered can be overwhelming, numbering hundreds of thousands of measurements. Most geophysicists record the data digitally so they can use computers to store, process, and display them.

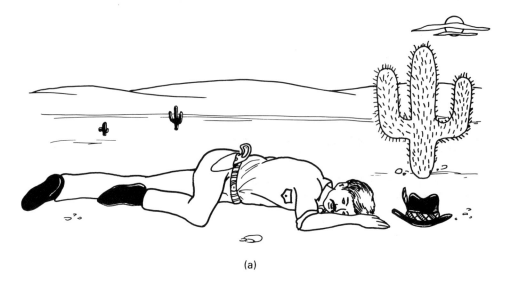

(a)

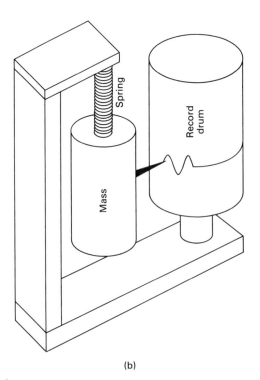

(b)

Figure 1 Seismic methods, fanciful and real. (a) One can ''put one's ear to the ground'' and listen for the thundering herd. Cowboys did this in the old Wild West movies. (b) Mechanical seismograph. A seismic signal causes the surface of the earth to move up and down and sideways. The seismograph instrument moves with the surface of the earth. The mass is suspended on a spring, and the inertia of the mass tends to keep it at a constant

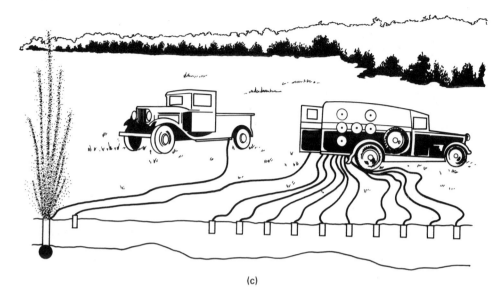

(c)

Figure 1 (continued) position as the earth moves. For simplicity, guides to constraint the mass motions to up and down are not shown. Here the seismograph is designed to record vertical components of motion. Actual amplitudes of motion are very small, and real mechanical seismographs use levers and optical magnification. The stylus marks the relative motion of the recording drum and the mass. The recording drum turns to give a record of motions versus time. (c) Seismic field party in the 1930s. The sketch is based on a photograph of a Conoco seismic field party from 1934.

Information or data processing is the crucial step, and geophysicists spend many people years developing programs to separate signals from noise and interpret the results in logical ways. Computers do not do much thinking but are very good at manipulating the data. Geophysicists do a lot of thinking and use much esoteric knowledge to instruct the computers. Eventually they deduce a solution and present it to a jury. If the employer is a petroleum company, the verdict is whether the geophysicist found oil or not. Academic geophysicists often get a nebulous verdict on their work. If a paper passes through the scientific journal system to publication, the author waits to see if anyone reads it, uses it, or damns it.

Our purpose is to give an elementary introduction to geophysical skills. Initially the situations are simple and display the basic physics. The situations become more complicated as the reader's skill improves. Geophysics is a quantitative science, so it often requires much numerical work. Many kinds of graphic displays are used to make the results more comprehensible. Although it is assumed that the reader has a computer or an advanced calculator available, all the problems can be done with paper and pencil, graphic techniques, and patience.

As geophysics has developed over the years, the common usage of terms has expanded. For example, *seismic* originally referred to earthquakes and related phenomena. Disturbances due to explosions and earth shakers are now called *seismic disturbances*. Disturbances traveling in solid or elastic media are commonly called *seismic waves*, while those traveling in gases and liquids are called *sound waves*. The term *disturbance* is used

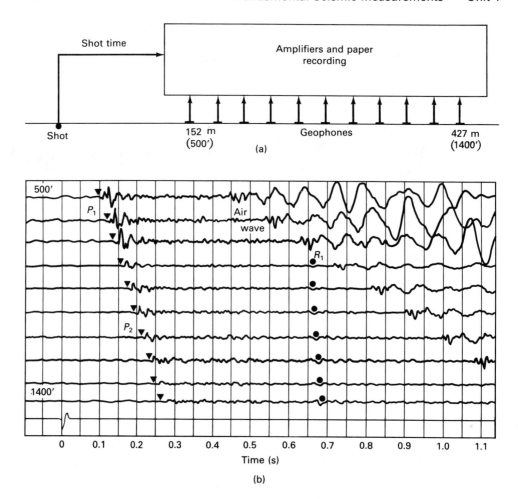

Figure 2 Seismic measurements in the 1950s. The figure shows a typical seismic reflection measurement in the 1950s. The data were recorded on photographic paper. If the geophysicist did not like the data, the party took another shot and made a new record. Since the 1960s, geophysicists have recorded their data on magnetic tape. The tape-recorded records are reprocessed back at the office. (a) Field arrangement of explosion and geophones. The source or shot was about 0.5 kg of TNT. The charge was buried 1 m deep. The data were taken along the beach on the Fire Islands of New York. (b) Paper record. The data are from M. Blaik, J. Northrop, and C. S. Clay, "Some seismic profiles onshore and offshore," *J. Geophys. Res.*, 64 (1959), 231–39. Copyright by the American Geophysical Union. (1 foot = 0.3048 m).

to describe a wave traveling in the medium without specifying whether the wave is being observed as a change in density, pressure, acceleration, velocity, or a displacement.

The fictional detective is not a chemist or a physicist or a psychologist, but he or she uses selected bits of information from each of these fields to solve the puzzle. Likewise, geophysicists select information from physics, mathematics, signal theory, geology, and chemistry. A few mathematical selections appear in Section A1.1 of Appendix A, and formulas are found in Appendix D. Thumb through these pages to see what is there.

1

Disturbances or Signals in Liquids

1.1 TRANSMISSION OF DISTURBANCES IN LIQUIDS

The measurement of the physical properties of a liquid by measuring the transmission of sound waves is very instructive. The measurements are simple and require a modest-sized tank of water, about 2 m^3 for laboratory experiments. The system is shown in Figure 1.1a. The experiment consists of transmitting from the source to a receiver at increasing separations. The challenge is to generalize the results as much as possible. Here generalize means to describe a phenomenon in as much detail as possible with the smallest number of rules and parameters, that is, to use the minimum number of equations and parameters.

We begin by doing sound transmission experiments, at least in thought. We can use a miniature microphone and a miniature loudspeaker and do the experiments in a large room. For experiments in a water tank the signal frequency must be high enough that the dimensions of the water volume are hundreds of sound wavelengths, and the sources and receivers should be small compared with a wavelength. Ferroelectric ceramics make fine transducers for the source and receivers. Depending on the signal generator, the source can be driven with impulsive signals or short gated sine waves or pings.

The trigger generator initiates the signal transmission and starts the sweep of the oscilloscope. If the trigger repeats after the echoes from the tank walls are over, it is easier to see the signals on the cathode-ray oscilloscope. The source transducer expands and contracts according to the driving signal, and the compressions and refractions travel outward from the source. For a source that is small in comparison with the sound wavelength, the disturbance moves outward with spherical symmetry. The pressure-sensitive receivers sense the changes of pressure as the sound wave passes them. We ignore the ambient pressure because it is constant. After amplification the pressure signal appears on the oscilloscope screen. A single receiver can be used and moved to measure the signals at a set of distances.

As sketched in Figure 1.1b, the trigger initiates the transmission at $t = 0$. The set

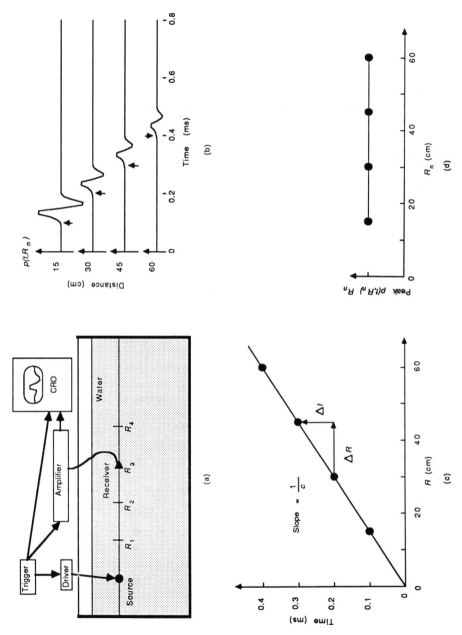

Figure 1.1 Transmission experiment. (a) Trigger, source driver, source, receiver, amplifier, and display (oscilloscope). In water the receivers are pressure-sensitive hydrophones. (b) Oscilloscope traces for distances R_1, R_2, R_3, and R_4. t_{max} is the maximum display time. (c) Travel times to the initial part of the transient signals. (d) Peak signal amplitude times the corresponding distance.

of transmissions shows that the signal arrives later as the separation between source and receiver increases. Except for the time delays and relative amplitude changes, the signals appear to have the same waveforms. By measuring the signals at constant distance and in all directions from the source, we find the same waveform and amplitude. Thus, the wave is spreading with spherical symmetry.

A plot of the times of the initiation of the arrivals versus R is shown in Figure 1.1c. The travel times fall on a straight line, that is, they increase linearly with distance. In an experiment the intercept of the travel time line at $R = 0$ is usually a little greater than zero because it takes time for the (electromechanical) transducer to respond to the driving voltage or current. The velocity of propagation is constant and equal to 1/(slope), or

$$c = \frac{\Delta R}{\Delta t} \tag{1.1}$$

We conclude that the velocity of propagation along the path is constant (within measurement error) because all data points fall on the line.

Notation for the velocities in the medium varies; the letters c, α, and V are commonly used for compressional (sound wave) velocity.

The source radiates an expanding spherical wave. With increasing range from the source, peak amplitudes of the signal appear to decrease as $1/R$. To test this hypothesis we multiply the peak amplitude p_n by the range and then plot the results versus R_n, Figure 1.1d. The results fall on a horizontal line. Within measurement errors the amplitude is proportional to $1/R$.

Since R_n and p_n are constant, we define ($\equiv$) a function $q(R, t)$ to express the signals, namely,

$$q(R, t) \equiv R\, p(t, R) \tag{1.2}$$

As sketched in Figure 1.2, the $q(R, t)$ have the same waveforms except for the time delays,

$$t_n = \frac{R_n}{c} \tag{1.3}$$

This is a traveling wave. We wish to develop an expression that has this characteristic. One way is to define a variable,

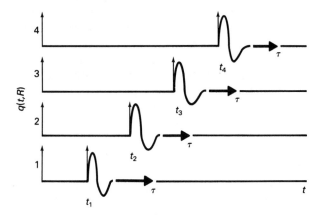

Figure 1.2 Traveling waves. $\tau = t - R/c$ and is a traveling coordinate.

$$\tau_n \equiv t - \frac{R_n}{c} \tag{1.4}$$

Another way is to let $\tau_n = t + R_n/c$ or in general to let

$$\tau = t \pm \frac{R}{c} \tag{1.5}$$

and then consider replacing the dependence on t and R with τ

$$q(\tau) = q(t, R) \tag{1.6}$$

For a test let $q(\tau)$ be zero for τ less than zero. At the range R_n this means that t must be greater than R_n/c for $q(t, R_n)$ or $q(\tau)$ to start. Thereafter, $q(t)$ follows its regular time dependence. This is the function we seek. From Figure 1.2 τ is a traveling coordinate that moves at the propagation velocity. For $\tau = t - R/c$ the wave travels in the positive R-direction. A wave traveling in the negative R-direction has $\tau = t + R/c$. The traveling wave expressions of Eq. 1.6 are

$$q_+(t, R) = q\left(t - \frac{R}{c}\right), \qquad +R \text{ direction} \tag{1.7}$$

$$q_-(t, R) = q\left(t + \frac{R}{c}\right), \qquad -R \text{ direction} \tag{1.8}$$

If we include spherical or geometrical spreading in Eq. (1.2), the outgoing experimental signal $p(t, R)$ is

$$p(t, R) = \frac{q(t - R/c)}{R} \tag{1.9}$$

Functional expressions of the traveling wave are very important and will appear in all the seismic chapters.

One important experiment remains: to determine whether the amplitude of the received signal is proportional to the amplitude of the transmitted signal. In a sense we have already shown this because the amplitudes are proportional to $1/R$. We can halve or double the source transducer drive and see if the received signals change corrspondingly. If the amplitude of the signal is proportional to the source drive, then the system is said to be linear. This is important because it means that the signals from several sources add algebraically.

When the power output of the transducer is very large and the signal is not proportional to the drive, then the transmission is nonlinear. Nonlinear effects cause additional losses and are usually negligible beyond a short distance from the source. Except very near a source, such as an explosion or a vibration, the transmission of seismic waves is linear.

PROBLEMS

1. Write the expression for a sine wave having frequency f and velocity c traveling in (a) the positive x-direction; (b) traveling in the negative x-direction.

2. Let $f = 10$ Hz and $c = 1500$ m/s (sound velocity in water). Use the results of Problem 1 to plot time-dependent graphs of $q(t, 0)$, $q(t, 30$ m$)$, $q(t, 60$ m$)$ for traveling in (a) the positive x-direction; (b) traveling in the negative x-direction. *Instructions*: Use the form of Figure 1.2 and plot only two or three cycles of the sine waves. You can plot the maxima, nulls, and minima of the sine wave and sketch the shape. The objective is to show graphically that the peaks move away from or toward the origin at velocity c.

1.2 RELATIONSHIPS OF THE DISTURBANCES TO PROPERTIES OF THE MEDIUM: THE WAVE EQUATION

A sound wave disturbs the medium in several ways. At the disturbance the local density may increase or decrease, the local (sound) pressure may increase or decrease, and particles of the medium move from their rest positions. We will need the relationship between the sound pressure and particle motion when we describe the reflection and transmission of seismic waves at boundaries. The derivation of these leads directly to a generalization known as the *wave equation*. The derivations require only a modest amount of physics and mathematics and are here because they demonstrate the fundamental properties of a propagating wave. Trusting readers can skip the details and accept the results.

1.2.1 Newton's Law and Hooke's Law

The derivations use Newton's law, $f = ma$ (force f, mass m, and acceleration a), conservation of material, and Hooke's law, a change of volume is proportional to a change of pressure. The first step is to relate $f = ma$ for an accelerating mass to the acceleration of a volume element in a continuous medium. In Figure 1.3a the mass of the volume element is $\rho \, \Delta x \, \Delta y \, \Delta z$, where ρ is density. The volume element accelerates when the force on one face is greater than the force on the opposite face. (We ignore forces due

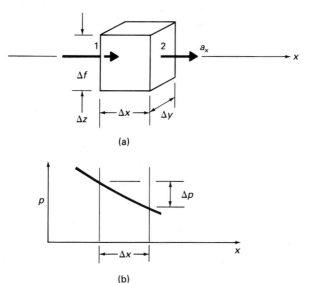

(a)

(b)

Figure 1.3 Force and pressure on a small volume in the medium. (a) The volume element has a difference of force along the x-direction. The acceleration (a_x) is in the x-direction. (b) The pressure dependence on x. Δp is the pressure difference across the volume element.

to the ambient pressure because the volume element is in static equilibrium.) In the figure the force on the left face is greater than the force on the right face, so the volume element accelerates to the right along the $+x$-direction. Newton's law becomes

$$\Delta f = (\rho \, \Delta x \, \Delta y \, \Delta z) \, a_x \qquad (1.10)$$

where Δf is the net force to the right, and a_x is the acceleration in the x-direction. The force on an area is the pressure times the area, so

$$\Delta f = \Delta p (\Delta y \, \Delta z) \qquad (1.11)$$

where $(\Delta y \, \Delta z)$ is the area and Δp is the pressure difference. From Figure 1.3b and calculus,

$$\Delta p = \frac{\partial p}{\partial x} \Delta x \qquad (1.12)$$

where $\partial p / \partial x$ is the *partial derivative*, that is, the derivative of p along the x-axis with all other variables held constant. For the elemental volume, $\partial p / \partial x$ is the slope. The combination of Eqs. (1.10), (1.11), and (1.12) gives

$$\frac{\partial p}{\partial x} = -\rho a_x \qquad (1.13)$$

By defining the displacement of the volume element as X in the x-direction and the velocity of the volume element as v_x, we obtain the time derivative or acceleration

$$a_x = \frac{\partial v_x}{\partial t} \qquad (1.14)$$

$$a_x = \frac{\partial^2 X}{\partial t^2} \qquad (1.15)$$

The use of $\partial / \partial t$ instead of the total derivative d/dt is valid when v_x is much less than c. X, v_x, and a_x are the particle displacement, velocity, and acceleration. They are vectors along the x-direction. Using the derivatives, we can write Eq. (1.13) as

$$\frac{\partial p}{\partial x} = -\rho \frac{\partial v_x}{\partial t} \qquad (1.16)$$

$$\frac{\partial p}{\partial x} = -\rho \frac{\partial^2 X}{\partial t^2} \qquad (1.17)$$

The $\partial / \partial x$ operation on p, a scalar pressure, gives the particle acceleration, a vector. These relations are important because they give the interrelations of the particle displacement, particle velocity, and pressure. After completing the derivations, we will return to these expressions.

The disturbance is a change of pressure, and the change of pressure causes the volume to change because the medium is compressible. The amount of material in the elemental volume is constant, so its density must change as the volume changes.

For a linear medium (recall the last experiment in Section 1.1), the change of volume, ΔV, is proportional to the change of the average pressure on the volume, V, and

$$\overline{\Delta p} = -B\frac{\Delta V}{V} \tag{1.18}$$

This is a form of Hooke's law, where B is the bulk modulus. From Figure 1.4 the initial volume is

$$V = \Delta x\, \Delta y\, \Delta z$$

The material in the volume is displaced by X, and the volume of the material changes. For a change of dimension in the x-direction only, $\Delta x + \delta x$, where δx is the amount of expansion, the change of volume ΔV is

$$\Delta V = (\Delta x + \delta x)\, \Delta y\, \Delta z - \Delta x\, \Delta y\, \Delta z \tag{1.19}$$
$$\Delta V = \delta x\, \Delta y\, \Delta z$$

and Eq. (1.18) becomes, after substitution of $\Delta V/V$,

$$\overline{\Delta p} = -B\frac{\delta x}{\Delta x} \tag{1.20}$$

Since the pressure, Δp, is a function of x, we use an integral to compute the average pressure

$$\overline{\Delta p} = \frac{1}{\Delta x}\int_0^{\Delta x} \Delta p\, dx \tag{1.21}$$

and write, using Eqs. (1.20) and (1.21),

$$\int_0^{\Delta x} \Delta p\, dx = -B\,\delta x \tag{1.22}$$

The operation of $\partial/\partial x$ on both sides of Eq. (1.22) gives

$$\Delta p = -B\frac{\partial \delta x}{\partial x} \tag{1.23}$$

δx is the change of the length of the volume. This change depends on the displacement and the pressure at the displaced position X. $\partial(\delta x)/\partial x$ gives the rate of change of δx as the material is displaced along x. By substitution of Eq. (1.12), Eq. (1.23)

$$\frac{\partial p}{\partial x}\Delta x = -B\frac{\partial \delta x}{\partial x} \tag{1.24}$$

To relate δx to the particle displacement X, we recall that δx is proportional to Δx and the change of X along x, Figure 1.4,

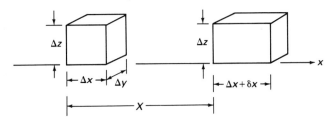

Figure 1.4 Change of volume of elemental volume and displacement X.

$$\delta x = \frac{\partial X}{\partial x} \Delta x \tag{1.25}$$

and Eq. (1.24) becomes

$$\frac{\partial p}{\partial x} = -B \frac{\partial^2 X}{\partial x^2} \tag{1.26}$$

Equating Eqs. (1.17) and (1.26) gives

$$\frac{\partial^2 X}{\partial x^2} = \frac{\rho}{B} \frac{\partial^2 X}{\partial t^2} \tag{1.27}$$

This is the one-dimensional wave equation, which gives the general relationship of the space dependence and time dependence of a traveling wave. By taking time derivatives of both sides we can use Eqs. (1.14) and (1.15) to write

$$\frac{\partial^2 v_x}{\partial x^2} = \frac{\rho}{B} \frac{\partial^2 v_x}{\partial t^2} \tag{1.28}$$

$$\frac{\partial^2 a_x}{\partial x^2} = \frac{\rho}{B} \frac{\partial^2 a_x}{\partial t^2} \tag{1.29}$$

Although this completes the derivation of the wave equation, a few tasks remain.

1.2.2 Plane Wave Solution

The first task is to show that the traveling waves of Section 1.1 and Eqs. (1.7) and (1.8) have the right form to be solutions of the wave equation. Let us write X_+ as follows:

$$X_+ = X\left(t - \frac{x}{c}\right) \tag{1.30}$$

where $(t - x/c)$ is the specific functional dependence of X. The quantity $(t - x/c)$ always appears in X_+ as the same combination of t and x. To take the partial derivatives of Eq. (1.30) it is convenient to use $\tau = t - x/c$ and then to use the implicit differentiations

$$\frac{\partial X_+}{\partial t} = \frac{\partial X_+}{\partial \tau} \frac{\partial \tau}{\partial t} \tag{1.31}$$

$$\frac{\partial X_+}{\partial x} = \frac{\partial X_+}{\partial \tau} \frac{\partial \tau}{\partial x} \tag{1.32}$$

The partial differentiations of τ are

$$\frac{\partial \tau}{\partial t} = \frac{\partial (t - x/c)}{\partial t} = 1 \tag{1.33}$$

$$\frac{\partial \tau}{\partial x} = \frac{\partial (t - x/c)}{\partial x} = -\frac{1}{c} \tag{1.34}$$

The partial differentiations of X_+ are

$$\frac{\partial^2 X_+}{\partial x^2} = \frac{1}{c^2}\frac{\partial^2 X_+}{\partial \tau^2} \tag{1.35}$$

$$\frac{\partial^2 X_+}{\partial t^2} = \frac{\partial^2 X_+}{\partial \tau^2} \tag{1.36}$$

and replacing $\partial^2 X_+/\partial \tau^2$ with $\partial^2 X_+/\partial t^2$, we have

$$\frac{\partial^2 X_+}{\partial x^2} = \frac{1}{c^2}\frac{\partial^2 X_+}{\partial t^2} \tag{1.37}$$

This looks like and is the wave equation. Comparison of Eqs. (1.27) and (1.37) shows that

$$c^2 = \frac{B}{\rho} \tag{1.38}$$

and gives the dependence of the velocity of propagation on the physical properties of the medium. Measurements of B made by measuring ρ and c yield a *dynamic* value of B. The *static* values of B come from compression measurements in a pressure vessel.

To relate the sound pressure and particle velocity, we recall Eq. (1.16):

$$\frac{\partial p}{\partial x} = -\rho\frac{\partial v_x}{\partial t} \tag{1.39}$$

Since p and v_x are traveling waves, implicit differentiations using Eqs. (1.31) to (1.34) give

$$-\frac{1}{c}\frac{\partial p}{\partial \tau} = -\rho\frac{\partial v_x}{\partial \tau} \tag{1.40}$$

for waves traveling in the $-x$-direction. Integration of both sides on τ for waves traveling in the $+x$- and $-x$-directions, Eqs. (1.7) and (1.8), gives

$$p = \pm \rho c v_x \tag{1.41}$$

where $+$ represents the $+x$-direction and $-$ represents the $-x$-direction. Later we use Eq. (1.41) to compute reflection and transmission coefficients.

Other forms of the wave equation follow directly by taking time derivatives of both sides of Eq. (1.37) or combining Eqs. (1.28), (1.29), and (1.38)

$$\frac{\partial^2 v_x}{\partial x^2} = \frac{1}{c^2}\frac{\partial^2 v_x}{\partial t^2} \tag{1.42}$$

$$\frac{\partial^2 a_x}{\partial x^2} = \frac{1}{c^2}\frac{\partial^2 a_x}{\partial t^2} \tag{1.43}$$

and since p is proportional to v_x,

$$\frac{\partial^2 p}{\partial x^2} = \frac{1}{c^2}\frac{\partial^2 p}{\partial t^2} \tag{1.44}$$

The pressure form of the wave equation is convenient because p is a scalar quantity. The three-dimensional derivation follows the same procedure using vectors for the particle displacements, and so forth. From Section A1.1 of Appendix A the result is

$$\frac{\partial^2 p}{\partial x^2} + \frac{\partial^2 p}{\partial y^2} + \frac{\partial^2 p}{\partial z^2} = \frac{1}{c^2}\frac{\partial^2 p}{\partial t^2} \tag{1.45}$$

The operation on the left side of Eq. (1.45) appears often, and the following symbol is used to represent it:

$$\nabla^2 \equiv \frac{\partial^2}{\partial x^2} + \frac{\partial^2}{\partial y^2} + \frac{\partial^2}{\partial z^2} \tag{1.46}$$

where ∇^2 is called the Laplacian and

$$\nabla^2 p = \frac{1}{c^2}\frac{\partial^2 p}{\partial t} \tag{1.47}$$

is the form that appears most often in the literature.

There are two ways to interpret the results of the derivation, depending on whether one is an experimentalist or a theoretician. The experimentalist might say that solutions of the wave equation give traveling waves like the results of experiments, which verifies the applicability of the theory to the situation. A theoretician might say that if the results of an experiment do not fit solutions of the wave equation, then the experiment is wrong. We have enough comparisons of experiment and theory to say that any experiment that fails to agree with solutions of the wave equation needs to be examined very carefully.

1.2.3 Spherically Spreading Waves

Most of the time the source used for geophysical measurements can be approximated as a point source. In a homogeneous medium the signal spreads spherically and, from our "experiment," keeps the same waveform with its decreasing amplitude as the wave travels away from the source. We verify this by solving the wave equation for spherical symmetry. The Laplacian formula for spherical symmetry is

$$\nabla^2 = \frac{\partial^2}{\partial R^2} + \frac{2}{R}\frac{\partial}{\partial R} \tag{1.48}$$

The wave equation for a spherically spreading function $u(t, R)$ in a homogeneous and thus spherically symmetrical medium is

$$\frac{\partial^2 u(t, R)}{\partial R^2} + \frac{2}{R}\frac{\partial\, u(t, R)}{\partial R} = \frac{1}{c^2}\frac{\partial^2 u(t, R)}{\partial t^2} \tag{1.49}$$

The substitution of

$$\frac{q(t, R)}{R} \equiv u(t, R) \tag{1.50}$$

in the wave equation gives

$$\frac{\partial^2 q(t, R)}{\partial R^2} = \frac{1}{c^2} \frac{\partial^2 q(t, R)}{\partial t^2} \tag{1.51}$$

The last equation has the form of the plane-wave equation. The function $q(t, R)$ travels without change of waveform. The spherically spreading wave $u(t, R)$ keeps the same waveform as its amplitude decreases.

PROBLEMS

3. Prove that $X(t + x/c)$ is a solution of the wave equation (1.37). Prove that $X(t, x) = a \sin[2\pi f(t - x/c)] + b \sin[2\pi f(t + x/c)]$ is a solution, where f is the frequency. What are the two components of $X(t, x)$?

4. *Purpose*: to show the relative phases and numerical values of the pressure and motions. $X = 10^{-6} \sin[2\pi f(t - x/c)]$ m, $c = 1500$ m/c, $\rho = 1000$ kg/m^3. Compute the peak amplitudes of X, v_x, a_x, and p. Sketch graphs of each to show the relative phase; two cycles are enough.
   (a) $f = 10$ Hz   **(b)** $f = 100$ Hz   **(c)** $f = 1000$ Hz

2

Complications

2.1 DISSIPATION AND SPREADING LOSSES

Measurements show that sound waves and seismic waves lose energy to the medium as the waves travel. We expect dissipation losses because the disturbances cause particles of the medium to move relative to each other and to have viscous losses in fluids and frictional losses in solids. The ions in seawater cause additional losses. Fluid-filled porous solids have both loss mechanisms. These losses cause decreases of signal amplitudes that are different from the decreases due to geometrical spreading of the waves.

2.1.1 Plane Waves and Absorption

For plane waves, or no geometrical spreading, the change of amplitude of a disturbance, da, is proportional to the distance, dx, and the amplitude, a. Using $-\alpha$ for the proportionality constant, we have

$$da = -\alpha\, a\, dx \tag{2.1}$$

Integration gives

$$\log\left(\frac{a}{a_0}\right) = -\alpha x \tag{2.2}$$

where we evaluate the constant of integration by letting $a = a_0$ at $x = 0$. The familiar exponential decay follows by exponentiating Eq. (2.2)

$$a = a_0 e^{-\alpha x} \tag{2.3}$$

where α is the amplitude decay coefficient and has the MKS units of m^{-1} or nepers/m. α depends on the material and frequency.

Seismologists commonly express α in terms of a quality or damping factor, Q. The factor Q is $2\pi W/\Delta W$, where W is the mean energy in the wave field, and ΔW is the energy lost per cycle.

$$\alpha = \frac{\pi f}{cQ} \tag{2.4}$$

or

$$Q = \frac{\pi f}{\alpha c} \tag{2.5}$$

Q depends on the type of rock and is nearly constant in the seismic frequency range for most rocks. Q is dimensionless and ranges from 10 to several hundred for common rocks. There is considerable interest in the dependence of α or Q on rock structure, pore fluid, and external forces such as the overburden pressure.

Fluids have viscous losses, which cause α to be approximately proportional to f^2. Dissipations due to ionic interactions in seawater depend on frequency and add to the loss. At high frequencies, the attenuation coefficients for fresh water and seawater are orders of magnitude less than the attenuation coefficients for rocks. Comparative graphs of attenuation coefficients are shown in Figure 2.1.

As seismic waves propagate away from the source, the amplitudes decrease because of geometric spreading. Two kinds of geometric spreading are particularly important, spherical spreading and cylindrical spreading. Our experiment in Section 1.1 is an example of spherical spreading where the signal amplitudes are proportional to $1/R$.

2.1.2 Conservation of Energy

Application of the principle of conservation of energy shows that spherical spreading of the wave is consistent with $1/R$ amplitude dependence when the dissipation is negligible.

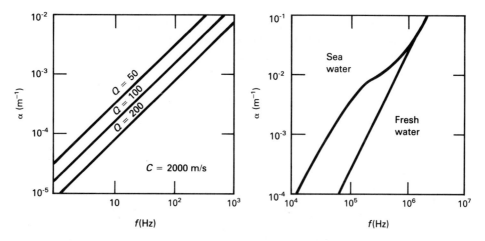

Figure 2.1   Frequency dependence of the attenuation coefficient α for sedimentary rocks, water, and seawater.

From mechanics, work energy is force times distance, and power is the rate of doing work or force times velocity, where the force is along the displacement. Within a continuous medium, we use the power or alternatively the energy passing through a surface normal to the direction of propagation of the propagating waves. In the continuous medium, force per unit area is pressure, and the velocity is the particle velocity. Thus, power per unit area is

$$j = p\, v_R \tag{2.6}$$

where v_R is particle velocity along the radius vector away from the source. Ignoring dissipation losses, the power passing through a spherical surface (area $= 4\pi R^2$) is the source power,

$$4\pi R^2 p v_R = \text{source power} \tag{2.7}$$

From Eq. (1.41), p and v_R are proportional (for a plane wave), so we can write

$$p \sim \frac{(\text{source power})^{1/2}}{R} \tag{2.8}$$

$$v_R \sim \frac{(\text{source power})^{1/2}}{R} \tag{2.9}$$

Since our experiment gives a R^{-1} amplitude dependence on R, we conclude that the spreading of the wave is spherical, and the dissipation losses are negligible.

Some of the main interferences and sources of noise in seismic measurements are the waves that travel in and on the thin layers on the surface of the earth. For reasons we develop later, part of the seismic energy from seismic sources is trapped in these layers when the sources are on the surface or in the layers. In general, only part of the source power of energy is trapped in these layers because the upper and lower boundaries may not be perfectly reflecting. Layers that trap seismic energy are also known as *wave guides* (Figure 2.2).

Transmission of signals in a thin layer is more complicated than in an homogeneous medium, and we use a sine-wave signal to drive the source here. The source is in the layer Δz thick. If we call the part of the source power that is trapped in the layer the *trapped power* and use conservation of energy, the trapped power is constant and

$$2\pi r\, \Delta z\, p\, v_r = \text{trapped power} \tag{2.10}$$

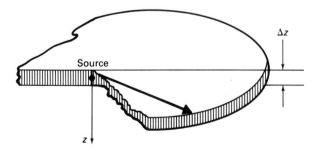

Figure 2.2 Source in thin layer or waveguide. The waveguide is a short cylinder, and the radiation from the source is said to spread cylindrically within the waveguide. A source on the surface of a half-space excites surface waves. Actually the surface wave has particle motions that decrease as a function of depth. With this understanding we can say that surface waves also spread cylindrically.

where Δz is the height of the short cylinder, v_r is the particle velocity along r, and dissipation is neglected. Again using the proportionality of p and v_r,

$$p \sim \frac{(\text{trapped power})^{1/2}}{r^{1/2}} \tag{2.11}$$

$$v_r \sim \frac{(\text{trapped power})^{1/2}}{r^{1/2}} \tag{2.12}$$

The amplitudes of trapped disturbances decrease as $r^{-1/2}$.

In addition to being trapped in a layer, waves can travel along an interface. *Boundary* or *interface waves* exist because of a discontinuity of the properties of the upper and lower media at the interface. A gravity wave on the surface of the ocean is a familiar example. The maximum amplitude of particle motion is on the surface, and the amplitudes decrease below the surface. Since the wave motion extends downward, we can regard boundary waves as being trapped in a short cylinder. The *Rayleigh wave* is a seismic example of a wave that propagates at the surface of an elastic solid half-space. The amplitudes of boundary waves decrease as $r^{-1/2}$.

The combination of spreading and dissipation losses gives amplitudes as

$$v_R \sim p \sim \frac{e^{-\alpha R}}{R} \tag{2.13}$$

for spherically spreading waves. Trapped waves and interface waves have amplitudes of

$$v_r \sim p \sim \frac{e^{-\alpha r}}{r^{1/2}} \tag{2.14}$$

for radially spreading waves. Of course, the numerical values of the attentuation coefficients depend on frequency and are different for spherically spreading waves and trapped waves.

PROBLEMS

1. *Purpose*: to show the effects of dissipation losses on signals having different frequencies and to determine why seismic signals have low frequencies. Assume plane wave transmission. For a sandstone assume $Q = 50$, $c = 2400$ m/s. The range is 1000 m. Compute α, αr, and relative signal amplitude $a = a_0 e^{-\alpha r}$ for the frequencies $f = 1, 10, 20, 50, 70, 100$, and 200 Hz. $a_0 = 1$. Graph the amplitudes versus frequency on both linear and log-log graph paper or plot $\log_{10} a$ versus $\log_{10} f$.

2. Repeat Problem 1 using $Q = 150$.

3. *Purpose*: to show graphically the relative losses due to spherical spreading of the waves and dissipation. The amplitudes and ranges cover several orders of magnitude, so use log-log graph paper or plot $\log_{10} a$ versus $\log_{10} R$. Let $a_0 = 1$ for all frequencies. Let $Q = 150$ and $c = 2400$ m/s. Do computations and graphs for $f = 1, 10$, and 100 Hz. Let R range from 1 to 10^4 m, and do the computations for $R = 1, 3, 10, 30, 100$, and so forth. (*Note*: $\log_{10} 3 \simeq 0.5$).

4. Let a be pressure with units of pascals (Pa) or N/m^2 and $a = a_0 e^{-\alpha R}/R$. Use dimensional analysis to determine the units of a_0.

2.2 TRANSMISSIONS IN AND ON ELASTIC SOLIDS

We do an experiment that is similar to ones routinely done by exploration geophysicists. We measure the transmission of seismic signals in the solid earth by initiating a disturbance at the surface and transmitting signals to receivers in a borehole (Figure 2.3a). The receivers that sense the particle motions in or on the solid earth are known as *geophones*, analogous to microphones in air and hydrophones in water. Locking arms press the geophone against the borehole wall to sense the motions of the earth (Figure 2.3b).

A velocity-sensing geophone consists of a coil of wire that moves in the magnetic field of a permanent magnet, Figure 2.4a. A spring supports the coil and holds the coil in alignment with the magnet. The combination of the coil mass and spring controls the mechanical resonant frequency of the geophone. Earth motion moves the case and magnet relative to the coil. The electrical output voltage of the coil is proportional to the velocity of the coil relative to the velocity of the magnetic field. The typical response of a geophone is sketched in Figure 2.4b. Well above the resonant frequency the coil is effectively motionless while the case and magnet moves up and down. At frequencies much less than the resonant frequency, the coil moves almost as fast as the magnet and has a small output voltage. Velocity-sensitive geophones are sensitive, simple, and rugged.

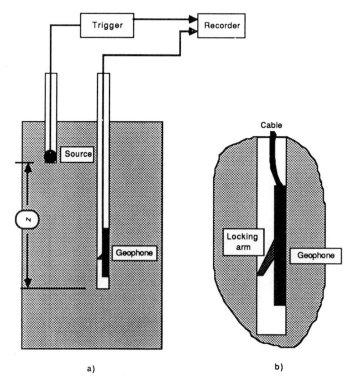

Figure 2.3 Transmission measurements to a receiver (geophone) in a borehole. (a) The source is in a shallow hole near the instrument hole. The geophone is pressed against the borehole wall by a locking arm. (b) Inset shows an enlargement of the geophone in the borehole.

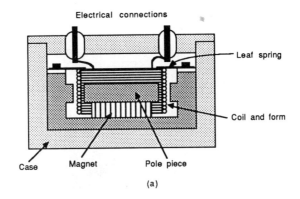

(a)

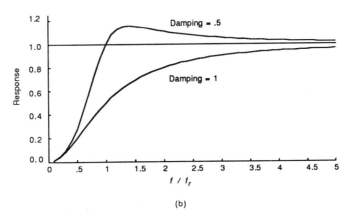

(b)

Figure 2.4   (a) Structure of a vertical-component geophone. The coil mass and leaf spring constant give the resonance frequency. (b) Response curves for critically damped geophone, damping = 1.0, and under damped response, damping = 0.5.

To illustrate the fundamental transmissions in solid (elastic) material, we simplify the measurements by placing the source and receiver at large depths so that the medium is effectively infinite. The first experiment uses an impulsive expanding source and a three-component geophone to measure v_x, v_y, and v_z. The results for a set of source-receiver separations are sketched in Figure 2.5. Figure 2.5a shows the geometry where the explosive source is above four geophones in the hole. Each geophone is a three-component instrument that senses x-, y-, and z-components of motion. The geophones are locked to the hole wall with locking arms. An enlargement of an instantaneous ''picture'' of the local densities is in part b of the figure. The grid lines indicate a coarse grid of particles. The lines from the undisturbed medium to the disturbed medium show how the longitudinal or compressional wave moves the particles of the medium downward and in the direction of propagation. The particles press together, increasing the densities. There are no motions of particles in the x- and y-directions. The signals observed on vertical-component geophones are sketched in Figure 2.5c. As one would expect, the horizontal-component geophones (Figures 2.5d, e) do not respond to the passing compressional wave.

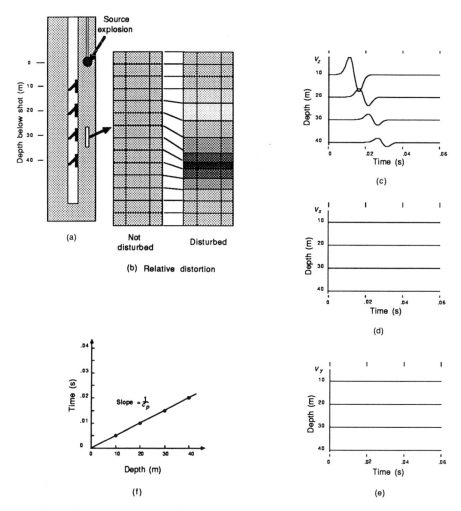

Figure 2.5 Explosive source. The source in an explosion. The source is in a hole near and above a string of geophones. The shot is put in a hole next to the instrument hole to prevent the waves that travel in the tube from getting to the geophones. The geophones have locking arms to get good contact to the wall of the hole. See Problems 5 and 6. (a) Source and string of three-component (x, y, z) geophones. (b) An enlarged snapshot of the relative densities and particle displacements in the medium as the impulsive wave front travels downward through the medium. (c) Signals observed by the vertical-component geophones v_z (d) and (e) Signals on horizontal-component geophones v_x and v_y (f) Travel time versus distance from the source.

In actual experiments, horizontal-component geophones may respond to the longitudinal signals because of imperfect geophones and poor coupling of the geophones to the ground.

A plot of the arrival times versus depth (Figure 2.5f) gives a straight line in a homogeneous medium. The velocity of the seismic wave is c_p. The subscript stands for primus or first arrival. The amplitudes are proportional to $1/z$, where z is the separation.

The source for the second experiment is a hammer (Figure 2.6). Ideally we would

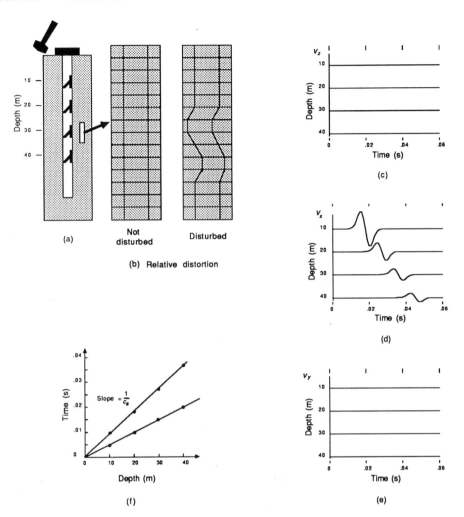

Figure 2.6 Horizontal hammer source. The source is a hammer blow on a plank. The source driver (not shown) stands on the plank and strikes the end of the plank to cause it to move in the x-direction. The source is above a string of geophones. The geophones have locking arms to get good contact to the wall of the hole. See Problem 7. (a) Source and string of three-component (x, y, z) geophones. (b) Enlarged snapshot of the relative densities and particle displacements in the medium as the impulsive wave front travels downward through the medium. (c) Signals observed by the vertical-component geophones v_z. (d) and (e) Signals on horizontal-component geophones v_x and v_y. (f) Travel time versus distance from the source. The travel time data from Figure 2.5 are plotted for comparison.

use a hammer to hit the side of the hole, and we would do an experiment that is parallel to the design of the explosive source experiment. The hammer hits the plank on the surface of the ground and initiates a seismic disturbance. The person swinging the hammer can stand on the plank to improve the coupling to the ground. The hammer blow causes the plank to move in the x-direction. As in Figure 2.5, geophones are locked to the wall of the hole.

Following the organization of the sketches in Figure 2.5, we show the instantaneous "picture" of the particles in the medium, both disturbed and undisturbed. Here the particle or slab motions are sideways or transverse and along the x-direction. The vertical-component geophones have very little response to this disturbance, whereas the x-component geophones give good signals. The y-component geophones have a small output. Although we do not display the results, we could rotate the plank and hammer to be along the y-direction. Then the y-component geophones would have good signals and the x-component geophones would not. Depending on the direction of the source, we can initiate x-component disturbances, y-component disturbances, or both. These disturbances are known as *shear waves* because the particles or slabs of the medium move transversely and tend to shear one slab from the slab next to it.

The graph of travel time versus separation gives a propagation velocity, c_s, that is less than c_p. Again, the amplitudes of the signals are proportional to $1/z$, where z is the separation. The subscript s stands for secundus and shear.

The experiments show the following: (1) The expanding source initiates a traveling wave having the propagation velocity c_p. (2) The particle displacements and velocities are along the direction of propagation, here along the z-axis. (3) The amplitude decreases as $1/z$ or $1/R$. (4) The horizontal x-displacement source initiates a wave that travels along the z-direction at the velocity c_s. (5) The displacements and particle velocities are along the x-axis for propagation along the z-axis. (6) The amplitudes decrease as $1/z$. The radiation from the shear wave source is more complicated than the radiation from an explosive source because the displacement source along the x-direction also starts a longitudinal wave that has a component of motion in the x-direction.

From elasticity theory the velocities of compressional and shear waves depend on the mechanical properties of the medium, as follows:

$$c_p = \left(\frac{B + 4\,\mu/3}{\rho} \right)^{1/2} \tag{2.15}$$

$$c_s = \left(\frac{\mu}{\rho} \right)^{1/2} \tag{2.16}$$

where ρ = density (kg/m^3)

μ = rigidity modules (N/m^2)

B = bulk modules (N/m^2)

When μ is zero, the medium does not transmit shear waves, and Eq. (2.15) becomes

$$c = \left(\frac{B}{\rho} \right)^{1/2} \quad \text{(fluids)} \tag{2.17}$$

which is the same as the result from the wave equation derivation, Eq. (1.38).

If we were to observe the magnitude and directions of motions of particles as the p- and s-wave fronts traveled outward from the hammer source, we would see the patterns shown in Figure 2.7. At the same radial distances, the shear wave amplitudes are maximum

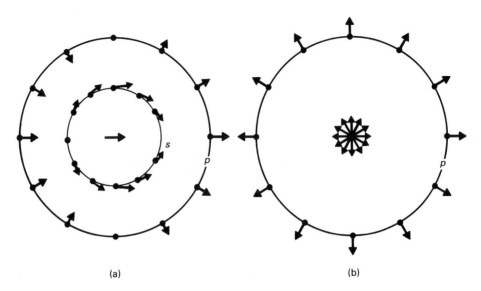

(a)   (b)

Figure 2.7   (a) Particle displacements for displacement and explosive sources in an infinite medium. The compressional and shear waves travel outward with the velocities c_p and c_s, where c_p is greater than c_s. (b) Explosive or volume expansion source. Only compressional waves are initiated by the source.

on the z-axis and minimum on the x-axis. The compressional wave amplitudes are maximum on the x-axis and minimum on the z-axis. The volume expansion or explosive source gives equal-amplitude compressional waves in all directions. Seismologists use worldwide observations of an earthquake and concepts such as shown in Figure 2.7 to determine the earth motions at the earthquake focus, that is, the source.

The third experiment is a measurement of the tramsmission on the surface of an elastic half-space (Figure 2.8a). The seismogram, that is, the signals at a set of geophones, is shown in Figure 2.8b. There are three arrivals, and the travel time-distance graphs show the velocities c_p, c_s, and c_R. We expect c_p and c_s on the basis of our previous experiments in the solid. The arrival having the velocity c_R is known as the *Rayleigh wave* [Lord Rayleigh described it in 1886. Lord Rayleigh (John William Strutt, 1842–1919) was a giant in physics. He was named a Nobel laureate in 1904.] The wave involves a coupling of shear and compressional components at the surface. The particle motion at the surface is elliptical, and the amplitudes of motion decrease with depth beneath the surface. Along the x-direction, the Rayleigh wave has components of motion in the x- and z-directions. The Rayleigh wave velocity depends on the elastic constants λ, μ, and ρ or the velocities c_p and c_s, Eqs. (2.15) and (2.16). The Rayleigh wave velocity is

$$c_R = 0.92c_s, \qquad \text{for } \lambda = \mu, \quad \lambda = B - \frac{2\mu}{3} \qquad (2.18)$$

Rayleigh wave amplitudes decrease at $r^{-1/2}$.

Geophysicists often refer to Rayleigh waves and other waves trapped in the near surface layers as *ground roll*. Characteristically, ground roll has low frequencies and low velocities of propagation.

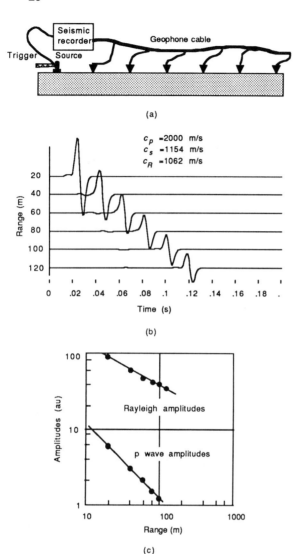

Figure 2.8 Hammer seismic measurements on a half-space. (a) Experiment. Distance *r* is along the array and in *m*. (b) Synthetic or theoretical seismogram. Amplitudes of the *p*- and *s*-arrivals were increased for display. (c) Amplitude versus range for the *p*-arrivals and Rayleigh waves. Although not shown, the *s*-arrival has $1/r$ amplitude dependence.

PROBLEMS

Purpose: to examine three fundamental papers in experimental seismology and to do wave propagation experiments. The first paper was written by one of the pioneers in exploration seismology, Norman Ricker (1896–1980). Figures 2.5 and 2.6 are based on these three papers; the original papers should be read. Two of the experiments were done in the Pierre shale near Limon, Colorado. The shale section is uniform from depths of 30 m to over 1200 m. Norman Ricker chose the area because he needed a uniform medium to test his seismic theory of wave propagaton in an absorbing medium. Many of Ricker's concepts entered the exploration seismic literature. The *Ricker wavelet* is an example. McDonal et al. repeated Ricker's experiments. Jolly did shear wave studies.

5. Read N. Ricker, "The form and laws of propagation of seismic wavelets," *Geophysics* 18 (1953), 19–40. The paper contains a detailed description of his experiments and the analysis of data. The experiments were done when all data were recorded on paper records and analyzed by hand. Compare the compressional wave signals as a function of travel distance. Can you see waveform changes? Do you think Ricker proved his ideas about seismic wave propagation?

6. Read F. J. McDonal, F. A. Angona, R. L. Mills, R. L. Sengbush, R. G. Van Nostrand, and J. E. White, "Attentuation of shear and compressional waves in Pierre shale," *Geophysics*, 23 (1958), 421–38. These authors repeated Ricker's experiments. The paper gives very nice examples of simple seismic signals. Why did these people repeat Ricker's experiment? Compare the signals they observed with those of Ricker. Also compare the hydrophone signal or pressure signal with Ricker's signals. Did they come to the same or different conclusions? Recall our discussion of attenuation in an earlier section.

7. Read R. N. Jolly, "Investigation of shear waves (Love elastic)," *Geophysics* 21 (1956), 905–38. Jolly used a cannon instead of a hammer for his shear wave studies. His experiments demonstrated the properties of a shear wave source. Why did Jolly do these experiments? What did Jolly learn about ground roll? How did he separate compressional waves from shear waves?

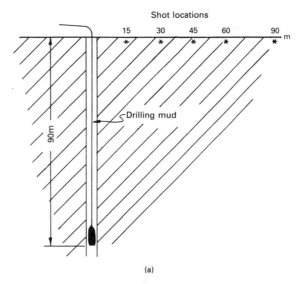

Figure 2.9 Comparisons of seismic arrivals on a vertical-component velocity geophone and a hydrophone (a sound-pressure sensor). The data were taken on the beach of the Fire Islands of New York. Figure 2.9a shows the arrangement. A vertical-component geophone was in an uncased hole at a depth of 90 m. A spring pressed the geophone against the side of the borehole. A hydrophone was at 87 m depth. From the surface to well below these depths, the materials were unconsolidated sandy-clay sediments. The hole was filled with drilling mud. The sources were 0.22 kg blocks of TNT. The shots were fired in shallow holes in the beach sand and were at the top of the water table, at about 1 m depths. Figure 2.9b shows the geophone signals, and Figure 2.9c shows the hydrophone signals. Additional information on the geophysical structure is in Problem 5 of Chapter 6 and Figure 6.10.

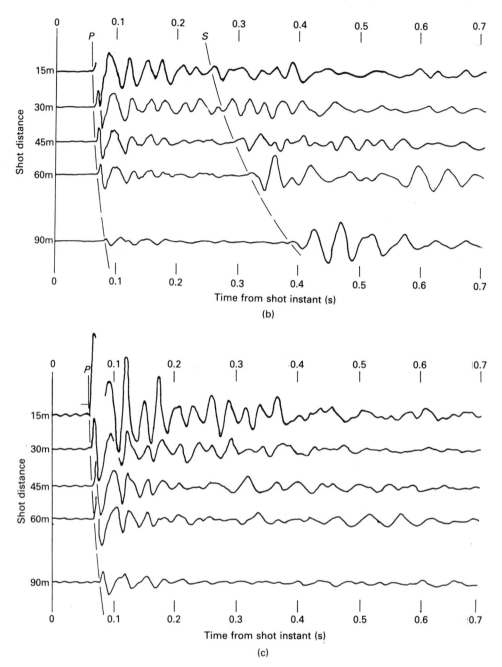

Figure 2.9 (continued)

8. Slinky® experiments. The slinky is a coil or long helix. Slinkys are usually sold in toy stores; the metal ones are preferable. When the slinky is stretched about 2 m, it makes a good medium for demonstration of one-dimensional waves. The problem assignment is to play with the slinky. Examples are the following:

Have the source person start waves by holding the end of the slinky in one hand and moving it right, left, up, down, forward, or back. Observe the dependence of the signal on the velocity and amplitude of hand motion. The receiving person can sense the different messages. Best viewing is usually along the slinky. Observe particle motions by putting a little tab of tape on a coil and then watching the tab move as the signal passes. Observe the reflected signals. Stretch the slinky and then let it lie on a smooth table. Transmit signals and compare the damping of the waves to the damping of a freely suspended slinky.

9. The data are shown in Figure 2.9. Use the data to compute the velocities of *p*- and *s*-waves. Do the first arrivals on the geophone and hydrophone give the same velocities? Give reasons why the hydrophone and geophone signals for the 90-m (300 ft.) shot distance are different in the time between 0.4 and 0.6. Give qualitative reasons why the amplitudes of the *p*-wave arrivals on the vertical-component geophone decrease as shot distance increases. Give qualitative reasons why the *s*-wave amplitudes on the geophone increase as shot distances increase. Commonly, people regard explosions as being sources of *p*-waves and not sources of *s*-waves. Give physical explanations why these shallow explosions also started shear or *s*-waves. Would it be different if the shots were fired deep in a uniform section?

3

Snell's Law
of Refraction,
Generalizations,
and Ray Paths

Quantitative optical refraction effects have been known for a long time. Claudius Ptolemaeus published a table of experimental refraction data for the water-air interface in A.D. 140. An accurate analytical expression was given by Willebrord Snell in 1621. Snell (1591–1626) was a professor of mathematics at Leyden. We can regard the refraction of light and Snell's law as the accurate description of an experiment. Some physics texts use constructions such as Huygens' wavelet construction to obtain Snell's law. However, this exercise does not prove Snell's law but shows that Huygens' construction gives the right answer.

Snell's law is also a consequence of the application of Fermat's principle of least time (about 1650). Pierre de Fermat (1601–65) was a lawyer and recreational mathematician. His idea was that light traveling from a point a to a point b follows the path having the shortest travel time. Richard Feynman (1918–1988), an outstanding teacher and a Nobel laureate in physics, stated that the ''least time principle'' was more fundamental because it showed a way of thinking that made Snell's law evident. In seismic geophysics, we measure travel times and apply Fermat's principle when we combine ray paths and travel times to interpret data. In his elementary text (R. P. Feynman, R. B. Leighton, and M. Sands, *The Feynman Lectures on Physics*, Reading, MA: Addison-Wesley, 1965), Feynman said that ''something'' being a minimum seems to be a fundamental principle in physics.

We will explore many applications of Snell's law in this chapter. A derivation of Snell's law that uses solutions of the wave equation can be found in Section A1.1.4 of Appendix A.

3.1 SNELL'S LAW OF REFRACTION

From Figure 3.1 Snell's law of refraction is

$$\frac{\sin \theta_1}{c_1} = \frac{\sin \theta_2}{c_2} \tag{3.1}$$

A derivation of Snell's law that uses solutions of the wave equation is given in Section A1.1 of Appendix A and Eqs. (A.67) to (A.76).

Snell's law is a deceptively simple relation because it is general. The formula describes the refraction of wave fronts for light, electromagnetic waves, sound waves, and seismic compressional, or p-, and shear, or s-, waves. Using the notation in Figure 3.2, we obtain the following formulas for reflection and refraction angles.

$p_1 p_1$ reflection:

$$\frac{\sin \theta_{1p}}{c_{1p}} = \frac{\sin \theta_{1p}}{c_{1p}} \tag{3.2}$$

$p_1 s_1$ and $s_1 p_1$ reflections:

$$\frac{\sin \theta_{1p}}{c_{1p}} = \frac{\sin \theta_{1s}}{c_{1s}} \tag{3.3}$$

$p_1 p_2$ refraction:

$$\frac{\sin \theta_{1p}}{c_{1p}} = \frac{\sin \theta_{2p}}{c_{2p}} \tag{3.4}$$

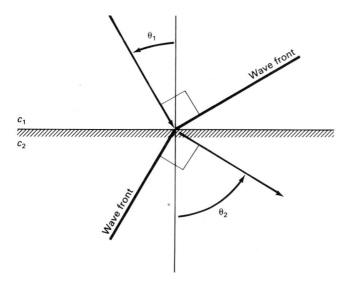

Figure 3.1   Refracted wave front. The arrows are the normals to the wave fronts in mediums 1 and 2. Each medium is isotropic.

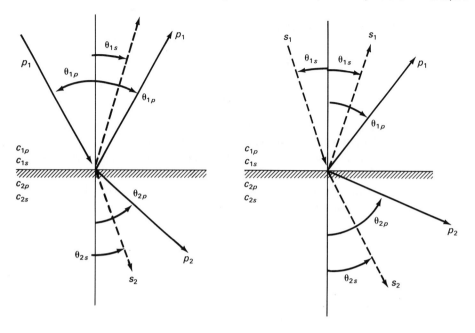

Figure 3.2   Incident, reflection, and refraction angles for *p*- and *s*-waves.

$p_1 s_2$ refraction:

$$\frac{\sin \theta_{1p}}{c_{1p}} = \frac{\sin \theta_{2s}}{c_{2s}} \tag{3.5}$$

$s_1 s_1$ reflection:

$$\frac{\sin \theta_{1s}}{c_{1s}} = \frac{\sin \theta_{1s}}{c_{1s}} \tag{3.6}$$

$s_1 p_2$ refraction:

$$\frac{\sin \theta_{1s}}{c_{1s}} = \frac{\sin \theta_{2p}}{c_{2p}} \tag{3.7}$$

$s_1 s_2$ refraction:

$$\frac{\sin \theta_{1s}}{c_{1s}} = \frac{\sin \theta_{2s}}{c_{2s}} \tag{3.8}$$

Snell's law has the same form for each situation. At each interface in a solid, the incident *p*-waves reflect and refract as both *p*- and *s*-waves. The same is true for incident shear waves. At vertical incidence, compressional waves do not convert shear waves and vice versa. Exploration geophysicists usually make seismic reflection measurements near vertical incidence and ignore shear waves.

3.1.1 Snell's Law and Minimum Travel Time

Although the energy travels in the wave front, seismic literature uses raypath analogies extensively. In isotropic media the rays are normal to the wave fronts. Rays are used because (1) it is easier to draw the normals to the wave fronts, and (2) the Snell's law travel path of the ray gives the minimum time for the wave front to travel from point a to point b, Figure 3.3.

We use calculus to determine the value of x that gives minimum travel time. The travel time from a to x to b is

$$t = \frac{(h^2 + x^2)^{1/2}}{c_1} + \frac{[z_1^2 + (x_1 - x)^2]^{1/2}}{c_2} \tag{3.9}$$

For the minimum travel time we set $dt/dx = 0$

$$\frac{dt}{dx} = \frac{x}{c_1 (h^2 + x^2)^{1/2}} - \frac{(x_1 - x)}{c_2 [z_1^2 + (x_1 - x)^2]^{1/2}} \tag{3.10}$$

or

$$\frac{1}{c_1} \frac{x}{(h^2 + x^2)^{1/2}} = \frac{1}{c_2} \frac{(x_1 - x)}{[z_1^2 + (x_1 - x)^2]^{1/2}} \tag{3.11}$$

By geometry $\sin \theta_1 = x/(h^2 + x^2)^{1/2}$ and $\sin \theta_2 = (x_1 - x)/[z_1^2 + (x_1 - x)^2]^{1/2}$. Snell's law is the result. This was Fermat's discovery. Of course, the travel time is minimum for the particular combination, here $p_1 p_2$. Other combinations have other minimum travel times. For example, the path $p_1 s_2$ has a larger travel time than $p_1 p_2$. Snell's law for $p_1 s_2$ gives the minimum travel time for the $p_1 s_2$ transmission.

For plane waves it does not make any difference whether we use wave fronts or their normals, the rays. The use of rays requires care when the wave fronts are expanding from a small source or a diffracting edge in the medium. *If the radius of curvature of the wave front is much greater than the seismic wavelength, then we can use the normal to a patch of wave front as it moves through the medium.* Remember, a single expanding wave front has an infinite number of rays, and we often choose to follow only a few of

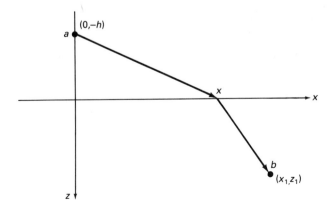

Figure 3.3   Least travel-time geometry.

these rays. With these warnings, we use ray paths as shortcuts because ray paths have the least travel times and are usually less work to draw.

3.2 LAYERED MEDIUM AND THE RAY PARAMETER

Snell's law extends directly to a layered medium having plane parallel interfaces. The seismic velocity and density are a function only of depth, z. It is possible to use any combination of p- and s-paths through the medium by choosing the appropriate velocities in each layer. For simplicity we drop the notation indicating p- or s-paths and list them as c_1, c_2, and so forth. Using Figure 3.4, we apply Snell's law from layer to layer

$$\frac{\sin\theta_1}{c_1} = \frac{\sin\theta_2}{c_2} = \frac{\sin\theta_3}{c_3} = \cdots \qquad (3.12)$$

or

$$\frac{\sin\theta_1}{c_1} = \frac{\sin\theta_i}{c_i}$$

For the angle of incidence, θ_1, $\sin\theta_1/c_1$ is a constant parameter for all layers and is the ray parameter p, where

$$p \equiv \frac{\sin\theta_1}{c_1} \qquad (3.13)$$

The seismic literature commonly uses p for the ray parameter. Confusion between the uses of p for ray parameter, p for compressional wave, and p for pressure should be minimal because they appear in such different ways. The ray parameter p has the dimensions time/distance, or s/m.

3.2.1 Phase Velocity and the Ray Parameter

In our one-dimensional solution of the one-dimensional wave equation, $X(t - c/x)$ moves along the x-axis at velocity c. If we pick a peak or valley of the waveform, it also moves

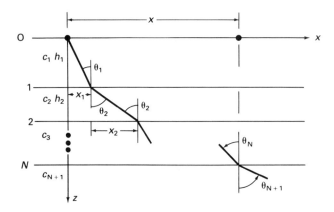

Figure 3.4 Multilayered medium. In the context of ray traces, x is the horizontal "displacement" of the ray path.

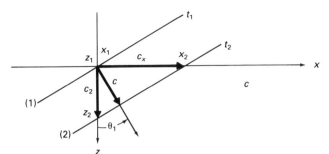

Figure 3.5 Phase velocities along x and z. The wave front moves from (1) to (2) in the time $(t_2 - t_1)$. $c_z = c/\cos \theta_1$, $c_x = c/\sin \theta_1$.

along the x-axis with velocity c. Because the peak or valley is a *phase of the waveform*, we can say that it moves with *phase velocity* c_x along x. Here, of course, c_x is the same as c because the wave is traveling in the x-direction.

The situation is different when the wave front of a peak or valley is incident at an angle θ_1 to the x-axis (Figure 3.5). Suppose we have detectors at x_1 and x_2 and observed the wave front at times t_1 at x_1 *and* t_2 at x_2. The distance the contact of the wave with the x-axis appears to move along the x-direction in the time between t_1 and t_2 is

$$\left.\begin{aligned}
\Delta x &= c_1 \frac{\Delta t}{\sin \theta_1} \\[6pt]
\Delta t &= t_2 - t_1 \\[6pt]
\Delta x &= x_2 - x_1 \\[6pt]
c_x &\equiv \frac{\Delta x}{\Delta t} = \frac{c_1}{\sin \theta_1}
\end{aligned}\right\} \tag{3.14}$$

where c_x is defined as the *phase velocity* along x. The phase velocity is the velocity of a contact moving along a coordinate. Passing to the limit and inverting c_x, we have

$$\left.\begin{aligned}
\frac{dt}{dx} &= \frac{\sin \theta_1}{c_1} = p \\[6pt]
p &= \frac{1}{c_x}
\end{aligned}\right\} \tag{3.15}$$

Equation (3.15) is useful for determining p and θ_1. The parameter p is the *wave slowness* along the x-direction. At vertical incidence ($\theta_1 = 0$), the phase velocity is infinite, and wave slowness is zero. By Snell's law, p and c_x are the same at all interfaces in a medium having parallel plane interfaces.

3.3 RAY-TRACE COMPUTATIONS

The following definitions and trigonometric relations simplify the numerical computations for multilayer ray traces.

$$\left.\begin{aligned}
a_i &\equiv p\, c_i \\[6pt]
b_i &\equiv (1 - a_i^2)^{1/2} \qquad \text{for } a_i < 1
\end{aligned}\right\} \tag{3.16}$$

$$\sin \theta_i = a_i \tag{3.17}$$

$$\cos \theta_i = b_i \tag{3.18}$$

$$\tan \theta_i = \frac{a_i}{b_i} \tag{3.19}$$

The case $a_1 \geq 1$ comes later. To compute the travel times and horizontal displacement of the ray along the x-axis, using Figure 3.4, we choose θ_1, compute p, and then compute the following:

$$t_i = \frac{h_i}{c_i \cos \theta_i} = \frac{h_i}{c_i b_i} \tag{3.20}$$

$$x_i = h_i \tan \theta_i = \frac{a_i h_i}{b_i} \tag{3.21}$$

The total travel time and displacement of the ray to the Nth interface are

$$t_N(\theta) = \sum_{i=1}^{N} t_i \tag{3.22}$$

$$x_N(\theta) = \sum_{i=1}^{N} x_i \tag{3.23}$$

$$z_N = \sum_{i=1}^{N} h_i \tag{3.24}$$

A brief discussion of computational techniques to evaluate these expressions appears in Section A1.4 of Appendix A.

The travel time and displacement for a reflection from the Nth interface are twice the travel time and displacement down to the interface.

The condition $p\, c_{i+1} = 1$ gives the critical angle for the refracted ray to travel parallel to the ith interface in the $(i + 1)$st layer, $\theta_{i+1} = 90°$. We use the subscript notation $\theta_{1,i+1}$ for the angle in layer 1 that gives critical reflection at the ith interface and a ray refracted parallel to it in the $(i + 1)$st layer. Thus,

$$\sin \theta_{1,i+1} = \frac{c_1}{c_{i+1}} \tag{3.25}$$

This case is very important and is in the next section.

The condition $p\, c_{i+1} > 1$ gives reflection beyond the critical angle, or the *total reflection*. It is the deepest penetration of the ray subject to the c_1/c_{i+1} velocity ratio. For fluid media this is the deepest penetration for a ray. In solid media we can have critical reflection for p-waves and penetration of a p-s converted wave through the interface. Total reflection in solid media requires that the angle of incidence be greater than the critical angle for incident p- to refracted s-waves.

PROBLEMS

1. *Purpose*: to become familiar with Snell's law for refraction and different ratios of c_1 and c_2.
   (a) For $\theta_1 = 30°$, compute θ_2 for $c_1 = 1500$ m/s, $c_2 = 2000$ m/s.
   (b) for $\theta_1 = 30°$, compute θ_2 for $c_1 = 2200$ m/s, $c_2 = 1100$ m/s.

2. Reflections and transmissions at a solid media interface. $c_{p1} = 1000$ m/s, $c_{s1} = 500$ m/s, $c_{p2} = 1500$ m/s. $c_{s2} = 1100$ m/s. For an incident p-wave
   (a) Compute the reflection angles of $p_1 p_1$ and $p_1 s_1$ waves. $\theta_1 = 10°$ and $60°$.
   (b) Compute the refraction angles of the $p_1 p_2$ and $p_1 s_2$ waves. $\theta_1 = 20°$.
   (c) Compute the critical angles for $p_1 p_1$ reflection (θ_{1p2p}) and $p_1 s_1$ reflection (θ_{1p2s}).
   (d) Can there be transmission into the lower medium for angles between θ_{1p2p} and θ_{1p2s}?

3. Write a computer program to compute θ_2 for an incident θ_1. Remember that practically all computers use θ in rad. Include an escape for reflection at or beyond the critical angle.

4. *Purpose*: to display the wave front as it propagates downward in a two-layer medium. Place the source at the 0 interface and let $c_1 = 800$ m/s, $h_1 = 800$ m, and $c_2 = 1600$ m/s. By traveling along the ray, calculate the position of the wave front at 0.8 s, .0 s, 1.2 s, and 1.4 s. Choose rays at $\theta_1 = 0°, 5°, 10°, 15°, 20°, 25°$, and $29.5°$. Graph the positions of the wave fronts.

5. Add the up-going wave front reflected at interface 1 to Problem 4. Choose times 1.2 s and 1.4 s.

6. Write a simple computer program to compute the travel time and x-displacement along the ray.

7. Use the program to calculate ray traces for structures of your choice. Graph the ray traces through the structure. Start with $N_1 = 3$, $c_1 = 800$ m/s, $h_1 = 800$ m, $c_2 = 1600$ m/s, $h_2 = 1600$ m, $c_3 = 1600$ m/s. Then let $N_1 = 3$, $c_1 = 1600$ m/s, $h_1 = 800$ m, $c_2 = 800$ m/s, $h_2 = 1600$ m, $c_3 = 800$ m/s. *Note*: The program requires the third interface to display the ray path in layer 2.

8. Compute and make a travel time–distance graph for direct arrival p_1 and the reflection $p_1 p_1$ for $c_1 = 1600$ m/s and $h_1 = 800$ m. Let x range from 0 to 1600 m. Notice that the travel times of $p_1 p_1$ approach p_1 at large range. You can use Eqs. (3.20) and (3.21). $t_{p1} = x/c$.

9. Repeat the graphing in Problem 8 using the p_1 and s_1 direct arrivals, $p_1 p_1$ reflection, and $p_1 s_1$ reflection. $h_1 = 800$ m, $c_{1p} = 1600$ m/s, $c_{1s} = 800$ m/s.

3.4 IMAGE SOURCE AND APPROXIMATIONS

Much of the older seismic literature approximates the actual subsurface structure as a homogeneous layer to the reflecting interface. The approximation is acceptable for ray paths near vertical incidence. We write the reflection times and displacements for the reflection ray paths from the bottom of the first layer. For layer 1, reflections are (Figure 3.6)

$$t_1(\theta_1) = 2 \frac{h_1}{c_1 \cos \theta_1} \tag{3.26}$$

$$x_1(\theta_1) = 2h_1 \tan \theta_1 \tag{3.27}$$

Equations (3.26) and (3.27) are suitable for calculations; however, we can cast them into a more convenient form by using the geometry of Figure 3.6a. $\cos \theta_1$ is

$$\cos \theta_1 = \frac{h_1}{[(x/2)^2 + h_1^2]^{1/2}} \tag{3.28}$$

and the time becomes

$$t_1(\theta_1) = \frac{(x_1^2(\theta_1) + 4h_1^2)^{1/2}}{c_1} \tag{3.29}$$

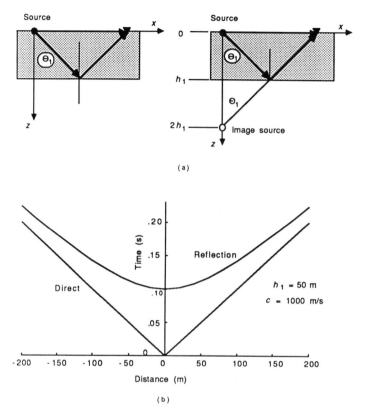

Figure 3.6 Ray-path construction, one layer: (a) ray path; (b) image construction; (c) travel time (t) versus distance (x) for a source at $x = 0$, $z = 0$, $h_1 = 50$ m, and $c = 1000$ m/s. The time is in seconds and the distance is in meters.

The reflection appears to come from an image source at depth $2h_1$ (Figure 3.6b). Changing notation to indicate the functional dependence of $t(x)$ on x, we often write Eq. (3.29) as follows:

$$t_1^2(x) = t_1^2(0) + \frac{x^2}{c_1^2}$$

$$t_1(0) = \frac{2h_1}{c_1}$$

(3.30)

It is easy to interpret seismic reflection data by graphing $t_1^2(x)$ versus x^2. The slope gives c_1^{-2}, and the intercept at $x = 0$ gives $4h_1^2/c_1^2$. An extension to multiple horizontal layers is given in Section A1.2 of Appendix A.

The ray-trace equations, (3.20) to (3.24), use an initial angle θ_1 or ray parameter p and compute the travel time and displacement. If we wish to know the travel time and θ_1 for a particular displacement x, we can try different values of θ until the answer is close enough. Alternatively, we can write $t_N^2(x)$ as a polynomial,

$$t_N^2(x) = a_0 + a_1 x + a_2 x^2 + a_3 x^3 + a_4 x^4 + \cdots \tag{3.31}$$

calculate a set of $t_N(\theta)$ and $x_N(\theta)$ for a set of θ_1, and then use the $t_N(\theta)$ and $x_N(\theta)$ to determine (by least squares) the coefficients a_0, a_1, and so forth. We suggest using at least a second-order polynomial in x. We can obtain p by differentiating Eq. (3.31) and solving for $d[t_n(x)]/dx$.

The $t^2(x)$ form of Eq. (3.28) is a good approximation for reflection ray paths near vertical incidence when the subsurface has plane parallel layers. From the derivation in Section A1.2 of Appendix A, we replace the velocity c_1 with the root mean square velocity (rms)

$$c_{n,\,rms}^2 = \frac{\displaystyle\sum_{i=1}^{n} c_i^2 \tau_i}{t_N(0)} \qquad \begin{matrix}(3.32)\\ \text{or (A.106)}\end{matrix}$$

where

$$\tau_i \equiv \frac{2h_i}{c_i} \qquad \begin{matrix}(3.33)\\ \text{or (A.103)}\end{matrix}$$

$$t_n(0) = \sum_{i=1}^{n} \tau_i \qquad \begin{matrix}(3.34)\\ \text{or (A.107)}\end{matrix}$$

The approximate travel time expression for reflection from the nth interface is

$$t_n^2(x) = t_n^2(0) + \frac{x^2}{c_{n,\,rms}^2} \qquad \begin{matrix}(3.35)\\ \text{or (A.105)}\end{matrix}$$

The Dix-Dürbaum formula for measurement of the velocity of a layer, Eq. (A.110) is given in Appendix A. The derivation uses these results.

Over horizontal layers the travel times to the receivers at $-x$ and x are equal; thus, the coefficients of the odd powers in Eq. (3.31) vanish. For moderately sloping interfaces geophysicists have devised ways to take data or process data that remove the odd powers of x. One technique is given in Problem 10. In the rest of this section we assume that the reflection data were taken and processed to have only even powers of x.

For interval velocity analysis or the determination of the seismic velocity in a layer, we need a set of reflection data for the nth interface and the $(n + 1)$th interface. The reflection data are (least squares) fitted to second-order polynomials in x,

$$t_n^2(x) = t_n^2(0) + \frac{x^2}{c_{n,\,rms}^2}$$

$$\tag{3.36}$$

$$t_{n+1}^2(x) = t_{n+1}^2(0) + \frac{x^2}{c_{n+1,rms}^2}$$

to determine the coefficients $t_n(0)$, $t_{n+1}(0)$, $c_{n,\,rms}$, and $c_{n+1,\,rms}$.

The values of these coefficients are substituted into the Dix-Dürbaum formula, Eq. (A.110),

$$c_{n+1}^2 = \frac{c_{n+1,\,rms}^2\, t_{n+1}(0) - c_{n+1,\,rms}^2\, t_n(0)}{t_{n+1}(0) - t_n(0)} \qquad \begin{matrix}(3.37)\\ \text{or (A.110)}\end{matrix}$$

to determine c_{n+1}. In effect, the method strips off all layers above n and uses data from only the nth and $(n+1)$st interfaces to give c_{n+1}. It gives excellent results when the data are of high quality and free of interfering multiple reflections.

PROBLEMS

10. A derivation of the travel time and distance equation for a dipping interface is given in Section A1.2 of Appendix A. The derivation is for a fixed source and a set of receiver stations. A *common midpoint* geometry is also used to make the seismic reflection measurements. In common midpoint measurements the source and receiver are set so that the source is at $-x/2$ and the receiver is at $x/2$. The source and receiver are at equal distances from the midpoint at $x = 0$. For the dipping layer, Figure A.3, derive an expression for t^2 as a function of x. Let the source be at $-x/2$ and the receiver at $x/2$. Show that $t^2 = 4h^2/c^2 + x^2 \cos^2 \phi_1/c^2$.

The dependence on dip ϕ_1 is still in the expression. Devise ways of determining c and ϕ_1. For these considerations and more complicated geometry see F. K. Levin, "Apparent velocity from dipping interface reflections," *Geophysics*, 36 (1971), 510–16.

11. Reflection data for two interfaces are given below. Assume that the interfaces are horizontal. Interpret the data to determine the seismic velocities and thicknesses of the layers.

Distance (m)	0	100	200	300	400	500
(1) Time(s)	1.000	1.002	1.005	1.011	1.020	1.030
(2) Time(s)	1.556	1.567	1.559	1.564	1.570	1.557

12. Read C. S. Clay and P. A. Rona, "Studies of seismic reflections from thin layers on the ocean bottom in the western North Atlantic," *J. Geophys. Res.*, 70 (1965), 855–69. They give coefficients through fourth order for Eq. (3.31). What are the major problems in using an interval velocity formula to compute (or estimate) the velocity from field data?

13. Read P. M. Shah and F. K. Levin, "Gross properties of time-distance curves," *Geophysics*, 38 (1973), 643–56. They give examples of seismic velocity analysis and compare the inclusion of second- and fourth-order terms in Eq. (3.31). When would you expect the errors to be large? Did Shah and Levin obtain the same theoretical expressions as Clay and Rona?

4

Head-Wave or Refracted Arrival

The use of head waves to determine subsurface structure has a history that involves academic scientists, wars, and entrepreneurs. In 1909, Zagreb was shaken by an earthquake. Andrija Mohorovičić (1857–1936), director of the Zagreb Meteorological Observatory, wrote a report on the earthquake in 1910. He identified two *p*-wave arrivals. One arrival was the direct arrival from the earthquake source or *focus*. It had a velocity of 5.6 km/s. The other arrival appeared to travel from the focus down to an interface at 50 km depth, travel along the interface, and then to travel up to the seismic station. The velocity of the wave as it traveled along the interface was about 7.7 km/s. The existence of this interface has been verified all over the earth and is known as the *discontinuity of Mohorovičić*, the *Moho*, or the *M-discontinuity*.

During World War I (1914–18) both armies fired big long-range guns. The German "Big Bertha" had a range of 100 km. The German, French, British, and American armies had research groups that were studying acoustic and seismic methods of gun location.

In 1917 J. C. Karcher (b. 1894) was working on gun location for the U.S. National Bureau of Standards. He built electrical geophones for observing seismic signals and air microphones for observing signals in air. His experimental source was an artillery field gun. Karcher observed that the seismic signals traveled much faster than the signals in air. He also noticed that the seismic waves traveled through deeper and deeper rock layers as the distance between the gun and the geophones increased. He recorded later arrivals and probably got a reflection from about 840 m depth.

On the battlefields most of the armies used air microphones to detect the airblast waves. They used an array of microphones to measure the travel-time differences and then used them to compute the location of the long-range guns. Actually, the sound location systems gave a better location of Big Bertha than airborne photography. The Germans had a decoy that photographed very well.

The Germans, mainly Ludger Mintrop (1880–1956), used mechanical seismographs to receive the signals initiated by the recoil of a gun. After the war Mintrop got a patent and organized the Seismos geophysical company. He used large explosives as sources and mechanical seismographs to record headwaves or refracted waves. In the 1920s, many Seismos crews worked in the Gulf coast.

After the war, J. C. Karcher completed his Ph.D. at the University of Pennsylvania in 1920 and returned to acoustic and seismic research. In 1921 Karcher, W. P. Haseman, I. Perrine, and W. C. Kite made reflection measurements in the Arbuckle Mountains, Oklahoma. Karcher had heard about Mintrop's success in finding salt domes with refraction measurements. Karcher also built refraction equipment and, in 1925, leased a crew to the Gulf Oil Company. He and his associates used electromagnetic geophones, moving-coil galvanometers in a recording camera, and radios for timing. Karcher's equipment was much better than Mintrop's mechanical seismic systems.

By the 1930s American geophysicists had started using electronic amplifiers, radios, and electromagnetic geophones. A few seismic reflection parties were in the field. The last Seismos refraction crew was dismissed. Because reflection techniques had much better resolution, the reflection seismograph became the principle geophysical method for the exploration of petroleum.

During World War II (1939–45) antisubmarine warfare was extremely important. Many American and European physicists worked on underwater sound problems in an effort to find submarines. They developed simple and robust equipment and techniques for making sound transmission and seismic measurements at sea. After World War II the scientists published several books and reports on their research. In particular, *Geological Society of America Memoir* 27 by M. Ewing, J. L. Worzel, and C. L. Pekeris (1948) had a profound influence on a generation of geophysicists. Ewing moved to Columbia University and organized the Lamont Geological Observatory. Ewing, his colleagues, and their students embarked on the task of measuring crustal thicknesses, that is, the depth to the Mohorovičić discontinuity in the world's oceans. Their main tool was refraction or head-wave measurements. In the process they used a lot of surplus World War II depth charges.

About 1947 M. A. Tuve of the Carnegie Institution of Washington started seismic crustal studies on land. Tuve and H. E. Tatel made an effort to record critical angle reflections and headwaves. University geophysicists entered the field, and crustal seismic refraction measurements became academic research. The refraction crustal studies on land and sea are the basis of our picture of the earth's structure to perhaps a 100-km depth.

4.1 AN EXPERIMENT

Almost all geophysicists do or have done the experiment shown in Figure 4.1a. A disturbance is initiated at the source and the transmissions to geophones along the surface are recorded. Crustal studies to depths of about 100 km require tons of explosives and ranges of hundreds of kilometers. Measurements of the weathered material over bedrock to depths less than 10 m require a few grams of explosive or hammer blows. The measurements can also be scaled down to laboratory models having layers a few centimeters thick.

A typical set of data from a short-range bedrock study is sketched in Figure 4.1b.

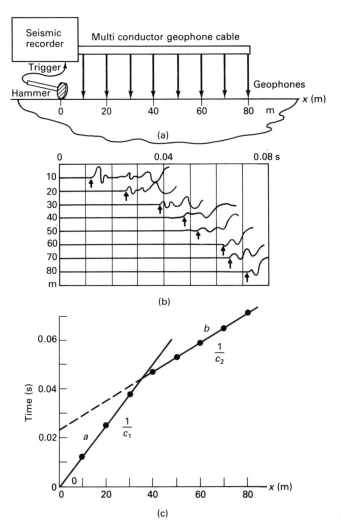

Figure 4.1   Seismic measurement. (a) Measurement. (b) Data. (c) Travel-time distance graph.

Geophysicists usually use only the *first arrivals* in these measurements because ground roll obscures later arrivals. The lines a and b are drawn through the points. The slopes of the travel time-distance graphs of the first arrivals indicate two velocities $c_1 = 800$ m/s, and $c_2 = 1600$ m/s. Extrapolation of the line b to $x = 0$ gives an intercept $t = 0.022$ s. The line a intercepts at $t = 0$. The data show the presence of two disturbances, one having the velocity c_1 and the other having the velocity c_2.

We reject the hypothesis that the two arrivals are shear and compressional waves traveling along the surface because the intercept of the c_2 arrival has a constant time delay (compare Figures 4.1 and 2.8). We test the hypothesis that the earth has two layers, where c_1 is the velocity of a disturbance in one layer and c_2 is the velocity of a disturbance in the other. This means that the arrivals are due to disturbances traveling horizontally in both layers. At short range the a arrival has velocity c_1, so we assume the top layer has velocity c_1. It is reasonable to assume that the time delay of the b arrival is the time required for the ray to go down to the c_2 layer and then up to the geophones. The hypothetical structure is shown in Figure 4.2.

At the critical angle the refracted ray travels parallel to the interface in the lower medium. Using the reverse of Snell's law, we see that the ray also comes out of layer 2 at the critical angle, θ_{12}, where

$$\sin \theta_{12} = \frac{c_1}{c_2} \tag{4.1}$$

If we assume compressional waves and use Figure 4.2a, the ray paths have the travel times

$$t_{P_1} = \frac{h_1}{c_1 \cos \theta_{12}} \tag{4.2}$$

$$x_1 = h_1 \tan \theta_{12} \tag{4.3}$$

$$L = x - 2x_1$$

$$L = x - 2h_1 \tan \theta_{12} \tag{4.4}$$

$$t_{P_1} = \frac{L}{c_2} \tag{4.5}$$

The total travel time is

$$t_{P_1 P_2 P_1} = \frac{2h_1}{c_1 \cos \theta_{12}} + \frac{x - 2h_1 \tan \theta_{12}}{c_2} \tag{4.6}$$

where the subscripts identify the path. Manipulation of Eq. (4.6) using the relations

$$\tan \theta_{12} = \frac{\sin \theta_{12}}{\cos \theta_{12}}$$

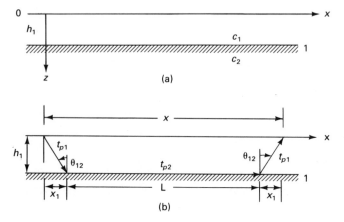

Figure 4.2 Hypothetical structure and ray paths. (a) Structure having parallel interfaces. (b) Ray paths for the p_2 ray parallel to interface 1.

$$c_2 = \frac{c_1}{\sin \theta_{12}}$$

$$\cos^2 \theta_{12} + \sin^2 \theta_{12} = 1$$

gives

$$t_{p_1 p_2 p_1} = \frac{2h_1 \cos \theta_{12}}{c_1} + \frac{x}{c_2} \qquad (4.7)$$

This equation is the usual form. The wave moves along the x-direction at velocity c_2 and has a fixed time delay. The time delay is the value of $t_{p_1 p_2 p_1}$ at $x = 0$ or the *intercept time*. Since the raypath requires distances greater than that for total reflection, the minimum distance for the refraction arrival is $2h_1 \tan \theta_{12}$. The implications of the seemingly small transformations from Eq. (4.6) to (4.7) are in Section A1.3 of Appendix A.

The p_1 arrival travels along the surface and

$$t_{p_1} = \frac{x}{c_1} \qquad (4.8)$$

This is the first arrival for short range.

The refraction method of seismic exploration uses this procedure. The $p_1 p_2 p_1$ arrival is known as the *refraction arrival*.

4.2 NUMERICAL EXAMPLE

In Figure 4.1c the slopes of lines a and b are $1/c_1$ and $1/c_2$. The velocities are

$$c_1 = 800 \text{ m/s} \qquad (4.9)$$

$$c_2 = 1600 \text{ m/s} \qquad (4.10)$$

The critical angle is $\arcsin(c_1/c_2)$, and

$$\theta_{12} = 30° \qquad (4.11)$$

The intercept at $x = 0$ of the line b is 0.022 s:

$$t_{p_1 p_2 p_1} \Big|_{x=0} = \frac{2h_1 \cos \theta_{12}}{c_1} \qquad (4.12)$$

$$0.022 = \frac{2h_1 \cos 30°}{800} \qquad (4.13)$$

$$h_1 = 10 \text{ m} \qquad (4.14)$$

where two significant figures for h_1 are consistent with two significant figures for the intercept time.

The minimum distance for the existence of the refracted wave is $2h_1 \tan \theta_{12}$, and

$$2h_1 \tan \theta_{12} = 11.6 \text{ m} \qquad (4.15)$$

This is also the distance for reflection at the critical angle.

Next we combine the travel times of the refraction arrival and the reflection. Equation (3.29) gives $t_{p_1 p_1}$ versus x, or $x_{p_1 p_1}$. The travel times versus x are shown in Figure 4.3. Multiple reflections, such as $p_1 p_1 p_1 p_1$, add additional curves. It is helpful to keep the form of this graph in mind when looking at data.

The mathematics of calculating the travel times for refraction ray paths is simple. However, there is a question about why the p_2 ray comes back up into medium 1. The answer is a difficult problem involving solution of the wave equation for a point source in a layered medium. We give the results of a laboratory experiment and an intuitive explanation that shows the nature of the wave.

Suppose the source is on the interface between c_1 and c_2, where $c_1 < c_2$. An impulsive disturbance from the source spreads as spherical waves in media 1 and 2 (Fig. 4.4). Since c_2 is greater than c_1, the p_2 wave front at the interface is always at greater x than the p_1 wave front. The letters (a) and (b) represent the contacts of the p_1 and p_2 wave fronts at the interface. A model of this experiment is shown in Figure 4.5. The experiment shows a wave front originating at (b) (Fig. 4.4) and extending as a line tangent to the p_1 wave front. The contact of the p_1 wave at (a) does not disturb the lower medium. The contact of p_2 at (b) initiates a wave in medium 1. Since the wave is tangent to the spherical p_1 wave front, we can calculate its angle easily. From Figure 4.6 the angle θ is given by

$$\sin \theta = \frac{R_1}{R_2} \tag{4.16}$$

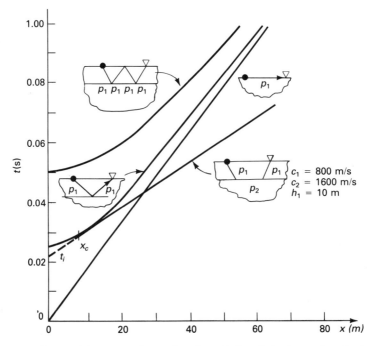

Figure 4.3   Travel time versus distance for a few arrivals. x_c is the distance for reflection at the critical angle. t_i is the intercept of $p_1 p_2 p_1$ at $x = 0$.

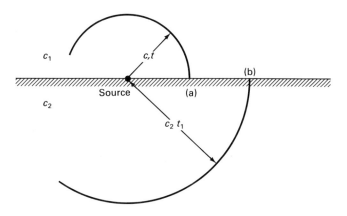

Figure 4.4 Spherically spreading wave fronts from a source at the interface.

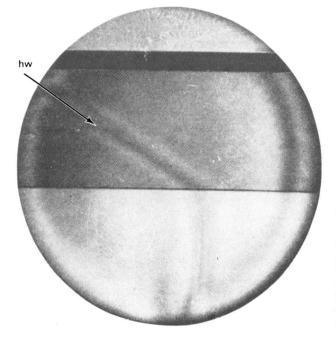

Figure 4.5 J. F. Evans, C. F. Hadley, J. D. Eisler, and D. Silverman, "A three-dimensional seismic wave model with both electrical and visual observation of waves," *Geophysics*, 19 (1954), 220–36. The photograph shows the contact of the head wave in the upper medium and the out-going wave in the lower medium. The moving contact is (b) in Figure 4.6.

$$R_1 = c_1 t$$

$$R_2 = c_2 t$$

$$\sin \theta = \frac{c_1}{c_2} \tag{4.17}$$

and θ is the critical angle. The wave front (hw) is the *refraction* wave and is also known as the *head wave*. Head wave refers to this particular wave, whereas refraction can mean this wave or simply the transmitted wave in medium 2.

A disturbance moving at a velocity greater than the velocity of propagation in the medium initiates this type of wave front. We can demonstrate the process by moving a

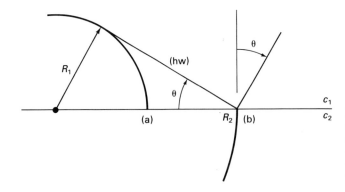

Figure 4.6 Geometry for head wave. The contact at (b) can be regarded as a source that moves along the interface at velocity c_2. The wave front (hw) originates at (b) and extends to its tangent contact on the p_1 wave front. It is the head wave.

fingertip along the surface of water. The angle of the wave front depends on the velocity of the finger. The bow waves from fast boats have the same appearance. By analogy, the head wave in medium 1 is due to a disturbance (b) in Figure 4.6 moving along the interface at a velocity c_2 greater than the velocity in medium 1. The disturbance in medium 1 at (a) is totally reflected.

Experiments and theory show that the amplitude of the head wave decreases as $(xL^3)^{-1/2}$, where x is horizontal range and L is the distance along the refractor, Figure 4.2b. The waveform of the head wave is the integral of the outgoing signal.

4.3 DIPPING INTERFACE: REVERSED PROFILES

The seismology literature contains the terms *unreversed* and *reversed profiles*. The example in Figure 4.1 is an unreversed profile because the source was at one end. Sketches of a reversed profile and the travel time–distance data for first arrivals are shown on Figure 4.7. It is customary to measure the geophone positions from one source, A here. The source at B gives the reversed profile. Each source position gives a set of p_1 and $p_1p_2p_1$ arrivals. We reject the hypothesis that the velocity of the second layer is different when it is measured in the opposite direction and propose the hypothesis that the second layer is dipping.

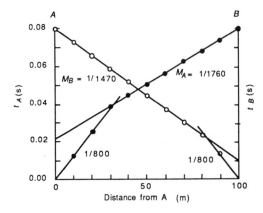

Figure 4.7 Reversed profile. Source at A gives times t_A (solid points). Source at B gives times t_B (circles).

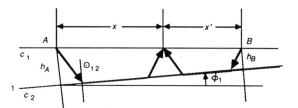

Figure 4.8 Dipping interface and reversed profiles.

The geometry of a dipping second layer is shown in Figure 4.8. The critical angles, θ_{12}, are relative to the first interface. h_A and h_B are normal to the interface. Before doing the problem analytically, we can test the dipping layer hypothesis. Consider path $p_{A1}p_2p_{Ax1}$. As x increases, the length of path p_{Ax1} decreases, and the travel time decreases. The head-wave velocity for path p_2 appears to be larger than c_2. From the other direction, with a source at B, as x' (measured from B) increases, path $p_{Bx'1}$ increases, and the travel time increases. The reversed profile gives a velocity that appears to be less than c_2. For small slopes we would expect the average of the two velocities to be approximately the actual velocity c_2, or

$$c_2 \simeq \frac{1470 + 1760}{2}$$

$$\simeq 1600 \text{ m/s}$$

(4.18)

The measurements are consistent with the dipping interface hypothesis. The depth at B should be smaller than the depth at A.

The theoretical travel time–distance formulas for dipping layers are given in Section A1.3 of Appendix A. From the geometry in Figure 4.8 the travel times are

$$t_{Ax} = \frac{2h_{A1}\cos\theta_{12}}{c_1} + \frac{x}{c_1}\sin(\theta_{12} - \phi_1)$$

(4.19)

$$h_{B1} = h_{A1} - (AB)\sin\phi_1$$

(4.20)

$$t_{Bx'} = \frac{2h_{B1}\cos\theta_{12}}{c_1} + \frac{x'}{c_1}\sin(\theta_{12} + \phi_1)$$

(4.21)

The slopes of the travel-time lines are

$$M_A = \frac{\sin(\theta_{12} - \phi_1)}{c_1}$$

(4.22)

$$M_B = \frac{\sin(\theta_{12} + \phi_1)}{c_1}$$

(4.23)

The intercept times for t_{Ax} and t_{Bx}, are the corresponding values for x and $x' = 0$.

Numerical example

The slopes and intercepts, $t_A(0)$ and $t_B(0)$, from Figure 4.7 are

$$c_1 = 800 \text{ m/s}$$

$$M_{A2} = \frac{1}{1760} \text{ s/m}$$

$$M_{B2} = \frac{1}{1470} \text{ s/m}$$

$$t_A(0) = 0.022 \text{ s}$$

$$t_B(0) = 0.010 \text{ s}$$

First, determine $\theta_{c_{12}}$ and ϕ_1 as follows:

$$\theta_{12} - \phi_1 = \arcsin(c_1 M_A)$$

$$\theta_{12} + \phi_1 = \arcsin(c_1 M_B)$$

$$\theta_{12} - \phi_1 = 27°$$

$$\theta_{12} + \phi_1 = 33°$$

$$\theta_{c_{12}} = 30°$$

$$\phi_1 = 3°$$

The velocity in layer 2 is $c_2 = c_1/\sin \theta_{12}$

$$c_2 = \frac{800}{\sin 30°} \text{ m/s}$$

$$= 1600 \text{ m/s}$$

The thickness computations use the intercept time $t_A(0)$

$$t_A(0) = \frac{2h_{A1}\cos \theta_{12}}{c_1}$$

$$h_{A1} = \frac{0.022 \cdot 800}{2 \cos 30°}$$

$$\simeq 10 \text{ m}$$

$$h_{B1} = h_{A1} - 100 \sin 3°$$

$$\simeq 4.8 \text{ m}$$

The computations carry two significant figures. Analysis of the data gives an interface that is shallower at B than at A. The results are consistent with the qualitative discussion.

This discussion is a bare introduction to the art of refraction measurements. There are techniques for analyzing data taken over nonplanar layers in addition to those for multiple dipping plane layers. References can be found in the Suggested Readings.

PROBLEMS

1. Construct the travel-time graph for the following crustal-type structure: $h_1 = 35$ km, $c_{1p} = 6$ km/s, $c_{1s} = 3.5$ km/s, $c_{2p} = 8$ km/s. The interface is horizontal. Determine the minimum distance for critical p_1p_1 reflection. Graph p_1, s_1, p_1p_1, p_1s_1, and $p_1p_2p_1$.

2. Construct the travel-time graph for the following structure. All interfaces are parallel and horizontal. Layer 1: $h_1 = 2$ m, $c_1 = 300$ m/s. Layer 2: $h_2 = 230$ m, $c_2 = 1740$ m/s. Layer 3: $h_3 = 353$ m, $c_3 = 2100$ m/s. Layer 4: $c_4 = 5020$ m/s. Ignore multiple reflections and shear waves. Make the time-distance graph to at least 2 km. Ignore all arrivals later than 1.4 s.

3. Use the travel-time data in Problem 2 for a multilayer head-wave interpretation technique, Section A1.3 of Appendix A. Assume that all interfaces are horizontal.

5

Amplitudes of Seismic Signals

Applications of Snell's law give the ray paths of seismic signals. Fermat's principle of least travel time assures us that the first disturbance to arrive followed the Snell's law ray path, where *disturbance* means the arrival of the very beginning of a signal. Of course, in a complicated geophysical structure, many possible paths exist, and each type of path has its minimum time and ray path. Computations of signal amplitudes require more theoretical developments.

The exact computation of signals due to a point source in a layered medium is a difficult theoretical problem that is beyond the level of this text. The difficulty comes from the interaction of a spherical wave front with a plane interface.

An approximation that combines ray paths, plane-wave reflection coefficients, and transmission coefficients can be used in many seismic problems. The approximation assumes that a patch of the spherical wave front is locally reflected and transmitted as plane waves. The amplitude of the incident wave at the patch is computed by using ray-path algorithms and conservation of energy. The approximation errors are small, of the order of $\lambda/(2\pi z)$, when the source is many wavelengths above or below an interface, and the incident angle is far from the critical angle for total reflection. The approximation is most accurate near vertical incidence. Near the critical angle of incidence, the local plane-wave approximation is poor because the curvature of the wave front must be included, and part of the energy goes into the head wave.

The chapter begins with the theoretical calculation of the plane-wave reflection and transmission coefficients in fluids. The second part of the chapter combines these coefficients, ray paths, and conservation of energy to compute signal amplitudes.

5.1 REFLECTION AND TRANSMISSION AT AN INTERFACE

Reflection and transmission phenomena are familiar to all. The magnitudes of the reflected waves depend on the contrast between the properties of the materials and the roughness of the interface. We let the interface be smooth and derive the transmission and reflection coefficients at the plane interface between fluid media having ρ_1, c_1 and ρ_2, c_2. The calculation of the reflection and transmission coefficients uses the fundamental technique known as the satisfaction of the boundary conditions.

5.1.1 Solution using Boundary Conditions

The boundary conditions at a fluid-fluid interface are as follows:

1. The pressures are the same on each side of the interface.
2. The media maintain contact. Vertical components of particle displacements and velocities on each side are the same.

Condition 1 gives

$$p_i + p_r = p_t \tag{5.1}$$

where p_i, p_r, and p_t are the sound pressures of the incident, reflected, and transmitted waves, Figure 5.1.

Condition 2 gives for the incident angle, θ_1, reflected angle, θ_1, and refracted angle, θ_2,

$$v_i \cos \theta_1 + v_r \cos \theta_1 = v_t \cos \theta_2 \tag{5.2}$$

where v_i, v_r, and v_t are the incident, reflected, and transmitted particle velocities. Equation (1.41) gives $p = \pm \rho c v$. The particle velocity is along the direction of propagation, and the $+$ sign is for propagation in the $+$ coordinate direction. Because the reflected wave is in the backward direction, the pressures are

$$p_i = \rho_1 c_1 v_i \tag{5.3}$$

$$p_r = - \rho_1 c_1 v_r \tag{5.4}$$

$$p_t = \rho_2 c_2 v_t \tag{5.5}$$

The substitution of Eqs. (5.3) to (5.5) in Eq. (5.1) gives

$$\rho_1 c_1 v_i - \rho_1 c_1 v_r = \rho_2 c_2 v_t \tag{5.6}$$

Equations (5.2) and (5.6) provide solutions for the ratios v_r/v_i and v_t/v_i, which are the definitions of the *velocity reflection coefficient*,

$$R_{12} \equiv \frac{v_r}{v_i} \tag{5.7}$$

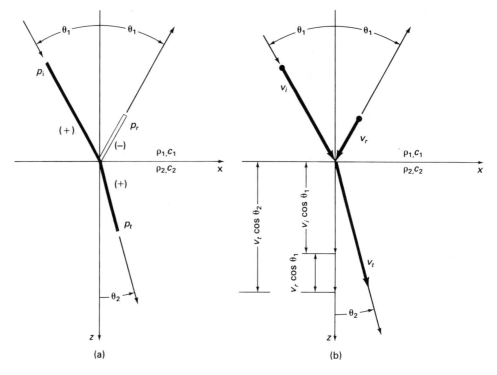

Figure 5.1 Incident reflected and transmitted components of pressure and particle velocity, $\rho_1 = \rho_2$, and $c_2 = c_1/2$. (a) Relative components of p_i, p_r, and p_t (scalars). (b) Relative components of v_i, v_r, and v_t (vectors).

and the *velocity transmission coefficient*,

$$T_{12} \equiv \frac{v_t}{v_i} \tag{5.8}$$

where the subscripts denote the 1-2 interface. Equations (5.2) and (5.6) become

$$\rho_1 c_1 - \rho_1 c_1 R_{12} = \rho_2 c_2 T_{12} \tag{5.9}$$

$$\cos \theta_1 + \cos \theta_1 R_{12} = \cos_2 T_{12} \tag{5.10}$$

The solutions for the *velocity* reflection and transmission coefficients are

$$R_{12} = \frac{\rho_1 c_1 \cos \theta_2 - \rho_2 c_2 \cos \theta_1}{\rho_1 c_1 \cos \theta_2 + \rho_2 c_2 \cos \theta_1} \tag{5.11}$$

$$T_{12} = \frac{2\rho_1 c_1 \cos \theta_1}{\rho_1 c_1 \cos \theta_2 + \rho_2 c_2 \cos \theta_1} \tag{5.12}$$

and

$$1 + R_{12} = T_{12} \frac{\cos \theta_2}{\cos \theta_1} \tag{5.13}$$

The order of the subscripts gives the directions of propagation, reflection, and transmission. The subscripts reverse for waves going in the other direction, so

$$R_{12} = -R_{21} \tag{5.14}$$

$$T_{21} = \frac{2\rho_2 c_2 \cos \theta_2}{\rho_1 c_1 \cos \theta_2 + \rho c_2 \cos \theta_1} \tag{5.15}$$

and

$$T_{12} T_{21} = 1 - R_{12}^2 \tag{5.16}$$

Snell's law relates θ_1 and θ_2. Beyond the critical angle θ_{12}, $\sin \theta_2$ is greater than 1, and the reflection is total:

$$|R_{12}| = 1 \qquad \text{for } \sin \theta_1 \geq \frac{c_1}{c_2}$$

$$T_{12} \rightarrow 0 \qquad \text{for } \frac{z}{\lambda} > 1 \tag{5.17}$$

where $|\ |$ is the absolute value. Details are given in Section A1.1.4 of Appendix A.

The reflection and transmission of compressional waves in solid media reduce Eqs. (5.11) and (5.12) at vertical incidence.

5.1.2 Reflection Sign Conventions

An alternative sign convention for reflection coefficients is convenient for following a signal along a ray path in a multilayered medium. We want a positive reflection coefficient to correspond to positive particle velocities in the direction of the reflected ray. For example, if $\rho_2 c_2 \cos \theta_1 > \rho_1 c_1 \cos \theta_2$, the particle velocity is positive in the direction of the reflected ray. An alternative definition of the reflection coefficient is

$$R_{12}^R \equiv \frac{\rho_2 c_2 \cos \theta_1 - \rho_1 c_1 \cos \theta_2}{\rho_2 c_2 \cos \theta_1 + \rho_1 c_1 \cos \theta_2} \tag{5.18}$$

where the superscript R indicates that the reflected particle velocity is along the direction of the ray.

Both sign conventions are useful. The sign convention in Eq. (5.11) refers the sign of all particle velocities to the common direction of the z-axis. This is desirable when the total signal is the vector sum of many upward- and downward-traveling waves in a multilayer medium. The sign convention of Eq. (5.18) is convenient for tracing a ray path through a multilayer medium, and the particle velocity is along the direction of the ray path.

The problem of sign conventions can be avoided in fluid media by using pressures and the pressure reflection coefficient, because pressure is a scalar. When many signal components from many directions arrive at the same time, the total pressure is the algebraic sum of pressures of components. Pressure or scalar computations are much simpler than vector computations of particle velocities. Equation (A.62) can be used to compute the acceleration or $\partial v/\partial t$ from the pressure.

$$\nabla p = -\rho \, \frac{\partial^2 \mathbf{R}}{\partial t^2} \tag{A.62}$$

PROBLEMS

1. Calculate the theoretical pressure reflection and transmission coefficients.

2. Write a computer program for computing reflection and transmission coefficients.

3. Calculate the reflection and transmission coefficients for $0 \leq \theta_1 \leq 90°$. Use $\rho_2 = 2\rho_1$. (a) $c_2 = 2c_1$; (b) $c_2 = c_1$; (c) $c_2 = c_1/2$; (d) other combinations of your choice.

5.2 AMPLITUDES OF SEISMIC WAVES: A RAY APPROXIMATION AND CONSERVATION OF POWER

Amplitude calculations for the seismic waves that travel through a multilayered medium are equivalent to a graduation exercise for Unit 1. The subject is practical because amplitude calculations are the first step in computing synthetic seismograms. All the concepts in earlier sections, have been discussed, namely, conservation of energy or power (Section 2.1), Snell's law and ray paths (Section 3.1), and the reflection and transmission of waves at a boundary (Section 5.1). Like the results of many conservation of energy computations, the final expressions are simple.

Ray approximations, whether they be in optics, acoustics, or seismology, are short-wavelength approximations. All interactions at an interface are many wavelengths from the source. All diffraction effects are ignored. We imagine that the ray traces the path of a small patch of a plane-wave front (Figure 5.2a). The minimum dimensions of a patch

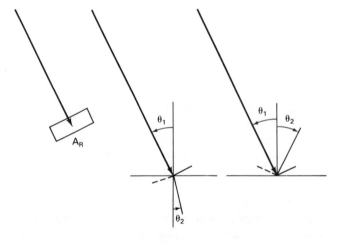

(a)   (b)

Figure 5.2 A ray and its patch of area. (a) The ray and patch of area A_R. (b) Side views of the incident, transmitted, and reflected patches. The dashed lines are the transmitted and reflected parts of the wave front.

are several wavelengths and roughly the dimensions of the first Fresnel zone of transmission by an aperture (discussed later in Section 14.1 where, we show that the dimension of a patch grow as the square root of the distance.) At an interface the wave-front patch is reflected and transmitted as if it were a plane wave (Figure 5.2b).

The ray approximation is accurate when the source is many wavelengths from an interface, and incident angles are not too close to critical angles. Near critical angles, part of the energy excites the head wave; then, amplitudes of the reflected wave, the transmitted wave, and the head wave depend on the curvature of the wave front. Analytical discussions of the reflection and transmission of spherical wave fronts and head waves are in more advanced texts.

Our procedure applies conservation of power to a tube of rays as the tube travels through a multilayered space. For simplicity, each layer is homogeneous, isotropic, and lossless. The interfaces are horizontal. The first part of the derivation defines a ray tube and the power in the tube. The second part is a calculation of power transmission and reflection coefficients. The third part gives examples of transmission and reflection of power tubes and computations of particle-velocity amplitudes.

5.2.1 Power in a Ray Tube

Figure 5.3 shows the geometry of a ray tube. In the ray approximation all the power entering a tube stays in the tube unless power is reflected or scattered out by interfaces or inhomogeneities. The ray tube is bounded by four rays. The differences between angles

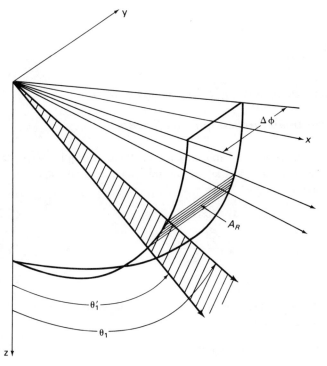

Figure 5.3 Geometry of source and a tube of rays. The ray tube is bounded by four rays. The shaded patch of area A_R is normal to a ray at the center of the tube. Power within the tube is conserved.

θ_1' and θ_1 are very small, and the patch of plane normal to a ray at the center is nearly normal to the bounding rays. Of course, the difference between θ_1' and θ_1 needs to be large enough that differences of the intercepts on an interface greatly exceed round-off error. From Eq. (2.6) and the power density j, in spherical coordinates, the power entering the tube is

$$\Delta\Pi = \int_{\theta_1'}^{\theta_1} j R^2 \sin\theta \, d\phi \, d\theta \tag{5.19}$$

$$j = pv_R \tag{5.20}$$

where p is the pressure and v_R is the particle velocity along the ray path. For j constant over the area, integration of Eq. (5.19) gives

$$\Delta\Pi = jA_R \tag{5.21}$$

$$A_R = R^2(\cos\theta_1' - \cos\theta_1)\,\Delta\phi \tag{5.22}$$

If j is constant in all directions, as for an omnidirectional source, the total power is ($\theta_1' = 0$, $\theta_1 = \pi$, $\Delta\phi = 2\pi$)

$$\Pi = 4\pi R^2 j \tag{5.23}$$

Equations (5.22) and (5.23) give ways of computing j for a known source power. Equation (5.23) is particularly useful for computing the pressure of particle velocity radiated by a simple omnidirectional source in water. Equation (1.41) relates p and v_R to j:

$$\left.\begin{aligned} p &= \rho c v_R \\ j &= \rho c v_R^2 \\ j &= \frac{p^2}{(\rho c)} \end{aligned}\right\} \tag{5.24}$$

where the subscript R indicates the particle velocity along the ray path.

5.2.2 Power Transmission and Reflection Coefficients

Computations of the power reflection and transmission coefficients require some care. Figure 5.4a shows the four bounding rays as the ray tube intercepts an interface. The intercepted area on interface A_1 is

$$L_1 = x_1 - x_1' \tag{5.25}$$

$$y_1 = \Delta\phi\frac{(x_1' + x_1)}{2} \tag{5.26}$$

$$A_1 = L_1 y_1 \tag{5.27}$$

The corresponding area of incident ray tube is the projection of A_1 on the plane normal to the center ray at the interface (Figure 5.4b),

$$A_{1,R}^i \simeq A_1\cos\overline{\theta}_1 \tag{5.28}$$

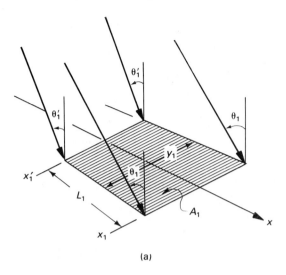

(a)

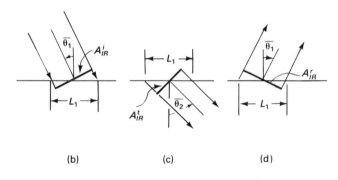

(b)   (c)   (d)

Figure 5.4 Geometry of ray tube areas: (a) intercept of the rays on an interface; (b), (c), and (d) side views of incident transmitted and reflected tube areas.

$$\bar{\theta}_1 \equiv \frac{\theta_1' + \theta_1}{2} \tag{5.29}$$

where the superscript i represents the incident ray tube and the subscripts 1 and R represent interface 1 and the normal to the ray, respectively. By Snell's law, the reflected angle is also $\bar{\theta}_1$, and

$$A_{1,R}^r \simeq A_1 \cos \bar{\theta}_1 \tag{5.30}$$

Using Snell's law for θ_2' and θ_2, we obtain the area of the transmitted tube,

$$A_{1,R}^t \simeq A_1 \cos \bar{\theta}_2 \tag{5.31}$$

$$\bar{\theta}_2 \equiv \frac{\theta_2' + \theta_2}{2} \tag{5.32}$$

The ray-tracing algorithms in Section 3.3 can be used to compute the θ_i and x_i intercepts at any interface.

We use the velocity transmission and reflection coefficients (Eqs. (5.12), (5.18), and (5.24)) to compute the transmitted component j_1^t and reflected component j_1^r at interface 1. From Eq. (5.24) the power-density components are

$$j_1^i = \rho_1 c_1 \, (v_{1,R}^i)^2 \tag{5.33}$$

$$j_1^t = \rho_2 c_2 \, (v_{1,R}^t)^2 \tag{5.34}$$

$$j_1^r = \rho_1 c_1 \, (v_{1,R}^r)^2 \tag{5.35}$$

After substitution of $v_{1,R}^i T_{12}$ for $v_{1,R}^t$, and so on, and from Eq. (5.33) the power densities become

$$j_1^t = \frac{\rho_2 c_2}{\rho_1 c_1} j_1^i \, T_{12}^2 \tag{5.36}$$

$$j_1^r = j_1^i R_{12}^2, \qquad R_{12}^2 = (R_{12}^R)^2 \tag{5.37}$$

At interface 1 the transmitted $\Delta\Pi_1^t$, the reflected $\Delta\Pi_1^r$, and $\Delta\Pi_1^i$ incident components of power are

$$\Delta\Pi_1^t = j_1^t \, A_1 \cos \bar{\theta}_2 \tag{5.38}$$

$$\Delta\Pi_1^r = j_1^r \, A_1 \cos \bar{\theta}_1 \tag{5.39}$$

$$\Delta\Pi_1^i = j_1^i \, A_1 \cos \bar{\theta}_1 \tag{5.40}$$

The power transmission and reflection coefficients follow from the ratios

$$T_{12}^\Pi \equiv \frac{\Delta\Pi_1^t}{\Delta\Pi_1^i} = \frac{\rho_2 c_2 \cos \bar{\theta}_2}{\rho_1 c_1 \cos \bar{\theta}_1} T_{12}^2 \tag{5.41}$$

$$R_{12}^\Pi \equiv \frac{\Delta\Pi_1^r}{\Delta\Pi_1^i} = R_{12}^2 \tag{5.42}$$

Substitution of Eqs. (5.12) and (5.18) for T_{12} and R_{12} and summation of Eqs. (5.41) and (5.42) gives

$$T_{12}^\Pi + R_{12}^\Pi = 1 \tag{5.43}$$

This demonstrates the conservation of power. Note that the power transmission coefficient depends on the densities of the upper and lower media, the directions of the ray tube, and the particle velocity transmission coefficient.

5.3 TRANSMISSION AND REFLECTION EXAMPLES

The transmission geometry is sketched in Figure 5.5a. The ray tube starts at a source on interface 0 and goes to an interface at depth z_4. Our purpose is to derive expressions for the incident power density and particle velocity at z_4. For numerical computations we use the ray-tracing programs in Section 3.1 and Section A1.4.2 of Appendix A to compute the angles and intercepts along the x-direction, and travel times. The parameters of the layers, initial angles θ_1', θ_1, and $\Delta\phi$ and power entering the ray tube are input data. If $\Delta\Pi$ is not known, we can set $\Delta\Pi$ to a convenient value, such as 1.

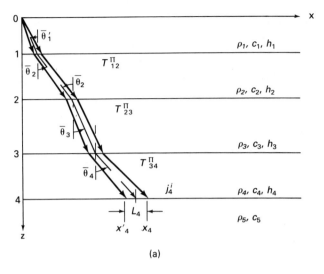

(a)

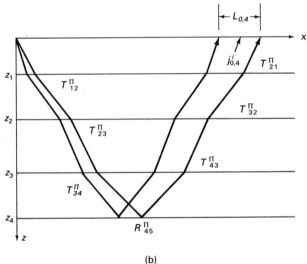

(b)

Figure 5.5 Ray tubes for transmission and reflection. The nomenclatures $L_{0,4}$ and $j^i_{0,4}$ indicate rays that have traveled to interface 4 and returned to the surface.

The fractions of power transmitted at each interface are given by the power transmission coefficients. The transmissions are in series, and the power in the ray tube that is incident on interface 4 is

$$\Delta\Pi^i_4 = \Delta\Pi \; T^{\Pi}_{12} \, T^{\Pi}_{23} \, T^{\Pi}_{34} \tag{5.44}$$

The substitution of Eq. (5.41) for the power transmission coefficients gives

$$\Delta\Pi^i_4 = \Delta\Pi \; \frac{\rho_4 c_4 \cos\overline{\theta}_4}{\rho_1 c_1 \cos\overline{\theta}_1} \, T^2_{12} T^2_{23} \, T^2_{34} \tag{5.45}$$

Perhaps this result is typical of power calculations. The output depends on the input parameters, output parameters, and transmission coefficients. Since our instruments usually measure pressure or the components of particle velocity, we use Eq. (5.40) to compute j_4^i and then use Eq. (5.33) to compute v_4^i. The substitution of the general form of Eq. (5.40), (where $\Delta\Pi = \Delta\Pi_1^i$)

$$\Delta\Pi_n^i = j_n^i A_n \cos \overline{\theta}_n \tag{5.46}$$

gives a power-density expression,

$$j_4^i = j_1^i \cdot \frac{A_1}{A_4} \frac{\rho_4 c_4}{\rho_1 c_1} T_{12}^2 T_{23}^2 T_{34}^2 \tag{5.47}$$

The ray-path intercept areas are

$$A_n = L_n \overline{x}_n \Delta\phi \tag{5.48}$$

$$L_n \equiv x_n - x_n' \tag{5.49}$$

$$\overline{x}_n \equiv \frac{x_n + x_n'}{2} \tag{5.50}$$

and Eq. (5.47) becomes

$$j_4^i = j_1^i \cdot \frac{L_1 \overline{x}_1}{L_4 \overline{x}_4} \frac{\rho_4 c_4}{\rho_1 c_1} T_{12}^2 T_{23}^2 T_{34}^2 \tag{5.51}$$

Reduction of Eq. (5.51) to an expression that uses particle velocities along the ray paths gives a simple result,

$$v_{4,R}^i = v_{1,R}^i \left(\frac{L_1 \overline{x}_1}{L_4 \overline{x}_4}\right)^{1/2} T_{12} T_{23} T_{34} \tag{5.52}$$

We have to compute the endpoints of the ray traces and the transmission coefficients to compute the particle velocity at the end of the path. The corresponding expression for pressure is

$$p_4^i = p_1^i \frac{\rho_4 c_4}{\rho_1 c_1} \left(\frac{L_1 \overline{x}_1}{L_4 \overline{x}_4}\right)^{1/2} T_{12} T_{23} T_{34} \tag{5.53}$$

While the computation is for the transmission from interface 0 to interface 4, the derivation is general, and we can take the ray tube from a source at any depth to any interface by any combination of reflections and transmissions.

Reflection example

The reflection geometry is sketched in Figure 5.5b. The reflection at the fourth interface introduces the reflection $R_{4,5}$, and then the transmission proceeds as before. The only complications are the notations, because the upward-traveling ray tube needs a notation that indicates it was reflected at the nth layer. We add a subscript, so that $L_{m,n}$ means length at the mth interface due to a reflection at the nth interface. Similarly, $j_{m,n}^i$ means incident power density at the mth interface of a reflection at the nth interface, and so forth, for all the terms. With these notational changes, the power transmission equation (5.45) becomes

$$\Delta\Pi_{0,4}^i = \Delta\Pi \; T_{12}^2 T_{23}^2 T_{34}^2 R_{45}^2 T_{43}^2 T_{32}^2 T_{21}^2 \tag{5.54}$$

Similarly, Eqs. (5.51) to (5.52) become

$$j_{0,4}^i = j_1^i \frac{L_1 \bar{x}_1}{L_{0,4} \bar{x}_{0,4}} \; T_{12}^2 T_{23}^2 T_{34}^2 R_{45}^2 T_{43}^2 T_{32}^2 T_{21}^2 \tag{5.55}$$

$$v_{0,4R}^i = v_{1R}^i \left(\frac{L_1 \bar{x}_1}{L_{0,4} \bar{x}_{0,4}} \right)^{1/2} T_{12} T_{23} T_{34} R_{45}^R T_{43} T_{32} T_{21} \tag{5.56}$$

$$p_{0,4}^i = p_1^i \left(\frac{L_1 \bar{x}_1}{L_{0,4} \bar{x}_{0,4}} \right)^{1/2} T_{12} T_{23} T_{34} R_{45}^p T_{43} T_{32} T_{21} \tag{5.57}$$

$$R_{45}^R = -R_{45} = R_{45}^p$$

It is easy to adapt the downward traveling transmission computation to the calculation of reflected signal amplitudes. The x-displacement of a reflected ray is twice the x-displacement of the downward traveling ray. The products of transmission coefficients also reduce, using Eq. (6.16)

$$T_{n,n+1} T_{n+1,n} = 1 - R_{n,n+1}^2 \tag{5.58}$$

Thus, we do not need to trace the ray paths back up to the surface. From Figure 5.5, the x-displacements of the reflected ray paths are twice the downward traveling ray paths and

$$\bar{x}_{0,4} = 2\,\bar{x}_4 \tag{5.59}$$

$$L_{0,4} = 2\,L_4 \tag{5.60}$$

These substitutions give

$$v_{0,4R}^i = \frac{v_{1R}^i}{2} \left[\frac{L_1 \bar{x}_1}{L_4 \bar{x}_4} \right]^{1/2} (1 - R_{12}^2)(1 - R_{23}^2)(1 - R_{34}^2)\, R_{45}^R \tag{5.61}$$

Similar expressions can be written for $j_{0,4}^i$ and $p_{0,4}^i$.

As promised at the beginning of Section 5.2, the results are simple and what one would expect from conservation of power. This detailed derivation shows how the ray intercepts, transmission coefficients, and reflection coefficients enter at each interface as the ray tube travels in the medium. A ray trace amplitude program can be found in Section A1.7 of Appendix A.

PROBLEMS

4. The formulation in this section is not very good for a vertically incident ray tube. Derive transmitted and reflected amplitudes for a vertically incident ray tube.

Data for Problems 5 and 6

The structure is a simplified version of a marine geophysical structure. The surface is interface 0. Layer 1, the water, is 1000 m deep. The sound velocity and density are 1500 m/s and 1030 kg/m^3. Layer 2 is unconsolidated marine sediment and has a thickness of 500 m, seismic velocity of 2000 m/s, and density of 1800 kg/m^3. Beneath interface 2 a sandstone layer has a seismic velocity of 3000 m/s and density of 2500 kg/m^3. Let the layer be infinitely thick.

5. Compute the incident power density, particle velocity along the ray, and pressure at interface 3, the top of the sandstone layer. The source is at 100 m depth and radiates omnidirectionally. The source transmitts a 0.05-s sinusoidal signal or ping. The peak signal power during a transmission is 10000 W. This permits measurement of the direct arrival and lets you ignore the path from the source up to the water surface and down. The transmission is from the source to the sediment sandstone interface 2 at 1500 m depth. For a set of incident angles θ_1 out to 20°, compute and graph the following versus the mean x-intercept distances:
   (a) x-intercept distance
   (b) Travel time
   (c) Incident power density
   (d) Incident particle velocity along the incident ray path
   (e) Incident pressure
   (f) Considering the reflection at interface 3, compute the pressure and the vertical component of particle velocity.

6. This is a reflection problem. The source is the same as in Problem 5. The receiver is at a depth of 100 m. Both the source and receiver are far enough from the water surface to ignore surface-reflected signals. For incident angles θ_1 out to 20° compute and graph the following data for reflections from interface 1 and interface 2:
   (a) Intercepts along x at 100 m depth
   (b) Incident power density
   (c) Incident particle velocity
   (d) Incident pressure

Unit I—Suggested Readings

AKI, K., AND P. G. RICHARDS, *Quantitative Seismology Theory and Methods*, Vols. I and II. New York: W. H. Freeman & Co., 1980. Gives an extensive discussion of seismic methods including the generalized inverse method.

CARMICHAEL, B. S., ed. *Physical Properties of Rocks*. Vol. I, Mineral Composition (1982); Vol II, Seismic Velocities (1982); Vol. III, Density of Rocks and Minerals. Boca Raton, Fla.: CRC Press, 1984.

CERVENY, V., AND R. RAVINDRA, *Theory of Seismic Head Waves*. Toronto: University of Toronto Press, 1971. Gives a careful development of ray theory for multilayered media. Their formulas for elastic wave reflection and transmission coefficients are convenient for computations. They develop ray theory and wave theory for head waves.

CLARK, S. P., *Handbook of Physical Constants Memoir 97*. New York: The Geological Society of America, 1966. Gives the data on the properties on rocks as compiled by experts in each specialized area.

DE BREMAECKER, J.-C., *Geophysics of the Earth's Interior*. New York: John Wiley, 1985. Gives an introduction to geophysics for the non-specialist. Uses calculus and vectors.

DOBRIN, M. B., *Introduction to Geophysical Prospecting*. Several editions. New York: McGraw-Hill, 1976. Covers the seismic exploration methods developed through the 1960s. This period was one of intense research and development effort in seismic techniques. Half of the text is seismic exploration. Requires knowledge of Fourier theory.

ELMORE, W. C., AND M. A. HEALD, *The Physics of Waves*. New York: McGraw-Hill, 1969. Gives an introduction to the propagation of waves in all kinds of media. The waves include acoustic, elastic, surface waves, and electromagnetic waves.

GARLAND, G. D., *Introduction to Geophysics*. Philadelphia: W. B. Saunders, 1971 and later for new editions. Gives basic theory for physics of the earth at a graduate level.

GRANT, F. S., AND G. F. WEST, *Interpretation Theory in Applied Geophysics*. New York: McGraw-Hill, 1965. Gives careful analytical developments of the methods of applied (measurement) geophysics. The text covers elastic media and seismic waves, gravity and magnetism, potential field theory, and electrical methods.

MUSGRAVE, A. W., ed., *Seismic Refraction Prospecting*. Tulsa, Okla.: Society of Exploration Geophysics, 1970. Contains an excellent collection of papers on the refraction method including model studies, theory, and practice.

PALMER, DERECKE, *The Generalized Reciprocal Method of Seismic Refraction Interpretation*. Tulsa, Okla.: Society of Exploration Geophysicists, 1980. Gives advanced applications of the refraction method.

STACEY, F. D., *Physics of the Earth*. New York: John Wiley, 1969, 1977. Gives basic theory for physics of the earth at a graduate level.

TELFORD, W. M., L. P. GELDART, R. E. SHERIFF, AND D. A. KEYS, *Applied Geophysics*. Cambridge: Cambridge University Press, 1976. Contains detailed discussions of all methods of geophysics.

TOLSTOY, IVAN, *Wave Propagation*. New York: McGraw-Hill, 1973. Uses the variational methods of classical physics to give a unified treatment of all kinds of waves in all media. He gives the normal coordinate method and uses it to derive the (Biot-Tolstoy) exact impulse solution for a point source and receiver near an ideal wedge.

TOLSTOY, I., AND C. S. CLAY, *Ocean Acoustics*, 2d ed. American Institute of Physics New York, NY, 1987. Gives an advanced treatment of wave propagation using normal modes. Includes arrays of sensors as mode filters in a waveguide. It includes the exact Biot-Tolstoy impulse solution for diffractions and reflections at a rigid wedge.

Unit II

DIGITAL OPERATIONS ON SIGNALS

Prior to 1960, most research in seismology and exploration geophysics used analog methods. Digital processing of data was very expensive and was rarely attempted. Then, most comparisons of theory and experimental data consisted of plotting the arrival times of events on distance-travel time curves or alternatively plotting the distance-travel time curves on experimental data. Other than trying to pick events having large amplitudes, wave forms and amplitudes were ignored. It took a lot of skill and imagination to use distance-travel time curves for the interpretation of real seismograms because the distance-travel time curves didn't look like the seismograms. Realistic theoretical seismograms or *synthetic seismograms* were needed. The required elements, seismic theory and signal theory, were well known and could be combined to make synthetic seismograms. Although geophysicists wrote elegant papers on how to compute synthetic seismograms, very few synthetic seismograms were actually calculated and displayed. The Fourier and convolution integrals were too laborious to evaluate by hand.

About 1960, the display and processing of seismic data began to change from largely analog methods to largely digital methods. Although the experimental geophysicists and engineers were very busy building more and more sophisticated analog equipment, a few engineers were beginning to build multichannel digital recording systems. The theoreticians were busy too. The theoreticians recognized that digital recording and digital computers would change everything because it would be practical to use powerful and advanced concepts from signal theory to process and interpret seismic data. The digital engineers and theoreticians were on the right course. Digital signal processing revolutionized geophysics.

Unit 2 gives an elementary discussion of the first stage of the digital revolution. Models of the subsurface structure and wave theory from Unit 1 are used to compute theoretical seismograms for impulsive sources. These are called the impulse response functions of the earth. For comparison to experimental data, the earth's impulse response and a signal function are convolved to give a synthetic seismogram. If the synthetic and experimental seismograms match, then the model earth is a good representation of the earth structure. The concepts involved in making synthetic seismograms help us to understand seismic measurements.

The exploration geophysical literature has many comparisons of the properties of rock layers penetrated by a borehole and seismic reflection data. Sometimes a complex sequence of rock layers gives a strong reflection signal, and sometimes the reflection signal is poor. After continuous seismic velocity logging instruments were introduced in 1952, geophysicists started measuring the seismic velocities of the rock layers. By 1960, digital technology enabled geophysicists to use seismic velocity logs and estimates of rock densities to compute theoretical earth response functions. These functions were convolved with signals to get synthetic seismograms. Comparisons of the synthetic seismograms and reflection data that were taken near the borehole gave much insight into how a complex sequence of layers could give large or small reflection signals.

But making synthetic seismograms wasn't the main reason that the petroleum industry's research laboratories each spent millions of dollars on digital systems and signal processing. Oil exploration had moved to offshore regions. Since there was much oil to be found offshore, the economic stakes were high. In areas such as the Persian Gulf, many seismic reflection records looked as if someone had turned on an oscillator and let it run. It was impossible to see reflection signals on these seismograms. The phenomenon was called ringing, singing, or reverberation. Experiments showed that the frequency and character of the ringing signals depended on the type of bottom and water depth. Geophysicists introduced the concept that the ringing signals could be modeled by a regular earth response function and a signal that rang like a bell. The task was to undo the ringing effect. In 1959, geophysicists and engineers used special analog filters to reduce the ringing and to get usable seismograms. Real improvements in data quality would require digitally recorded and processed data.

The solution to processing the ringing records from marine areas was to undo the convolution of the ringing signal and the earth response function. The undoing operation or inverse of convolution is known as *deconvolution*. Wiener's optimum filter theory became a basis for digital deconvolution processing. Digital deconvolution processing of seismic reflection data has become routine because it works. Unit 2 ends with examples of deconvolution operations.

6

Computations Using Digital Signals

We present the basic methods of using seismic signals in digital operations. Logically, the first task is to sample signal and create a sequence of numbers that accurately represent the original signal. Subsequent sections will introduce the elementary operations of time delay, multiplication by a constant, and addition. The procedures for doing these linear operations are in computational subroutines or algorithms.

6.1 DIGITAL REPRESENTATION OF A SIGNAL

A seismic signal is a continuous function of time, and we wish to represent it with a set of numbers. The simplest sampling operation is to measure the (almost) instantaneous voltages at equal time steps t_0. As shown in Figure 6.1, a clock controls the sampling instructions. The sampling period is t_0. Computer memory and electronic hardware set limits on the total number of samples and the sampling rate. *As a limiting condition*, t_0 *must be less than half the shortest period in the signals.* (The sampling frequency corresponding to t_0 is $f_0 = 1/t_0$. Corresponding to the shortest period being greater than $2t_0$, *the highest-frequency component of the signal should be less than* $f_0/2$. $f_0/2$ is also known as the *Nyquist*, or *folding*, *frequency*.) These are statements of the Nyquist sampling rule. It is customary to design the analog amplifiers to limit the frequency range of signals to less than $f_0/2$ prior to analog/digital (A/D) conversion.

We want the digital representation to be a reasonably accurate representation of the analog signal without using elaborate interpolation procedures. We can sketch a sine-wave signal then sample the signal at different sampling periods (see Problem 1). Samples taken with t_0 a little less than half of the sine-wave period give a poor representation. Samples taken with t_0 less than a quarter of the sine-wave period give an acceptable

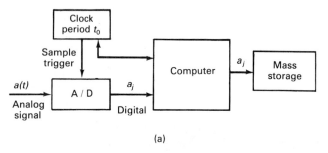

(a)

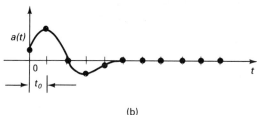

(b)

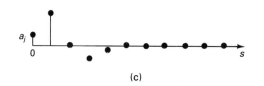

(c)

Figure 6.1   Digitization of a signal. (a) The clock sends a sample trigger instruction to the analog to digital converter (A/D). The computer receives the numbers a_j and stores them sequentially on a storage medium, such as magnetic tape or disks. (b) $a(t)$ is the analog signal. (c) a_j are the sampled data points.

representation. Samples taken with t_0 greater than a half-period give a representation having a *lower* frequency than the signal.

Rule: t_0 must be less than half the shortest period in the signal regardless of the kind of the data of interest.

The digitized data are a sequence of numbers. The computer stores them on magnetic tape, disks, or in the computer memory as data files. To operate on the signals, we instruct the computer to find the file and place it in an array so that we can read individual numbers. Writing the data files, retrieving the signals, and preparing them for digital processing can be difficult. Experienced people test the recording and retrieving programs on synthetic data before doing field work.

Three warnings:

1. Improper sampling of a physical phenomenon gives garbage.
2. Conceptual and programming errors give nonsense.
3. The creation of perfect computer programs can become an obsession.

PROBLEMS

1. *Purpose*: to show the effect of different sampling frequencies on the digital representation of the signal. Let $a(t) = \sin(2\pi\, t/T)$. Sketch a graph of $a(t)$ for $T = 2.4$ ms. Graphically sample $a(t)$ at the time steps $t_0 =$ (a) 1 ms; (b) 4 ms; (c) 6 ms; (d) 11 ms; (e) 12 ms; (f) 13 ms; (g) 23

ms. Display each graph separately. Connect the sampled data points with a smooth curve for comparison with the original $a(t)$. Use the same time scale and graph at least 200 periods of $a(t)$.

6.1.1 Digital Operations On Signals

The first five chapters give the basic seismic methods. Since the data sets are large, geophysicists do as much as possible in a computer. Our tasks are to change the algebraic formulas into computational algorithms, write the code, and display the results of calculations. All these operations are illustrated on seismic examples, although these operations are general and can be transferred to other geophysical data by changing the names of the variables and coordinates. To aid in the technology transfer, we give computational methods and algorithms in some detail.

Algorithms

If we ignore geometric spreading, $a(t - R/c)$ is a traveling wave, and we need concise digital expressions for this kind of signal. At $R = 0$, the digital amplitudes of $a(t)$ are a_0, a_1, a_2, and so forth, where the subscript represents the number of time steps. Arrays are a convenient way to handle subscripted numbers in a computer (see Section A1.4 of Appendix A).

6.1.2 Load Arrays

Before we can use or load data into arrays, the computer must reserve memory for the arrays. In our examples we use $J = 50$ as the maximum subscript and dimension the arrays. BASIC systems usually use subscripts that range from 0 to a positive integer, here 50. A BASIC dimension statement is DIM A(50) for the A(J) array, FORTRAN 77 is more flexible. For example, an array of real numbers can be dimensioned by writing real $A(-50:50)$. (FORTRAN uses *real* for decimal numbers, *complex* for complex decimal numbers, *integer* for integer numbers, and *character* for strings of letters. All these can be put in arrays.) In FORTRAN 77 the array subscripts can range from -50 to 50. Of course, the maximum array dimensions of 0 to 50 and -50 to 50 are given for illustration. Continuing in BASIC, assume that the arrays A(J), B(J), and C(J) are dimensioned and ready to use. The next step is to load the data into the arrays using a FOR-NEXT loop

```
FOR  J = 0 TO 50
     A(J) = a(j)
     B(J) = b(j)                                    (6.1a)
     C(J) = 0
NEXT J
```

The statement, $C(J) = 0$, initializes $C(J)$ to zero.

```
c     FORTRAN do loop

      do 10 j = - 50, 50                              (6.1b)
         A(j) = a(j)
1 0   continue
```

6.1.3 Multiplying a Signal by a Constant

The algebraic statement is

$$c_j = ka_j \tag{6.2}$$

and the corresponding program instructions are, where $K = k$,

```
FOR J = 0 TO 50
   C(J) = K * A(J)                                    (6.3)
NEXT J
```

For each J the computer reads the number at A(J), multiplies it by the number named K, and stores the product in the location named C(J). It erases the old value of C(J) when it stores the new value. A graphic example is shown in Figure 6.2.

6.1.4 Time of Shift of A(J)

The algebraic statement of a traveling wave corresponds to $a(t - R/c)$:

$$c_j = 0 \qquad \text{for } j < n \tag{6.4}$$

$$c_j = a_{j-n} \qquad \text{for } j \geq n \tag{6.5}$$

where j is the time step t and n is the time step corresponding to R/c. The first set of instructions set $C(J) = 0$ before the arrival of the signal. For the delay $N = n$, the instructions to program (6.4) are

```
FOR  J = 0 TO N
   C(J) = 0                                           (6.6)
NEXT J
```

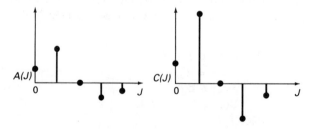

Figure 6.2   Multiplication, $K = 2$; $C(J) = K * A(J)$.

The next set of instructions starts at J = N. The instructions to do (6.5) are

$$
\begin{aligned}
&\text{FOR J = N TO 50}\\
&\qquad \text{C(J) = A(J - N)}\\
&\text{NEXT J}
\end{aligned}
\qquad (6.7)
$$

The instructions store the number A(0) = a_0 in the location C(N), the number A(1) = a_1 in the location C(N + 1), and so forth. A graphic example is shown in Figure 6.3.

6.1.5 Linear Combination of A(J) and B(J)

The algebraic equation is

$$
c_j = k_1\, a_{j-m} + k_2\, b_{j-n}
\qquad (6.8)
$$

The operation is linear because the components of the signal add. The parameters are K1 = k_1, K2 = k_2, M = m, and N = n. A procedure to add time-delayed signals is to enter the first into C(J) and then add the second signal to C(J). The instructions for a_{j-m} are

$$
\begin{aligned}
&\text{FOR J = 0 TO M}\\
&\qquad \text{C(J) = 0}\\
&\text{NEXT J}
\end{aligned}
\qquad (6.9)
$$

$$
\begin{aligned}
&\text{FOR J = M TO 50}\\
&\qquad \text{C(J)= K1 * A(J - M)}\\
&\text{NEXT J}
\end{aligned}
\qquad (6.10)
$$

Next, enter the second signal b_{j-n}:

$$
\begin{aligned}
&\text{FOR J = N TO 50}\\
&\qquad \text{C(J) = C(J) + K2 * B(J - N)}\\
&\text{NEXT J}
\end{aligned}
\qquad (6.11)
$$

The instructions begin at J = N because the array C(J) is loaded by instructions (6.9) and (6.10). In (6.11) the computer reads C(J), adds the product, and then stores the result in C(J). Instruction sets like (6.11) repeat for each arrival. A more compact algorithm can be found in Section 6.3. A graphic example is shown in Figure 6.4.

Figure 6.3 Time shift. C(J) = A(J − N), N = 3.

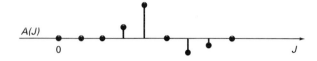

Figure 6.4 Time shift and add, $K_1 = 1$, $K_2 = 1$; $N = 3$. $C(J) = K_1 A(J - N) + K_2 B(J)$.

6.2 NUMERICAL EXAMPLE

We use the seismogram sketched in Figure 1.2 as an example. Since we are introducing new instructions, we limit the example to a single arrival. The algebraic equation is

$$s_i(t) = \frac{a(t - R_i/c)}{R_i} \tag{6.12}$$

where s is the signal measurement (pressure, velocity, or displacement), and the subscript i represents the ith receiver. The input parameters are

```
REM   C IN M/S, T IN S, AND R IN M.
REM THE : STARTS A NEW STATEMENT
```

$$
\begin{aligned}
&\text{C = 1500 :} \quad \text{T0 = 0.001} \\
&\text{A(0) = 5 :} \quad \text{A(1) = 15 :} \quad \text{A(2) = 0} \\
&\text{A(3) = -7:} \quad \text{A(4) = -2} \\
&\text{R(0) = 15:} \quad \text{R(1) = 30} \\
&\text{R(2) = 45:} \quad \text{R(3) = 60}
\end{aligned}
\tag{6.13}
$$

where the units of a_j are arbitrary. Fifty time steps are enough to display the seismogram because the travel time to $R(3)$ is 0.04 s or 40 time steps, and 40 plus the signal duration of 4 gives 44.

A two-dimensional array S(I,J) stores the output or seismogram. One dimension, I, gives the receiver channel, and the other, J, gives the time step. The dimension instructions are

$$DIM\ S(\ 3,50),\ R(3),\ A(50)$$

c   FORTRAN 77   (6.14)

$$real\ S(0:3,\ 0:50),\ R(0:3),\ A(0:50)$$

The time delay of the arrival R_i/c is an integer number of time steps, n. The BASIC instruction is

$$N = INT(\ R(I)/(C\ ^*\ T0) + 0.5\ )$$ (6.15)

where INT drops all decimal fractions, and the $+ 0.5$ causes round up. The computational equivalent of (6.12) is

$$S(I,J) = A(J - N)\ /\ R(I)$$ (6.16)

If necessary, the following instructions set $S(I,J) = 0$ (zero arrays on the Macintosh):

```
REM ZERO ARRAY

FOR I = 0 TO 3
  FOR J = 0 TO 50
    S(I,J) = 0                      (6.17)
  NEXT J
NEXT I
```

The next statements calculate the signal in each channel.

```
FOR I = 0 TO 3
  N = INT( R(I) / (C * T0) + 0.5 )       (6.18)

  FOR J = N TO 50
    S(I,J) = A(J - N) / R(I)
  NEXT J                                 (6.19)

NEXT I
```

The travel-time number is computed for each R(I) and then used in a FOR-NEXT loop on J. The seismogram is shown in Figure 6.5.

PROBLEMS

2. *Purpose:* to construct an idealized seismogram for the signals on the surface of a solid. Actual amplitudes and waveforms depend on the source, so use the a_j in Eq. (6.13). Let the P-wave amplitude be $0.7\ a_j$, the s-wave amplitude be a_j, and the Rayleigh wave amplitude be $5\ a_j$.

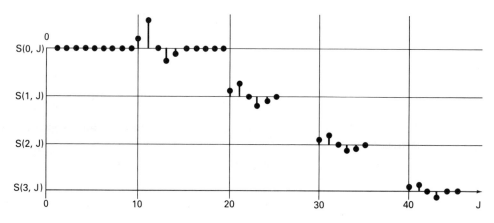

Figure 6.5 Seismograms for a spherically diverging wave front. The data are in Eq. (6.13).

$c_p = 2000$ m/s, $c_s = 1155$ m/s, and $c_R = 1060$ m/s. $t_0 = 1$ ms, $R_i = 20, 40, 60, 80, 100$ m. Use the amplitude versus range dependence for each type of wave. *Hint*: Compute the signal for each wave separately and then add them together. Compare your result with Figure 2.8.

6.3 LINEAR OPERATIONS ON DIGITALLY SAMPLED SIGNALS

We need powerful analytical methods to study the propagation of signals in the earth. Specifically we want mathematical techniques that are easy to use on digitally sampled signals. The sampling operation changes $a(t)$, for example, into a sequence of values,

$$a(t) \rightarrow a_0, a_1, a_2, \ldots, a_n \qquad (6.20)$$

for the times $0, t_0, 2t_0, \ldots, nt_0$. We want a simple way to represent Eq. (6.20). The so-called generating function is our choice. These functions were used by Laplace in *Théorie Analytique des Probabilités*, 1812. The same functions will serve our needs.

6.3.1 Generating Functions and Properties

The generating function is an algebraic way of writing the amplitudes of a sequence of operations and keeping the sequence in order. The generating function for the sequence in Eq. (6.20) is

$$A(z) \equiv a_0 + a_1 z + a_2 z^2 + \cdots + a_n z^n \qquad (6.21)$$

The sampling times at $0, t_0, 2t_0, \ldots, nt_0$ are represented by $z^0, z^1, \ldots, z^n$. The polynomial has the sequence of amplitudes in order of increasing times. There are a few conditions on $A(z)$. First, all the a_j are finite. Second, there are limits on allowed values of z, namely,

$$0 \le |z| < |z_0| \qquad (6.22)$$

where $|z_0|$ is finite. The absolute value signs mean that z can have real and complex values. If n tends to infinity, then z_0 must be chosen so that the series converges. Otherwise, z

may have any value within the allowed range. Properties of the generating-function representation of signals follow from the analysis in Section B.1.1 of Appendix B.

Property 1: Coefficients

If the generating functions $A(z)$ and $B(z)$ are equal, then the coefficients of like powers of z are equal:

$$A(z) = B(z)$$
$$a_0 + a_1z + \cdots + a_jz^j + \cdots + a_nz^n$$
$$= b_0 + b_1z + \cdots + b_jz^j + \cdots + b_nz^n \tag{6.23}$$
$$a_j = b_j$$

This property is demonstrated in Section B1.1 of Appendix B. The proof also shows that the coefficients of unlike powers are independent, that is, b_i is independent of b_j for $j \neq i$.

Property 2: Time shifts

Multiplication of a generating function by z^j gives a time shift of j time steps. For example, $A(z)z^j$ gives

$$A(z)z^j = a_0z^j + a_1z^{j+1} + \cdots + a_nz^{j+n} \tag{6.24}$$

The factor z^j is equivalent to a time delay of jt_0. The factor z^{-j} is equivalent to a time delay of $-jt_0$, or a time advance. Figure 6.6 shows examples of time shifts of z^2 and z^5. The signals are

$$\left.\begin{array}{l} A(z) = 1 + 0.5z \\ A(z)z^2 = z^2 + 0.5z^3 \\ A(z)z^5 = z^5 + 0.5z^6 \end{array}\right\} \tag{6.25}$$

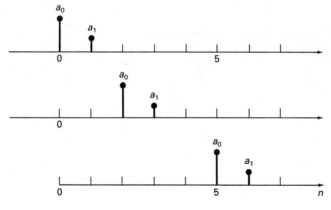

Figure 6.6 Time shifts or delays of $a_0 + a_1z$.

Property 3: Algebraic rules

The sums of generating functions follow the usual algebraic rules. Multiplication by a constant multiplies all coefficients by the same constant, e_0

$$e_0 A(z) = e_0 a_0 + e_0 a_1 z + \cdots + e_0 a_n z^n \qquad (6.26)$$

Let $C(z)$ be the sum of $A(z)$ delayed by m time steps and $B(z)$ delayed by n time steps

$$C(z) = A(z)z^m + B(z)z^n \qquad (6.27)$$

On the right side we gather the coefficients having the same powers of z and then equate them to the coefficient having the same power of z on the left side. An example is shown in Figure 6.7. The signals and computations are as follows:

$$\left. \begin{array}{l} A(z) = 1 + 0.3z \\ B(z) = -0.6 + 0.7z \end{array} \right\} \qquad (6.28)$$

$$\left. \begin{array}{l} C(z) = A(z)z^2 + B(z)z^3 \\ C(z) = (z^2 + 0.3z^3) + (-0.6z^3 - 0.7z^4) \\ C(z) = z^2 - 0.3z^3 - 0.7z^4 \end{array} \right\} \qquad (6.29)$$

Property 4: Convolution

Convolution is the operation of adding the same signal $A(z)$ at various time delays and amplitude factors. Convolution is also the algebraic multiplication of a pair of generating functions. To show this, we start by constructing the sum

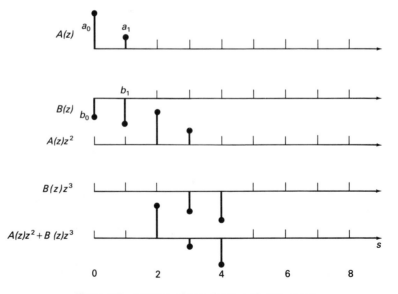

Figure 6.7 Addition of two signals with time delays.

$$C(z) = e_0 A(z) + e_1 A(z)z + e_2 A(z)z^2 + \cdots$$

$$
\begin{aligned}
c_0 + c_1 z + c_2 z^2 + \cdots &= e_0(a_0 + a_1 z + \cdots) \\
&\quad + e_1(a_0 + a_1 z + \cdots)z \\
&\quad + e_2(a_0 + a_1 z + \cdots)z^2 + \cdots
\end{aligned}
\tag{6.30}
$$

or in the factored form,

$$c_0 + c_1 z + c_2 z^2 + \cdots = (e_0 + e_1 z + e_2 z^2 + \cdots)(a_0 + a_1 z + \cdots) \tag{6.31}$$
$$C(z) = E(z) A(z)$$

where again we use $(a_0 + a_1 z + \cdots)$ to represent a signal of any length and e_0, $e_1, \ldots$ are amplitude factors.

Generating functions entered the geophysical and engineering literatures from different sources. As a result, the same name means something different to different people. Since some groups may be dogmatic in the terminology and applications, some explanations and connections are given in Section B1.1 of Appendix B.

6.4 IDEAL SEISMOGRAM

The computation of an ideal seismogram for two arrivals shows how the generating functions are used. The geometry is shown in Figure 6.8. The signals are

$$s_d(t) = \frac{a(t - x/c_1)}{x} \tag{6.32}$$

$$s_r(t) = R_{12}\frac{a(t - R/c_1)}{R} \tag{6.33}$$

where

$$R = \frac{2h}{\cos \theta_1} \tag{6.34}$$

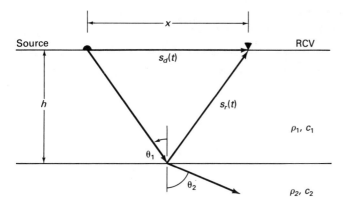

Figure 6.8 Two arrivals. The direct is $s_d(t)$, and the first reflection is $s_r(t)$.

and R_{12} is the reflection coefficient. For simplicity θ_1 is less than critical. Use the INT instruction, for example, Eq. (6.15), to compute the time-delay steps m and n for the direct and reflected arrivals, respectively. The input signal is $a(t)$, and the output signal is the sum of the two traveling waves

$$s(t) = s_d(t) + s_r(t) \tag{6.35}$$

Using generating functions for the digital signals, we have for $s(t)$

$$s_0 + s_1 z + s_2 z^2 + \cdots = \left(\frac{z^m}{x} + \frac{R_{12} z^n}{R} \right) (a_0 + a_1 z + \cdots) \tag{6.36}$$

where the time step is t_0 and the travel times are $m t_0$ and $n t_0$. By comparison of Eqs. (6.31) and (6.36), the coefficients e_i are

$$\left. \begin{aligned} e_m &= \frac{1}{x} \\[2mm] e_n &= \frac{R_{12}}{R} \\[2mm] m &= \text{INT} \left(x/(C1*T0) + 0.5 \right) \\[2mm] n &= \text{INT}(R/(C1*T0) + 0.5) \end{aligned} \right\} \tag{6.37}$$

and all other e_i are zero. The generating-function expressions are

$$s_0 + s_1 z + \cdots = (e_m z^m + e_n z^n)(a_0 + a_1 z + \cdots) \tag{6.38}$$

or

$$S(z) = E(z)A(z) \tag{6.39}$$

Equations (6.38) and (6.39) are statements of the convolution. Other common equivalent forms of the convolution are given in Section B1.2 of Appendix B.

6.4.1 Impulse Response

The *impulse response* is the signal at the receiver when the source is an *ideal impulsive function*. The unit impulsive signal at $t = 0$ is $a_0 = 1$, and all other a_i are 0. The substitution of $a_0 = 1$ in Eq. (6.38) gives

$$\begin{aligned} s_0 + s_1 z + \cdots |_I &= e_0 + e_1 z + e_2 z^2 + \cdots \\ &= e_m z^m + e_n z^n \end{aligned} \tag{6.40}$$

or

$$S_I(z) = E(z) \tag{6.41}$$

where the subscript I indicates impulse response. The impulse response is the earth transmission function. Figure 6.9 shows the impulse response and the result of convolving the impulse response with the signal $A(z)$.

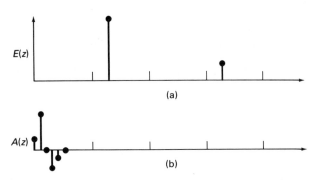

(a)

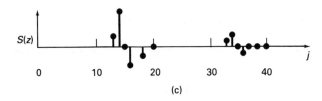

(b)

(c)

Figure 6.9 Ideal seismogram. $h = 23.3$ m, $x = 18.8$ m, $c = 1500$ m/s. $t_0 = 1$ ms. (a) Impulse response; (b) Signal set a_i, Eq. (6.13); (c) Transmission s_j.

From measurements of $S(z)$ and $A(z)$ we can solve Eq. (6.39) for $E(z)$ and get the earth impulse response,

$$E(z) = \frac{S(z)}{A(z)} \tag{6.42}$$

This is a deconvolving operation.

6.4.2 Algorithms

Convolution instructions in BASIC

The algorithm uses the form of Eq. (6.30). The input data are

```
FOR M = 0 TO M1
    E(M) = e(m)
NEXT M
```
$$\tag{6.43}$$
```
FOR N = 0 TO N1
    A(N) = a(n)
NEXT N
```

The convolution output is

$$C(I) = c(i) \tag{6.44}$$

The first step is to initialize $C(I) = 0$ (if necessary)

```
FOR I = 0 TO M1 + N1
    C(I) = 0
NEXT I
```
$$\tag{6.45}$$

The convolution computation uses the fact that $c_i z^i = e_m a_n z^{m+n}$ where $i = m + n$. The two loops step through all combinations of M and N. The algorithm is

FOR M = 0 TO M1

    FOR N = 0 TO N1
    I = M + N
    C(I) = C(I) + E(M) * A(N)   (6.46)
    NEXT N

    NEXT M

Compare the compactness of this algorithm with the sequence of operations for each arrival in Eqs. (6.9) to (6.11).

Convolution is a general operation and can include more than two functions. For example, the convolution of four functions is

$$G(z) = A(z)B(z)C(z)D(z)$$ (6.47)

Evaluation uses algorithms (6.45) and (6.46) by creating a temporary holding array $H(z)$ and reusing $G(z)$. The order is arbitrary, but the following is a possibility:

$$\left. \begin{aligned} G(z) &= C(z)D(z) \\ H(z) &= B(z)G(z) \\ G(z) &= A(z)H(z) \end{aligned} \right\}$$ (6.48)

where $G(z)$ is the output at the third step.

Instead of calling $A(z)$ and $E(z)$ the source and earth generating functions or z-transforms, geophysicists often call them the source and earth filters, in analogy with the filters used in electronic and communication systems. The electronic and electrical engineering literature on the use of filters is very large and has solutions to many problems of interest to geophysicists.

PROBLEMS

3. Write a program that convolves $E(z)$ $A(z)$ using algorithms (6.43) to (6.46). You will need to add output and graphics statements. Initially test the program on the source function $A(z) = 0.5 + 1z + 0.5z^2$ and earth function $E(z) = 1 + 3 z^5$. Multiply the polynomials and check your results by hand.

4. Use your program to compute a set of examples that are similar to the figures in Sections 6.1 to 6.4.

5. *Purpose*: to make synthetic seismograms. The earth model $E(z)$ can be held fixed. The output seismogram depends on the input signal function $A(z)$. The Ricker wavelets, Eqs. B.64 and B.65 of Appendix B, are very convenient source functions. You can use different widths, w_0, to study the effect of signal frequency on the "reflection seismogram." The data for the earth model are given in the table. The first step is the conversion of a layer thickness and velocity to an integral number of time steps. Round off to the nearest integer by using the INT( ) function. Use the

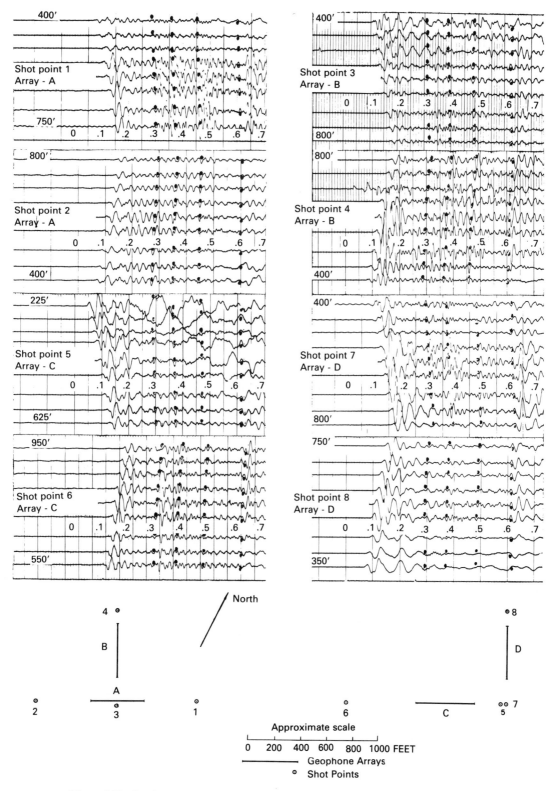

Figure 6.10 Beach reflection records and shot plan. The figure is from M. Blaik, J. Northrop, and C. S. Clay, ''Some seismic profiles onshore and offshore Long Island, New York,'' *J. Geophys. Res.*, 64 (1959), 231–39. Copyright by the American Geophysical Union. (1 foot = 0.3048 m).

values of velocity and density to compute reflection and transmission coefficients at each layer. *Ignore multiple reflections.* Start with one or two layers and get your program to work first. Remember to include the effect of the transmission coefficients for the passages through each layer.

SIMPLIFIED EARTH MODEL

Layer	Thickness (m)	Interval velocity (m/s)	Density (g/cm^3)
1	90	1620	2
2	150	1740	2.2
3	270	1950	2.3
4	300	2100	2.4
5	—	5000	3.0

Figure 6.10 shows several reflection profiles taken on the Fire Islands beach (M. Blaik, J. Northrup, and C. S. Clay, "Some seismic profiles onshore and offshore Long Island, New York," *J. Geophys. Res.* 64 (1959), 231–39. The earth model is simplified from this paper.

7

Multiple Reflections

Multiple reflections are usually a nuisance. The transmission of a signal through a layer always causes a signal to become more complicated because reflections within a layer add extra components to the rear and decrease the energy in the leading part of the signal. Multiple reflections interfere with signals that come later. Still worse, a multiple reflection can look like the reflection from a deep interface.

But multiple reflections are not all bad. Suppose a formation is a sequence of thin layers where the impedence contrasts between adjacent layers are small. The ensemble of reflections and multiple reflections may add to form a nice signal from the formation. One special case caused a flurry of theoretical and experimental work. Some shallow-water areas (depths of tens of meters) have large impedance contrasts between the water and sediments below the water. Here, multiple reflections within the water layer can ruin reflection records. A few areas on land have similar problems.

7.1 THIN-LAYER REVERBERATIONS

In the 1950s geophysicists were making exploration seismic measurements in water-covered areas such as the Persian Gulf (Figure 7.1a). In some areas they got good records (Figure 7.1b), in others the records looked as if someone had turned on an oscillator, and it was impossible to identify reflections from subsurface interfaces. (Compare no-singing with high-singing records.)

On the basis of their experience with ground roll on land, geophysicists initially thought that the long-singing records were due to the horizontal transmission of waves trapped in the water layer. However, much careful field research showed that the long-singing records were due to another cause. The nearly vertically incident reflection signals had multiple reflections in the water layer, and these added to cause the singing, ringing,

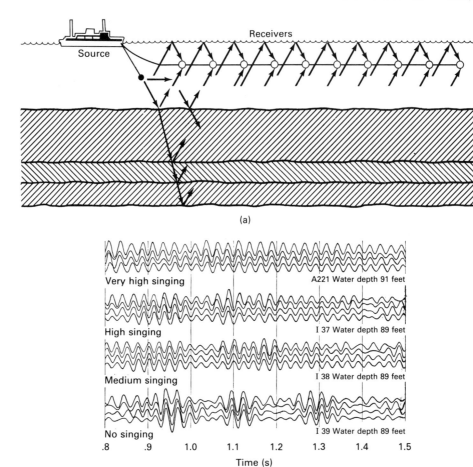

Figure 7.1 Exploration seismic measurements over a water-covered area. (a) Multi-channel marine seismic system. Commonly, a single ship tows the source and the receiving hydrophones in a long streamer. (b) Sections of seismic records with and without singing. The records were taken in an area having 27 m water depth. The receiving array was fixed, and the shot source was moved away. Each set of traces is for different shot distances. All shots are within a 1200-m distance. The automatic gain amplifiers give an average constant output signal level. The events at 0.95, 1.1, and 1.3 s are probably reflection signals. (From G. C. Werth, D. T. Liu, and A. W. Trorey, "Offshore singing-field experiments and theoretical interpretation," *Geophysics*, 24 (1959), 220–32, Fig. 1.)

or reverberation. Geophysicists and electronic engineers created special (analog) filters to remove most of the reverberation. The economic pressures to get good data were great, and they spurred a digital revolution in data recording, signal theory, and data processing.

Reverberation affects all seismic measurements. The magnitude of the effect depends on the structures at the source and receivers. The problem and solutions apply to exploration geophysics and seismological studies of natural events (earthquakes).

7.1.1 Receiver in a Thin Reverberating Layer

A thin layer is thin compared with the distances to deep reflecting interfaces and is usually less than a few acoustic wavelengths thick. The reverberations are troublesome when the upper and lower reflection coefficients for the layer are large, and the amplitudes of the multiple reflections decay slowly. The sequence of multiple reflections due to a single up-traveling plane-wave front in the layer is shown in Figure 7.2. We assume, for simplicity, that the detector is a vertical-component velocity geophone at the top of layer 1 and at the 0 or air-earth (or -water) interface. Since the waves are traveling vertically, it is convenient to spread the display of the multiple reflections by plotting time along the abscissa. The number of time steps for a one-way trip through the layer is k. We arbitrarily choose the arrival time of the signal at the top of the layer to be zero.

The geophone is at the top of layer 1, and the amplitude of the signal is proportional to $1 + R_{10}$, where 1 is the amplitude of the incident wave, and R_{10} is the reflected wave. (At a free surface $R_{10} = 1$ for particle velocity.) We call the transmission of the layer $T_{21}(z)$ and use Figure 7.2 to write the terms of the generating function or z-transform,

$$T_{21}(z) = (1 + R_{10})(T_{21})(1 + R_{10}R_{12}z^{2k} + (R_{10}R_{12}z^{2k})^2 + \cdots)$$

or

$$T_{21}(z) = T_{21} \cdot (1 + R_{10}) \sum_{n=0}^{\infty} (R_{10}R_{12}z^{2k})^n \tag{7.1}$$

where T_{21} and R_{12} are plane-wave reflection and transmission coefficients. Equation (7.1) is the sum of an infinite geometric progression having the form

$$s = \sum_{n=0}^{\infty} r^n = \frac{1}{1 - r} \tag{7.2}$$

and the reduced form of Eq. (7.1) is, for $r = R_{10}R_{12}z^{2k}$,

$$T_{21}(z) = \frac{T_{21} \cdot (1 + R_{10})}{1 - R_{10}R_{12}z^{2k}} \tag{7.3}$$

The compact form of Eq. (7.3) actually represents the infinite number of arrivals in Eq. (7.1). The rate of decay of the reverberation depends on the magnitude of $|R_{10}R_{12}|$. The

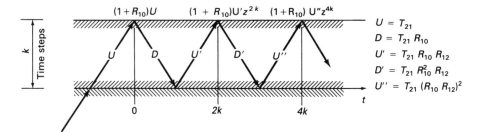

Figure 7.2 Upcoming waves incident on a near-surface thin layer. The incident wave is vertically incident. The abscissa shows time steps. The reflection coefficient at interface 1 is R_{21}, and the transmission coefficient is T_{21}.

decay is rapid for $|R_{10}R_{12}| < 0.5$ and is slow for $|R_{10}R_{12}|$ a little less than 1. The absolute value $|R_{10}R_{12}|$ is large and nearly 1 in areas where the reverberation is troublesome.

The reverberation function, $T_{21}(z)$, convolves the upcoming signal. Let the upcoming signal be the convolution of the earth $E(z)$ and signal $A(z)$. Then the signal $E(z)A(z)$ convolves with $T_{21}(z)$ to give

$$S(z) = E(z)A(z)T_{21}(z) \tag{7.4}$$

There are two ways to evaluate Eq. (7.4). The brute-force method requires multiplication of the corresponding power series in z for each generating function. In Section 7.2 we give a compact feedback technique that uses Eq. (7.3).

Examples of the dependence of the reverberation impulse response on the magnitude of $|R_{12}|$ are shown in Figure 7.3. To some extent we can identify the decaying sequences of reverberation spikes. The convolution of $E(z)\,T_{21}(z)$ with a signal $A(z)$ gives a very singing record.

So far we have limited our discussion to the reverberations at the receiver. Multiple reflections of the signal from the source give source reverberations. The layer thickness may be different. Using Eqs. (7.1) to (7.3) as models, we obtain the source generating function,

$$T_{12}(z) = \frac{T_{12}}{1 - R_{10}R_{12}z^{2h}} \tag{7.5}$$

where the source is at the top of layer 1, and h is the layer thickness in time steps. Inclusion of the source alters Eq. (7.4) to

$$S(z) = T_{12}(z)T_{21}(z)E(z)A(z) \tag{7.6}$$

An extension of this discussion to the reflection and transmission for multiple layers is given in Section B1.3 of Appendix B.

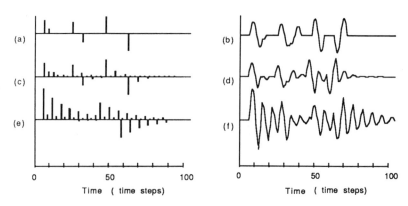

Figure 7.3 $S(z)$ for the reverberations of signal traveling up. The thickness of reverberating layer h is 6 time steps. The transmission in the reverberating layer at the receiver is $T_{21}(z)$. (a) Impulse display $E(z)$ for up-traveling signal, no reverberation. $R_{10}R_{12} = 0$. (b) $S(z)$ for a sine wave signal $A(t)$ having a 16 time-step period. $S(z) = A(z)E(z)$. (c) $E(z)T_{21}(z)$, $R_{10}R_{12} = 0.3$. (d) $S(z) = A(z)E(z)T_{21}(z)$, $R_{10}R_{12} = 0.3$. (e) $E(z)T_{21}(z)$, $R_{10}R_{12} = 0.7$. (f) $S(z) = A(z)E(z)T_{21}(z)$, $R_{10}R_{12} = 0.7$.

PROBLEMS

1. Read G. W. Werth, D. T. Liu, and A. W. Trorey, "Offshore singing: field experiments and theoretical interpretation," *Geophysics*, 24 (1959), 220–32. Concentrate on the physical descriptions of singing signals.

2. Read M. M. Backus, "Water reverberations: their nature and elimination," *Geophysics*, 24 (1959), 233–61. Backus uses both time and frequency domains to describe the cause of reverberation. As you read the paper, make an effort to understand the physical causes of the reverberation and how the reverberation affects the seismograms. Backus uses analog filters to deconvolve the seismic signals. R. F. Seriff et al. have a comment in *Geophysics* 26 (1961), 242. Do you think it is important?

3. Read F. K. Levin, "The seismic properties of Lake Maracaibo," *Geophysics*, 27 (1962), 35–47. Levin reports the results of a large experimental effort. One result, gas bubbles in the sediments, can cause the velocity reflection coefficient at the water-sediment interface to approach 1.

7.2 CONVOLUTIONS USING FEEDBACK

Sections 6.3 and 6.4 develop the convolution expression for the ideal seismogram. Section 7.1 shows what happens when the geophone and source are in reverberating surface layers. Combining the results from both sections, we begin with Eq. (7.6)

$$S(z) = E(z)T_{12}(z)T_{21}(z)A(z) \tag{7.7}$$

where

$$T_{12}(z) = \frac{T_{12}}{1 - R_{10}R_{12}z^{2h}} \tag{7.8}$$

$$T_{21}(z) = \frac{(1 + R_{10}) T_{21}}{1 - R_{10}R_{12}z^{2k}} \tag{7.9}$$

where $T_{12}(z)$ and $T_{21}(z)$ have local values of R_{10}, R_{12}, h, and k for the source and receiver positions. As mentioned earlier, the brute-force technique for computing $S(z)$ requires the expansion of $T_{12}(z)$ and $T_{21}(z)$ into infinite series and then multiplying all the series. This is a tedious task that we wish to avoid.

The *feedback* calculation simplifies the task. To demonstrate the basic technique, we use the following equation:

$$c_0 + c_1z + c_2z^2 + \cdots = \frac{1}{1 + bz} \tag{7.10}$$

where b is a constant. We multiply both sides by $(1 + bz)$ and gather the coefficients of z^n.

$$c_0 + (c_1 + bc_0)z + (c_2 + bc_1)z^2 + \cdots = 1 \tag{7.11}$$

The values of the coefficients c_n follow by equating like powers of z

$$c_0 = 1 \tag{7.12}$$

and all other coefficients are zero; thus

$$
\left.
\begin{aligned}
c_1 &= -bc_0 = -b \\
c_2 &= -bc_1 = b^2 \\
c_i &= -bc_{i-1} = (-1)^i b^i
\end{aligned}
\right\} \tag{7.13}
$$

Each coefficient depends on the value of the preceding coefficient, starting at $c_0 = 1$. The resulting series of $c_i z^i$ shows that

$$
\frac{1}{1 + bz} = \sum_{i=0}^{\infty} (-bz)^i \tag{7.14}
$$

The result is correct because this is the sum of the geometric progression of Eq. (7.2). The algorithm to compute the terms is short. Letting $B = b$, where $|b| < 1$, the loop is, using Eqs. (7.12) and (7.13),

```
B = b
I1 = i1
C(0) = 1

FOR I = 1 TO I1                                              (7.15)
    C(I) = - B * C(I-1)
NEXT I


END
```

where I1 is the maximum I of interest. A polynomial division algorithm is in Section B1.4 of Appendix B.

Evaluation of convolution (7.7) requires computation of $T_{12}(z)$ and intermediate storage in a holding array, $H(z)$. Using $S(z)$ as temporary storage, we have

$$
S(z) = E(z) T_{12}(z) \tag{7.16}
$$

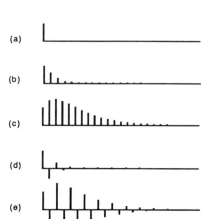

(a)

(b)

(c)

(d)

(e)

Time   (20 time steps per tic)

Figure 7.4 Impulse response of a reverberating layer. Time-step layer thickness = 6 time steps. (a) Impulse response $h = k = 6$ with no reverberation, $R_{12} = 0$. (b) $R_{10}R_{12} = +0.3$. (c) $R_{10}R_{12} = +0.7$. (d) $R_{10}R_{12} = -0.3$. (e) $R_{10}R_{12} = -0.7$.

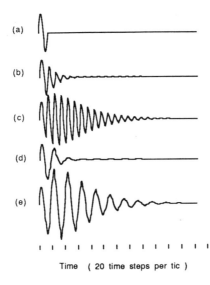

Figure 7.5 Reverberation for a single wavelet. Layer thickness $k = 6$ time steps. (a) No reverberation. (b) $R_{10}R_{12} = +0.3$. (c) $R_{10}R_{12} = +0.7$. (d) $R_{10}R_{12} = -0.3$. (e) $R_{10}R_{12} = -0.7$.

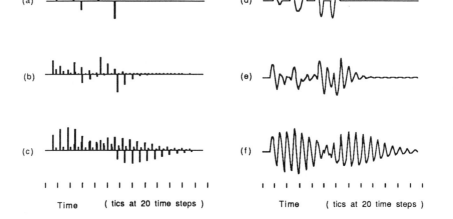

Figure 7.6 Reverberations. $h = k = 6$, and $R_{10}R_{12} =$ various values. (a) $R_{10}R_{12} = 0$. (b) $R_{10}R_{12} = 0.3$. (c) $R_{12} = 0.7$. The next set have the signal $A(z)$ in Figure 7.3 and show (d) $R_{10}R_{12} = 0$. (e) $R_{10}R_{12} = 0.3$. (f) $R_{10}R_{12} = 0.7$.

The next step is to convolve $S(z)$ and $T_{21}(z)$. The result is stored in the holding array

$$H(z) = S(z)\, T_{21}(z) \tag{7.17}$$

This is the impulse response, including source and receiver reverberations. Convolution with the source $A(z)$ gives the signal

$$S(z) = H(z)\, A(z) \tag{7.18}$$

where $S(z)$ is now the output.

Examples of the impulse response with and without $T_{12}(z) \, T_{21}(z)$ reverberation are shown in Figure 7.4 on page 90. The same responses with a signal are shown in Figure 7.5 on the preceding page.

Figure 7.6 shows the impulse response of a reflection signal for the same sequence of $R_{10}R_{12}$ coefficients as in Figure 7.5. The effects of convolving the impulse responses with the source function are also shown.

PROBLEMS

4. *Purpose*: to demonstrate the use of the feedback algorithm. Write a program to compute $A(z)/B(z)$ using Section B.1.4 of the Appendix B. Let $A(0) = 1$, $m = 1$, and $b = 0.5$. Use the program to compute the geometric sequence $(-\frac{1}{2})^i$. Let $b = -0.5$, and obtain the sequence $(\frac{1}{2})^i$. This is a program test.

5. Test your program for the impulse responses for $m = 12$ and $b = 0, 0.7, -0.7, 0.9$ and -0.9. Check your results by computing a few terms of the corresponding geometric series.

6. Write a program that does the reverberation convolution twice, following Eqs. (7.16) and (7.17).
   (a) Test the program for $R_{10}R_{12} = 0.3, 0.7, -0.3, -0.7$, and $h = k = 6$. *For simplicity*, let $T_{12}T_{21}(1 + R_{10}) = 1$ in Eqs. (7.8) and (7.9), since it is a constant multiplying factor. Compare your results with Figure 7.4 for $E(z) = z^{50}$.
   (b) What happens if you let $k = 5$ and $h = 7$? *Hints*: You may use the program of Problems 4 or 5 and enter b or $b_m = -R_{10}R_{12}$, $m = 2h$ for the receiver. On the repeated convolution $m = 2k$.

7. Advanced effort. Write and test a program to do the ratio of polynomials in z using Section B1.4 of Appendix B. Multiply $T_{12}(z)T_{21}(z)$ to obtain a single polynomial in the denominator. Run for the parameters in Problem 6 and compare for the same results.

8

Deconvolution

Deconvolution of seismic signals is an almost routine processing procedure. *Deconvolution* or *inverse filtering* means undoing a convolution. In this section we use a known structure to do what might be called "known deconvolution." The deconvolution is known in the sense that we know the generating function or z-transform of the convolving reverberation, and the deconvolution consists of undoing the reverberation.

Predictive deconvolution is a more powerful procedure that works when the convolving functions such as $T_{12}(z)$ and $T_{21}(z)$ are unknown. Predictive deconvolution uses Norbert Wiener's optimum filter theory and the statistical properties of signals. Section B1.5 of Appendix B contains a brief description of predictive deconvolution and examples.

8.1 DECONVOLUTION OPERATIONS

The previous section showed that the signal at the geophone $S(z)$ is the combination of many travel paths, including surface-layer reverberations. We express these as the convolutions of the generating functions, Eq. (7.6)

$$S(z) = T_{12}(z)T_{21}(z)E(z)A(z) \qquad (8.1)$$

where the earth $E(z)$ and the shallow-layer transmission functions for the source $T_{12}(z)$ and receiver $T_{21}(z)$ are indicated as separate convolution operations.

A deconvolution operation means undoing any one of or combination of the convolution operations. For example, solution of Eq. (8.1) for $E(z)$ gives

$$E(z) = \frac{S(z)}{T_{12}(z)\,T_{21}(z)\,A(z)} \qquad (8.2)$$

where $S(z)$, $T_{12}(z)$, $T_{21}(z)$, and $A(z)$ are known. In the real world, deconvolution operations are approximate because we do not know exact values of the functions. The effects of noise on deconvolution operations are given later in this section.

8.1.1 Deconvolution of Reverberation

Reverberation in the surface layer is a major problem, and its removal is an early major success in the art of deconvolution. When we are satisfied to remove only reverberation, the output signal is $A(z) E(z)$ and, using Eq. (8.1), is

$$A(z) E(z) = \frac{S(z)}{T_{12}(z) T_{21}(z)} \tag{8.3}$$

The substitution of Eqs. (7.3) and (7.5) for $T_{12}(z)$ and $T_{21}(z)$ gives

$$A(z)E(z) = S(z) F(z) \tag{8.4}$$

$$F(z) \equiv \frac{(1 - R_{10}R_{12}z^{2k})(1 - R_{10}R_{12}z^{2h})}{T_{12} T_{21} (1 + R_{10})} \tag{8.5}$$

where k and h are the time step equivalents of the water depths at the receiver and source. $F(z)$ is known as the *Backus filter*. The deconvolution algorithm, using the Backus filter, has the same form as (6.46). Using the following notation,

$$C(z) = F(z) S(z) \tag{8.6}$$

we obtain the expansion of Eq. (8.5),

$$F(z) = f_0 + f_{2k}z^{2k} + f_{2h}z^{2h} + f_{2k+2h}z^{2k+2h} \tag{8.7}$$

$$f_0 = \frac{1}{T_{21} T_{12} (1 + R_{10})} \tag{8.8}$$

$$f_{2k} \equiv \frac{-R_{10}R_{12}}{T_{12} T_{21} (1 + R_{10})} \tag{8.9}$$

$$f_{2h} = f_{2k} \tag{8.10}$$

$$f_{2h+2k} = \frac{(R_{10} R_{12})^2}{T_{12} T_{21} (1 + R_{10})} \tag{8.11}$$

The dereverberation algorithm is, assuming that $C(I) = 0$ initially and $S(I) = s_i$ is in memory,

$$
\begin{array}{c}
\text{M1 = 2 * H + 2 * K} \\
\text{REM N1 = LENGTH OF S(Z)} \\
\\
\text{FOR I = 0 TO M1} \\
\text{F(I) = f(i)} \\
\text{NEXT I}
\end{array}
\tag{8.12}
$$

```
FOR M = 0 TO M1
  FOR N = 0 TO N1
    I = M + N
    C(I) = C(I) + F(M) * S(N)                    (8.13)
  NEXT N
NEXT M
```

Figure 8.1 shows (a) a seismic signal without reverberation, (b) the result of reverberation in a surface layer, and (c) deconvolution using the same parameters as used for the reverberation calculation.

In experimental situations we do not know the exact depths and reflection coefficients. When $h = k$, we can guess k by measuring the time steps between peaks. From Figure 7.4 the reverberation repeats as positive impulses for $R_{10}R_{12} < 0$. Thus, the time between peaks is $2k$, and from Figure 8.1b $2k \simeq 12$. We must estimate $R_{10}R_{12}$, and we choose $R_{10}R_{12} = 0.6$. We may ignore $T_{12}T_{21}(1 + R_{10})$ because it is a constant amplitude factor, and our purpose is to simplify the waveform of the signal. The approximate deconvolution of Figure 8.1b using $k = 6$ and $R_{10}R_{12} = 0.6$ is shown in Figure 8.1d. (The exact value is $R_{10}R_{12} = 0.7$.) We can iterate other choices of k and $R_{10}R_{12}$ to improve the output.

8.1.2 Deconvolution of Source

The purpose of source function deconvolution is to determine the impulse response of the earth $E(z)$. Omitting the reverberation terms, we write

$$S(z) = E(z) A(z) \qquad (8.14)$$

The deconvolution gives

$$E(z) = \frac{S(z)}{A(z)} \qquad (8.15)$$

The evaluation uses an extended version of the feedback algorithm, Section B1.4 of Appendix B.

A source deconvolution example for three arrivals is shown in Figure 8.2. The signal is above, and the deconvolved impulse response is below. This is an exact calculation,

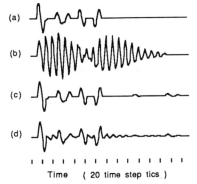

(a)

(b)

(c)

(d)

Time   (20 time step tics)

Figure 8.1 Reverberation deconvolution of synthetic seismograms. See Figure 7.6f. (a) The up-traveling arrivals. (b) Reverberation, $h = k = 6$ and $R_{10}R_{12} = 0.7$. (c) Deconvolution of (b) with $h = k = 6$ and $R_{10}R_{12} = 0.7$. (d) Deconvolution of (b) with $h = k = 6$ and $R_{10}R_{12} = 0.6$.

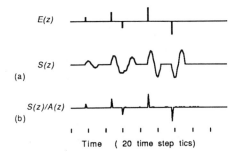

Figure 8.2 Signal deconvolution of synthetic seismograms. (a) Synthetic seismogram for several arrivals, $E(z)$ in Figure 7.6a. (b) Deconvolution for impulse response. The impulses correspond to the impulse response $E(z)$, Figure 7.6a.

and errors are due to limitations of the numerical computation. In practice, we do not have exact values for $A(z)$ and must make approximations. A general procedure uses the *Wiener least squares optimum filter theory*. A full exposition of the Wiener theory is beyond our level, but we give a brief discussion for a special case in Section B1.5 of Appendix B.

Noise

Noises from unwanted and unknown sources contaminate the signal. We can measure the noise by recording the geophone output without operating the source. The noise is

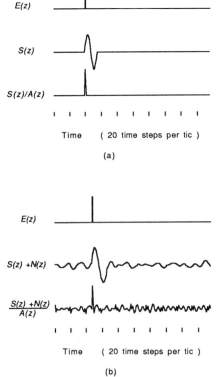

Figure 8.3 Signal deconvolution. (a) Signal and deconvolution without noise. (b) Signal and deconvolution with signal about seven times the noise amplitudes.

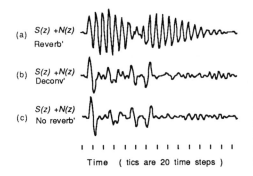

(a) $S(z) + N(z)$
    Reverb'

(b) $S(z) + N(z)$
    Deconv'

(c) $S(z) + N(z)$
    No reverb'

Time   (tics are 20 time steps)

Figure 8.4 Deconvolution of a noisy reverberating seismogram. (a) Noise added to the seismogram, Figure 8.1b. (b) Deconvolution of (a) using $h = k = 6$ and $R_{10}R_{12} = 0.7$. (c) Comparison of the same noise added to Figure 8.1a.

usually random, and although different samples have the same appearance, they are not alike. A particular record of noise is $N(z)$. For analysis we add $N(z)$ to the signal to obtain

$$S_{S+N}(z) = T_{12}(z)T_{21}(z)E(z)A(z) + N(z) \tag{8.16}$$

Deconvolution of $S_{S+N}(z)$ operates on both terms on the right side of Eq. (8.16). Since $N(z)$ comes from other sources and paths, the source and reverberation deconvolutions of $N(z)$ give noise out. If the noise amplitudes are small compared with the signal, we expect the convolutions to give good results.

Basic programming languages usually have an instruction such as RND(N) for generating a sequence of random numbers. These number sequences can be convolved or filtered to make imitations of seismic noise. For example, let $RN(z)$ be a sequence of random numbers and $W(z)$ be the Ricker wavelet, Section B1.3 of Appendix B. A filtered random signal is $N(z) = W(z)RN(z)$. For the examples we add ''seismic noise'' to the synthetic seismograms prior to deconvolving them.

An example of the signal deconvolution for a single arrival is shown in Figure 8.3 with and without added noise. For this small amount of noise, signal amplitude about seven times the noise amplitude, the arrival pulse is clearly evident. The output signal is about three to four times the noise.

For another example we add noise to the reverberating seismogram of Figure 8.1b to obtain Figure 8.4a. The deconvolution of Figure 8.4a gives Figure 8.4c. For comparison, Figure 8.4c shows the same noise added to the seismogram of Figure 8.1a.

PROBLEMS

1. Problems 4, 5, and 6 of Chapter 7 are forward problems for convolving signals with reverberation. Deconvolution undoes the convolution. Write a deconvolution program to remove the reverberation due to the surface layer. Test the program by multiplying the results of Problem 4 by $1 + bz$. It should give 1.

2. Test your deconvolution program on Problem 5. Again it should yield unity.

3. Deconvolve the results of problem 6. You may multiply the z-polynomials $T_{12}(z)$ and $T_{21}(z)$ into one polynomial.

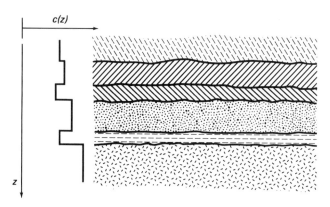

Figure 8.5 Formation of layers between thick uniform formations.

8.2 REFLECTION AT A THIN LAYER: ELEMENTS OF SEISMIC STRATIGRAPHY

The structures in sedimentary basins often consist of layers of sandstones, shales, limestones, and the like. Commonly, the interfaces are approximately parallel for distances of the order of a kilometer or more and many seismic wavelengths. The thicknesses of the layers range from a fraction of a wavelength to many wavelengths.

Almost all the events we call "reflections" on reflection records are actually the unresolved sum of reflections from a formation of thin layers between thick uniform layers, Figure 8.5. We call this the *formation reflection*. Since the formation may be the same over distances of several kilometers, the sequence of peaks and troughs of the formation reflection has an identifiable character. The identification and mapping of a formation reflection having a particular character is the beginning of seismic stratigraphy. The correlation of a formation reflection to a geological formation can require a lot of work.

We use a thin layer between two thick layers for examples. The derivation of the reflection coefficient of the layer is essentially the same as the reverberation derivation in Section 7.1 and is given in Section B1.3 of Appendix B. Using the notation in Figure 8.6, we derive the thin-layer reflection coefficient (note, $-R_{21} = R_{12}$),

$$R_{13}(z) = \frac{R_{12} + R_{23}z^{2k_2}}{1 + R_{12}R_{23}z^{2k_2}} \qquad (8.17)$$

where k_2 is the thickness of the layer in time steps. Algorithms for computing $R_{13}(z)$ appear in Sections B1.3 and B1.4 of Appendix B. As with reverberations in a surface layer, the magnitude of $|R_{12}R_{23}|$ controls the decay of the multiple reflection terms.

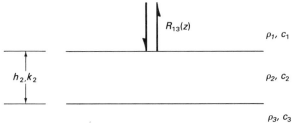

Figure 8.6 Thin-layer geometry. The thickness is k_2 time steps.

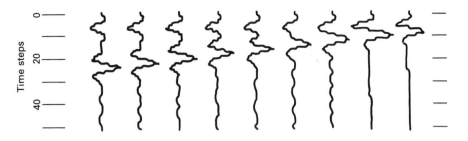

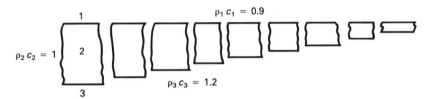

Figure 8.7 Effect of changing thickness of a layer. The up-going signal is observed at the top of layer 2. The signal is a Ricker wavelet. The thickness varies linearly from 1 to 9 time steps. For simplicity we give the relative values of the seismic impedances $\rho_1 c_1$, $\rho_2 c_2$, and $\rho_3 c_3$.

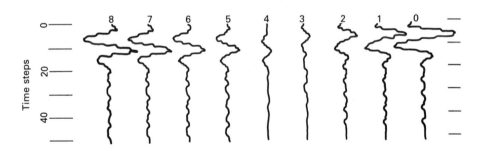

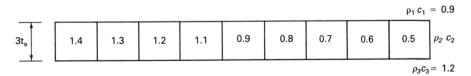

Figure 8.8 Effect of changing impedance in a constant time-step thickness layer. The up-going signal is observed at the top of layer 2. The signal is a Ricker wavelet. The relative impedance of the middle layer varies from 1.4 to 0.5.

Our first example is the *pinch out* or thick layer that thins as one makes a seismic section. The thinning of the layer, Figure 8.7, is gradual, and the interfaces are nearly parallel at each source-receiver location. The reflections from the bottom and top of the layer are resolved on the left side of the figure and unresolved on the right side. If the layer were to pinch out and end, the ending or discontinuity would cause diffraction arrivals. These arrivals are not included in Figure 8.7, for simplicity. Diffractions are discussed in Chapter 15.

The second example is shown in Figure 8.8. The impedance of the formation changes from a high impedance on the left side to a low impedance on the right. This simulates a formation that changes from one of low porosity $\rho_2 c_2 = 1.4$ to gas-filled pores at high porosity $\rho_2 c_2 = 0.5$. The reflections from the top and bottom interfaces are not resolved. A simple interpretation that ignores amplitude gives a down-dipping interface from locations 8 to 4, a discontinuity between 4 and 3, and a level interface between 3 and 0. This is a difficult problem because the eye follows a phase, that is, a peak or valley across the section. The large change of amplitude and change of phase hint that the impedance of the formation may be changing along the section.

These two examples are from an infinite number of geological possibilities.

PROBLEMS

Purpose: to demonstrate the dependence of the character of a reflection on formation parameters.

4. Write a program that computes the vertical incidence reflection coefficient given $\rho_1 c_1$, $\rho_2 c_2$, ρ_3, c_3, and h_2 or k_2 for a thin layer. The details and an algorithm are given in Section B1.3. Note that the reflection coefficients depend on the ratios of $\rho_1 c_1$, $\rho_2 c_2$, and $\rho_3 c_3$.

5. Study the effect of wavelet "breadth" by convolving your reflection algorithm with the Ricker wavelet. Use the parameters in Figure 8.7 and test your program by letting $w_0 = 3$ and $I_0 = 6$. Let the thickness vary from 9 time steps to 1 time step. Make runs with $w_0 = 2$ and $I_0 = 6$, $w_0 = 6$ and $I_0 = 12$.

6. Repeat the type of computations in Problem 5 for $\rho_1 c_1 = 1$, $\rho_2 c_2 = 0.8$, and $\rho_3 c_3 = 1$.

7. Repeat the type of computations in Problem 5 for the case in Figure 8.8.

8. Repeat the computations in Problem 7 for other cases such as $\rho_1 c_1 = 1$, $\rho_2 c_2$ variable, $\rho_3 c_3 = 1$.

9. Do the computations in Problem 8 for $k_2 = 9$ time steps. Are the reflections resolved? Do multiple reflections become important?

10. For advanced students, reflection from two thin layers.
    (a) Substitute Eq. (B.52) in Eq. (B.53) and clear the algebra to obtain $R_{14}(z)$ as the ratio of two polynomials.
    (b) Write a program using the ratio of polynomials algorithm in Appendix B1.4 to calculate $R_{14}(z)$. Find seismic velocity well logs in the geophysical literature for a pair of thin beds and compute synthetic reflections for $R_{14}(z)$ using Ricker wavelets having different breadths.

9

Frequency Response and Filters

The specifications of practically all analog equipment are given as input-output measurements using sine-wave signals. The basic measurement is shown in Figure 9.1. The source is a sine-wave oscillator, and the oscilloscope (or voltmeter and phase meter) measures the output as a function of frequency. It is customary to treat the instrument as a "black box" and to ignore how components are connected inside. Linear devices have an output amplitude that is proportional to the input amplitude. For sine-wave input, the output has the same frequency as the input sine-wave.

Many black boxes are filters that pass the signals in one frequency range and attenuate signals in other frequency ranges. Examples of filter responses are shown in Figure 9.1. The frequency response must be chosen to pass the seismic signals.

9.1 SIGNAL SPECTRA

Although it is natural to use signals in the time domain, (i.e., generating functions), as in the preceding sections, the alternative spectral or frequency description is convenient in discussions of equipment response and theoretical derivations. The details of how to replace a seismic signal with the sum of the harmonics of sine and cosine waves are given in Section B1.2 of Appendix B. The time-domain descriptions and the frequency-response descriptions are interchangeable because they are (Fourier) transformations of each other. The two descriptions are analogous to different languages. A frequency-response description is known as the *spectral representation* or the *signal spectrum*.

Two rules for approximating the spectrum are sufficient in our discussion. The rules come from the comparisons of seismic signals and their spectra. The first rule is the relationship of the frequency bandwidth of the signal to its duration.

Rule 1:

$$f_{BW} t_{\min} \geq 1 \tag{9.1}$$

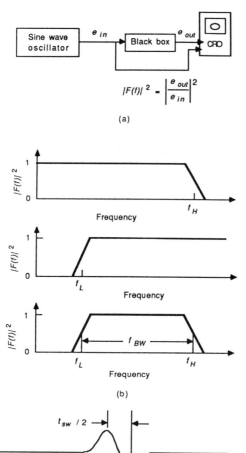

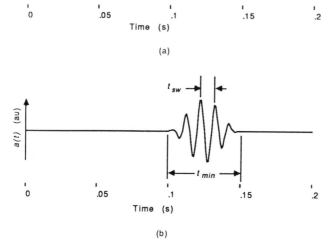

Figure 9.1 Frequency-response measurement. (a) "Black box" represents the instrument under test. Black box is the traditional name of electronic devices that do something to a signal. (b) Low-, high-, and band-pass filter responses. f_{BW} is the bandwidth. The exact definitions of f_L and f_H as the frequencies corresponding to half-amplitude, half-power, and so forth, vary.

Figure 9.2 Periods and durations of seismic waves, two examples. (a) Duration = 0.05s, period = 0.05s. (b) Duration = 0.05s, period = 0.01s.

where

$$f_{BW} \equiv f_H - F_L \qquad (9.2)$$

f_H is the highest-frequency component (sine wave) having appreciable amplitude, and f_L is the lowest-frequency component having appreciable amplitude. t_{min} is the minimum duration of the signal (Figure 9.2). We can use the measurements of signal duration to estimate the bandwidth, f_{BW}.

The second rule uses an estimate of the period of the seismic signal t_{SW} to estimate the peak frequency of the spectrum, f_{avg}, Figure 9.2.

Rule 2:

$$f_{avg} \simeq \frac{\text{number of cycles}}{t_{min}}$$

and $\qquad\qquad\qquad\qquad\qquad\qquad\qquad\qquad\qquad\qquad\qquad (9.3)$

$$f_{avg} \simeq \frac{(f_H + f_L)}{2}$$

The combination of the two rules gives

$$f_H \simeq f_{avg} + \frac{f_{BW}}{2}$$

$$f_L \simeq f_{avg} - \frac{f_{BW}}{2} \qquad (9.4)$$

Comparisons of the crude spectral estimates and numerical computation for a seismic wavelet are shown in Figure 9.3.

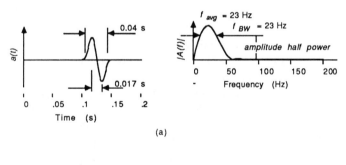

(a)

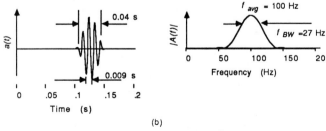

(b)

Figure 9.3 Seismic wavelets: the spectrum and crude spectral estimates. The $|A(f)|$ are modulus of the spectral components.

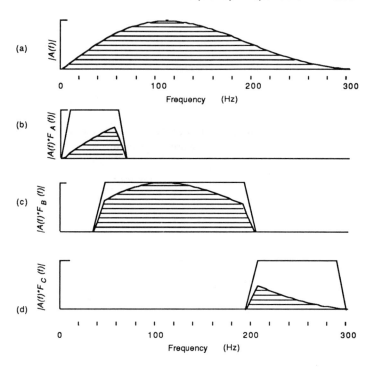

Figure 9.4   Signal $A(f)$ and three filters. The shaded areas are the outputs $A(f) \cdot F_A(f)$, $A(f) \cdot F_B(f)$, and $A(f) \cdot F_c(f)$.

9.2 FILTERS

The filter passes or attenuates the frequency components of the signal as shown in Figure 9.4. Filter B passes most of the spectral components of the signal, whereas filters A and C do not. The output of the filter is the multiplication of the signal and filter responses.

      Band-pass filters are important in seismology because they can sometimes separate one kind of arrival from another. For example, ground roll often has much lower frequencies than the reflection signal, Figure 9.5. The band-pass filter almost eliminates the ground roll and passes the reflection signal nicely.

PROBLEMS

1. Recall Problem 5 of Chapter 2. Ricker derived an expression for a seismic wavelet. He related the frequency spectrum of the wavelet to its breadth.
   (a) Using Ricker's data, give a relationship between the peak of the spectrum and wavelet breadth.
   (b) Using the approximate expressions Eqs. (9.1) and (9.2), use Ricker's seismograms of wavelets to estimate the peak and widths of the spectra of wavelets.
2. Recall Problem 6 of Chapter 2 and the data of McDonald et al. Using expressions (9.1) and (9.2), use their seismograms to estimate the peaks and widths of the wavelet spectra.

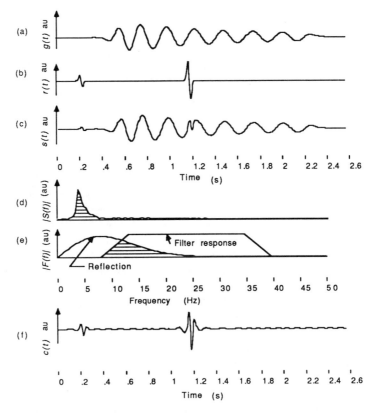

Figure 9.5 Use of a filter to reduce ground roll, $g(a)$, and enhance a reflection $r(t)$.

Unit II—Suggested Readings

FELLER, W., *An Introduction to Probability Theory and Its Applications*. New York: John Wiley, 1957. Has a chapter on the use of generating functions in probability.

KANASEWICH, E. R., *Time Sequence Analysis in Geophysics*, 3d ed. Edmonton, Alberta, Canada: University of Alberta Press, 1982. Gives an excellent overall discussion of modern geophysical data processing and interpretation. The techniques apply to all kinds of geophysical measurements.

RICE, R. A., "Inverse convolution filters," *Geophysics*, 27 (1962), 18. Uses Laplace generating functions in a least squares computation of inverse filters. Considers realizability of digital filters.

ROBINSON, E. A., "Predictive decomposition of seismic traces," *Geophysics*, 22 (1957), 767–78. Gives the philosophy behind predictive filter research and statistical estimation. Equation (1) in this paper is a Laplace generating function.

ROBINSON, E. A., *Multichannel Time Series Analysis with Digital Programs*. San Francisco: Holden-Day, 1967. Contains analytical discussions of the application of Wiener's optimum filter theory to seismic exploration.

ROBINSON, E. A., "Predictive decomposition of time series with application to seismic exploration," *Geophysics*, 32 (1969), 418–84. Use z-transforms in an elegant and complete development of optimum filter techniques.

ROBINSON, E. A., AND S. TREITEL, *Geophysical Signal Analysis*. Englewood Cliffs, N.J.: Prentice Hall, 1980. Gives a summation of Robinson and Treitel's numerous papers on the analysis of seismic reflection signals.

WADSWORTH, G. P., E. A. ROBINSON, J. G. BRYAN, AND P. M. HURLEY, "Detection of reflections on seismic records by linear operators," *Geophysics*, 18 (1953), 539–86. Explains the philosophy behind the use of prediction filtering.

WEBSTER, G. M, ED, *Deconvolution*. Tulsa, Okla.: Society of Exploration Geophysics, 1978. Contains Robinson (1969), Wadsworth et al. (1953), Rice (1962), and many other excellent papers.

Unit III

SIGNAL ENHANCEMENT

Signal enhancement is the art of making what we want to see more apparent than everything else. The fictional detective solves the crime by discarding the red herrings and selecting the real clues. Geophysicists have the same problem. We have to guess the signal or structure and choose or invent enhancement procedures to better display the signal or structure. The need for better enhancement methods has been a driving force in seismic research for decades.

Starting in the late 1920s and through the 1940s, geophysicists and engineers developed reliable geophysical amplifiers and recording cameras. The number of channels went from 6 to 24. People found that the simple increase from 6 to 24 channels greatly improved their ability to interpret the data. Reflection methods got most of the research effort because high-resolution reflection techniques were needed to locate oil-bearing structures. By 1950 amplifier and filter improvements still did not enable geophysicists to get usable records in many areas. Geophysicists added a new domain to their filters by using arrays of geophones on the surface. The purpose of the arrays was to receive the vertically traveling reflection signals and reject the horizontally traveling surface waves. Multiple geophone arrays worked very well and tremendously improved the quality of the records. Throughout this period most seismic data were recorded on paper and were not electrically reproducible. The selected readings contain a list of papers with examples of the history.

The development of multichannel magnetic tape recorders for seismic systems gave the geophysicist electrically reproducible records. The analog recording and replay systems were replaced by digital recorders and computers. Since the 1960s, geophysicists have been using digital computers to process their data in both time and space domains.

Most of this unit is about signals and noise, signals and interferences, and ways to combine the signals in time and space, that is, *stacking*, to enhance what we hope to see.

10

Random Signals and Sources

Random signals are present in all measurements on the earth or in the laboratory. We use the term *random signals* to describe the outputs of sensors that are not instrument noise. For example, random signals may be due to nearby stomping mules, distant earthquakes, highway noise, wind, or rain. Random signals come from many directions, have different characteristics, and arrive at seemingly unrelated times. Collectively, random signals are often called *ambient noise* to distinguish them from instrument noise. Random signals are discussed in the literature as the analysis of random time series and are the subject of many texts. This section is a brief introduction.

The simplest random sequence is a series of coin tosses. The amplitudes are 1 or 0 for heads or tails. For an unbiased, or "honest," coin, both 1 and 0 are equally likely. In this sequence the result of the next toss is completely independent of how many heads or tails have been tossed previously.

10.1 RANDOM SIGNALS

Computers usually have an instruction or subroutine to produce a sequence of random numbers. Some random number algorithms repeat the sequence after a great number have been given. The particular instruction sets vary, but most algorithms can be set at the same place in the sequence or at random locations, as desired. Program debugging is simpler when a sequence of random numbers can be repeated on command.

A common instruction in BASIC language is

$$x = RND(1) \tag{10.1}$$

It gives a new random number each time it is called. Commonly, the random number algorithm generates a random number x, where $0 \le x < 1$. The probability density function is uniform,

$$P(x, x + \Delta x) = \text{constant} \tag{10.2}$$

where P is the probability that x is between x and $x + \Delta x$. The mean value of a very large sample of random numbers tends to 0.5. The statement

$$x = \text{RND}(1) - 0.05 \tag{10.3}$$

shifts the set of x to a mean of zero, and the mean of x is

$$\langle x \rangle = \frac{1}{n_1} \sum_{n=0}^{n_1-1} [\text{RND}(1) - 0.5]$$

$$\langle x \rangle \to 0 \qquad \text{as } n_1 \to \infty \tag{10.4}$$

The mean square of x^2 is

$$\langle x^2 \rangle = \frac{1}{n_1} \sum_{n=0}^{n_1-1} [\text{RND}(1) - 0.5]^2$$

$$\langle x^2 \rangle \to \sigma^2 \qquad \text{as } n_1 \to \infty \tag{10.5}$$

where the mean is 0.5 and σ is the standard deviation for the infinite data set. For other distributions we write the mean value of x as follows:

$$\langle x \rangle = \frac{1}{n_1} \sum_{n=0}^{n_1-1} x_n \qquad \text{as } n_1 \to \infty \tag{10.6}$$

and the mean square as

$$\langle x^2 \rangle = \frac{1}{n_1} \sum_{n=0}^{n-1} x_n^2 \qquad \text{as } n_1 \to \infty \tag{10.7}$$

The standard deviation or fluctuation about the mean σ is

$$\sigma^2 = \frac{1}{n_1} \sum_{n=0}^{n_1-1} (x_n - \langle x \rangle)^2 \qquad \text{as } n_1 \to \infty \tag{10.8}$$

Squaring operation and reducing gives

$$\sigma^2 = \frac{1}{n_1} \sum_{n=0}^{n_1-1} x_n^2 - \langle x \rangle^2 \qquad \text{as } n_1 \to \infty \tag{10.9}$$

Although Eqs. (10.8) and (10.9) are the same mathematically, computational round-off errors can be much larger for Eq. (10.9) than for Eq. (10.8) if $\langle x \rangle$ is large compared to the fluctuations of x_n.

The choice of finite n_1 gives estimates of $\langle x \rangle$ and σ. If we choose an n_1 and make repeated estimates of $\langle x \rangle$ and σ, the estimates fluctuate. An example is shown in Figure

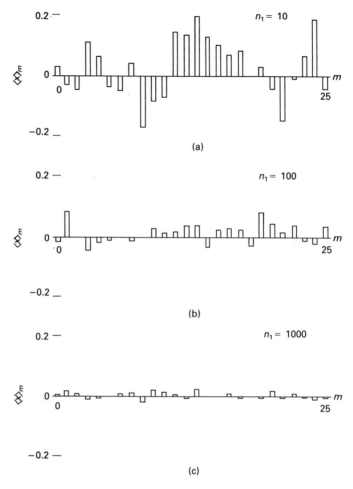

Figure 10.1   Averages of sets of random numbers. $-0.5 < x < 0.5$, $\langle x \rangle = 0$ for very large sample. Each value of $\langle x \rangle_m$ is the average of n_1 numbers. $n_1 = 10, 100, 1000$.

10.1. The estimates are labeled $\langle x \rangle_m$, where 0 is the zeroth trial, 1 is the first trial, and so forth. The figure shows the decrease of the fluctuations of the estimate of $\langle x \rangle$ as the averaging number n_1 increases. Elementary statistics texts show that the fluctuations decrease as $n_1^{-1/2}$. Fluctuations of the estimate of σ also decrease as $n_1^{-1/2}$ increases.

PROBLEMS

Purpose: to experiment with the properties of random numbers and incidentally the random number instruction. Commonly, the BASIC instruction RND(1) gives a new sequence each time it is called. The program

```
        REM Microsoft BASIC Complier: Macintosh
        REM INITIAL RANDOMIZATION STEP

        RANDOMIZE TIMER

        X = RND(1)

        FOR N = 0 TO N1
          X(N) = RND(1)
          PRINT X(N)
        NEXT N

        END
```

$$(10.10)$$

gives a new set of X(N) each time it is run. The instruction (APPLE computer) RND(-1) resets the random number algorithm to give the same sequence of random numbers each time the application program is run. The statements are

```
        X = RND(-1)

        FOR N = 0 TO N1
          X(N) = RND(1)
          PRINT X(N)
        NEXT N
```

$$(10.11)$$

In APPLE II BASIC, RND(-1) is equivalent to using the same seed number. Each time the program is run, it gives the same sequence of numbers. Let N1 $= 10$ and test these instructions before proceeding. The problems are intended to suggest experiments, not to limit or exhaust you.

1. Write a program to make a sequence of estimates of the means $M(m) \equiv \langle x \rangle_m$, where $\langle x \rangle_m$ is the mth average of n_1 random numbers. Let the number of estimates $m_1 = 25$. Compute sets of $M(m)$ for $n_1 = 4, 16, 64,$ and 256. Use Eq. (10.1), $x = $ RND(1), so that the mean for infinite n_1 tends to 0.5.

2. Use the set of $M(m)$ from Problem 1 to compute the standard deviation or σ_m of the fluctuations of $M(m)$ about $\langle M(m) \rangle$, where the average is over m_1. Plot a graph of σ_m as a function of n_1. Is σ_m proportional to $n_1^{-1/2}$?

3. Let $x = $ RND(1) $- 0.5$ and repeat Problems 1 and 2. The purpose is to demonstrate the effects of shifting the mean. Is σ_m nearly the same as in Problem 2? Does σ_m decrease as $n_1^{-1/2}$?

4. Examine the fluctuations of estimates of σ^2 by making a sequence of estimates using either Eq. (10.5) or Eq. (10.8) and computing the standard deviation of the σ^2 about their mean. Do this for $n_1 = 4, 16, 64,$ and 256. Does the standard deviation σ_s decrease as $n_1^{-1/2}$?

10.2 RANDOM-SIGNAL OR TIME-SERIES ANALYSIS

Random signals or time series are often frustrating to look at because sometimes they appear to make sense and sometimes they do not. Random signals are usually the result of a random source driving some kind of a process or in a general sense a filter (Figure

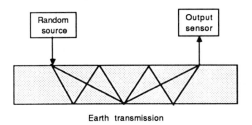

Figure 10.2 Random source driving an "earth transmission filter." Each ray represents numerous arrivals.

10.2). Since the filter does not change, it is deterministic and within limits of resolution can be measured. Occurrences of earthquakes all over the earth are unpredictable, and collectively they can be considered random-signal sources. The reverberations of large earthquakes and the numerous small earthquakes contribute to the random signal or ambient noise. Random signals are a combination of random sources and deterministic transmissions. A purpose of random-signal analysis is to measure or more properly estimate the deterministic attributes of random signals. Our next task is to give an analysis method for random signals.

10.2.1 Covariance

The *covariance* measurement is a traditional and powerful way to analyze random signals or time series. This part of the discussion is general and applies to many kinds of sequences of data points (or samples) in time or space domains. For generality, the variables are x_n, which may be replaced by any convenient symbols. A summation expression for the covariance is

$$\langle x \rangle_m = 0$$

$$c_j = \frac{1}{n_1} \sum_{n=-n_1/2}^{n_1/2-1} x_n\, x_{n-j} \tag{10.12}$$

$$n_1 \to \infty$$

where x_n are members of an infinite sequence having a mean value of zero, and the summation has n_1 terms. c_j is the covariance for the "lag" j. As in the preceding section, finite n_1 gives an estimate of c_j. The equivalent integral is

$$c(\tau) = \frac{1}{T_1} \int_{-T_1/2}^{T_1/2} x(t)\, x\,(t - \tau)\, dt \tag{10.13}$$

As necessary for a particular problem, the limits of Eqs. (10.12) and (10.13) may be changed to span any sequence of n_1 terms or duration of T_1.

The *normalized covariance* is also called the *correlation*. The correlation is

$$c_{Rj} = \frac{1}{n_1 c_0} \sum_{n=-n_1/2}^{n_1/2-1} x_n\, x_{n-j} \tag{10.14}$$

$$c_0 = \frac{1}{n_1} \sum_{n=-n_1/2}^{n_1/2-1} x_n^2 \tag{10.15}$$

To some extent the terms covariance and correlation are used interchangeably. The concept is much more important than the particular normalization.

When we measure the covariance of a sequence of numbers, we are asking the question, Are there any relationships of this number to any preceding numbers? Normally a random number generator gives a sequence of numbers that presumably do not have

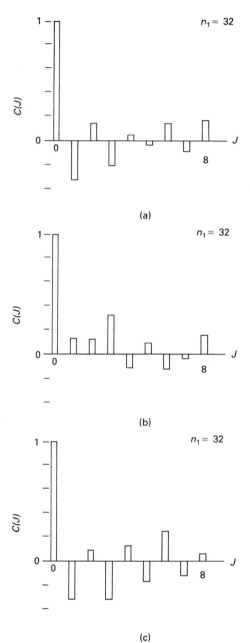

Figure 10.3 Covariances of sequences of random numbers. The covariances are normalized. All sequences have $n_1 = 32$ and a maximum lag of $j_1 = 8$.

any relationship. Thus, the correlation of a number to any preceding numbers ought to be zero. Examples of three trials of correlation computations for $n_1 = 32$ are shown in Figure 10.3. The normalization gives $c_0 = 1$, and the rest of the c_j fluctuate in the range of ± 0.2. Computations of c_j for finite n_1 have fluctuations of the order of $n_1^{-1/2}$. Here, n_1 is 32, and $n_1^{-1/2}$ is about 0.18. The three trials show that the fluctuations for any particular j are different.

Many signal-processing computations involve the sums of short sequences of random numbers and the root mean square (rms) fluctuations of the sum. We know that the sum of random numbers having a mean value of zero tends to zero as n_1 tends to infinity. The mean square of the sum of n_1 random numbers is

$$\frac{1}{n_1} \left(\sum_{m=0}^{n_1-1} x_m \right)^2 = \frac{1}{n_1} \sum_{m=0}^{n_1-1} \sum_{n=0}^{n_1-1} x_m x_n \tag{10.16}$$

where the square is written as the product of summations. The right side has the form of the covariance expression. Since the covariance terms for $j \neq 0$ or $m \neq n$ tend to zero, we write Eq. (10.16) as two summations,

$$\frac{1}{n_1} \left(\sum_{m=0}^{n_1-1} x_m \right)^2 = \frac{1}{n_1} \sum_{m=0}^{n_1-1} x_m^2 + \frac{1}{n_1} \sum_{n \neq m} x_m x_n \tag{10.17}$$

The second summation on the right, the fluctuation term, tends to zero for large n_1. The first term on the right is the major part of the estimate. If we drop the second summation and recall the definition of σ^2 (Eqs. (10.8) and (10.9)), the estimate of the mean square of the sum of random numbers is

$$\sigma^2 \simeq \frac{1}{n_1} \left(\sum_{m=0}^{n_1-1} x_m \right)^2$$

$$\sigma^2 \equiv \frac{1}{n_1} \sum_{m=0}^{n_1-1} x_m^2 \tag{10.18}$$

We often see the statement that "random numbers add as the sums of squares." They do so in the context of eqs. (10.17) and (10.18). This type of addition is also known as *incoherent addition*.

Covariance using generating functions

A procedure for computing covariances can be demonstrated by equating the product of $X(z^{-1})$ and $X(z)$ to $C(z)$, where, as in Eq. (10.12), the x_m are members of an infinite sequence and

$$X(z) \equiv \cdots + x_0 + x_1 z + x_2 z^2 + \cdots \tag{10.19}$$

$$X(z^{-1}) \equiv \cdots + x_0 + x_1 z^{-1} + x_2 z^{-2} + \cdots \tag{10.20}$$

$$C_X(z) \equiv c_{-j_1} z^{-j_1} + \cdots + c_{-1} z^{-1} + c_0 + c_1 z^1 + \cdots + c_{j_1} z^{j_1} \tag{10.21}$$

and then writing

$$C_X(z) = \frac{X(z)\, X(z^{-1})}{n_1} \tag{10.22}$$

The substitution of Eqs. (10.19) and (10.20) in the right side of Eq. (10.22) gives

$$C_X(z) = \frac{1}{n_1} \sum_{m=-n_1/2}^{n_1/2-1} x_m z^m \sum_{n=-n_1/2}^{n_1/2-1} x_n z^{-n} \tag{10.23}$$

Moving the summation symbols and changing the summation indices to j and m so that

$$j = m - n$$

gives

$$C_X(z) = \frac{1}{n_1} \sum_j \sum_{m=-n_1/2}^{n_1/2-1} x_m\, x_{m-j}\, z^j \tag{10.24}$$

The substitution of Eq. (10.21) on the left side of Eq. (10.24) and equating coefficients of like powers of z gives Eq. (10.12). This proves that we may use Eq. (10.22) to write the covariance.

Proof of the symmetry of the covariance, $c_j = c_{-j}$, uses the same technique. We replace z with z^{-1} in Eq. (10.22),

$$C_X(z^{-1}) = \frac{X(z^{-1})\, X(z)}{n_1} \tag{10.25}$$

The multiplications $X(z^{-1})\, X(z)$ and $X(z)\, X(z^{-1})$ give the same result on the right side, so

$$C_X(z^{-1}) = C_X(z) \tag{10.26}$$

Substituting Eq. (10.21) on the right side of Eq. (10.23), a similar expansion for $C_X(z^{-1})$ on the left side, and equating the coefficients of like powers of z gives

$$c_j = c_{-j} \tag{10.27}$$

A convenient algorithm for calculating c_j follows:

```
FOR J = 0 TO J1
   S = 0

   FOR M = J TO N1 + J
      S = S + X(M) * X(M - J)            (10.28)
   NEXT M

   C(J) = S / N1
NEXT J
```

The algorithm assumes that the $n_1 + j_1$ terms are in the time series. This completes our general discussion of covariance and correlation functions.

10.2.2 Covariance of Surface-Layer Reverberations

The generating function of the covariance is a powerful way to examine random seismic signals. From Section 6.4 and Eq. 6.39 the seismic signal is

$$S(z) = E(z)A(z) \tag{10.29}$$

Let $A(z)$ be a random-source signal having a mean value of zero and a mean square value of σ^2

$$\langle A(z) \rangle = 0 \tag{10.30}$$

$$\langle A^2(z) \rangle = \sigma^2 \tag{10.31}$$

$E(z)$ is the earth response function, and $S(z)$ is the signal at the geophone. The purpose of our development is to use the random signal $S(z)$ to make some measure of $E(z)$.

The covariance of $S(z)$ is, on replacing $X(z)$ by $S(z)$,

$$C_s(z) = \frac{S(z)S(z^{-1})}{n_1} \tag{10.32}$$

The substitution of $A(z)E(z)$ for $S(z)$ gives

$$C_s(z) = \frac{E(z)E(z^{-1})A(z)A(z^{-1})}{n_1} \tag{10.33}$$

after reordering the products of the z-polynomials. Both $E(z)E(z^{-1})$ and $A(z)A(z^{-1})/n_1$ are covariances.

$$C_E(z) = E(z)E(z^{-1}) \tag{10.34}$$

$$C_A(z) = \frac{A(z)A(z^{-1})}{n_1} \tag{10.35}$$

Substitution of Eqs. (10.34) and (10.35) in Eq. (10.32) gives

$$C_s(z) = C_E(z)C_A(z) \tag{10.36}$$

The *covariance* of the random seismic signal has the form of the *convolution* of covariances of the earth response and source functions.

The construction of a synthetic seismogram using covariance signal processing follows the same technique as used earlier in Chapter 6. We assume an incident signal $A(z)$, a structure $E(z)$, and convolve them to get the signal $S(z)$ (Figure 10.4). Here $A(z)$ is a random signal and correspondingly $S(z)$ is also a random signal. Figure 10.4 shows the first 0.5 s of the random signals that have 16 s duration. The rest of the signals have the same character as the first 0.5 s. The normalized covariance of $S(z)$ is $C_s(z)$ (Figure 10.4e). The covariance looks like a reflection record. The filled peaks and valleys correspond to the reflection terms in $E(z)$.

The quality of covariance signal processing depends on the number of independent trials or duration of the signal relative to width of the covariance function. This width is a measure of the non-zero duration of the covariance. In Figure 10.5 we show synthetic seismograms for a signal having 1 s duration. All other parameters, including the covariance

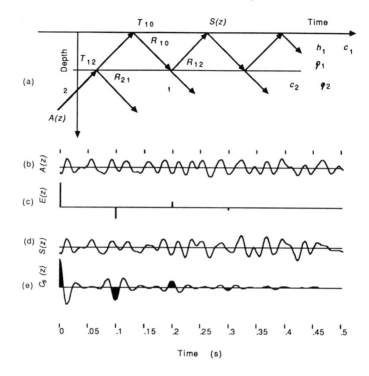

Figure 10.4 Synthetic example of covariance signal analysis for a time series, 16 s duration. The random signal $A(z)$ is vertically incident on a layer. The section is $h_1 = 25$ m, $c_1 = 500$ m/s, $\rho_1 = 1.5 \times 10^3$ kg/m^3, $c_2 = 1000$ m/s, $\rho_2 = 2.0 \times 10^3$ kg/m^3. The sampling interval is $t_0 = 0.0025$ s. The wavelet breadth for the Ricker wavelet is 0.025 s. For display, only the first 0.5 s of the signals are shown. (a) Section. The waves are vertically incident and the abscissa indicates time. (b) The incident random signal $A(z)$ is the convolution of a Ricker wavelet and a random sequence of numbers. (c) Impulse response for the direct and first 3 multiple reflections, $E(z)$. (d) Signal observed at the top of the section, $S(z)$. (e) $C_s(z)$ is the normalized covariance or autocorrelation of $S(z)$. The first 0.05 s can be used as an approximation for the covariance of the incident signal $C_A(z)$.

width, are the same. Comparisons of Figure 10.4e and Figure 10.5d show a significant reduction in the quality of $C_s(t)$ when the signal duration is reduced to 1 s.

Recalling Eq. (10.9), the mean square fluctuation σ^2 is proportional to $1/n_1$, where n_1 is the number of independent trials. Since n_1 is proportional to signal duration, the change from 16 s duration to 1 s duration signals ought to make the fluctuations about 4 times larger in Figure 10.5. In Figure 10.4e, the first three events are clearly above the noise while only the first 2 events are above the noise in Figure 10.5d.

Often the signals and reverberations come from earthquakes or "events," and the durations of the signals are short compared with the time between earthquakes or events. Since each event is independent of the others, one can treat them as random sources and average the autocorrelation or covariances. Let the jth event have the covariance $C_{s,j}(z)$. The expansion is

$$C_{s,j}(z) = c_{s,j,0} + c_{s,j,1}\, z + \cdots \qquad (10.37)$$

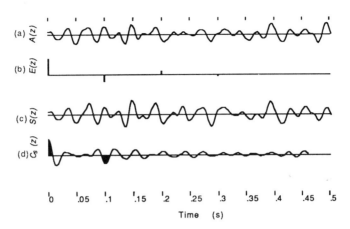

Figure 10.5 Synthetic example of covariance signal analysis of a time series 1 s duration. All parameters are the same as in Figure 10.4. (a) Incident random signal $A(z)$. (b) Impulse response $E(z)$. (c) Signal $S(z)$. (d) Normalized covariance $C_s(z)$.

The average over a set of j_1 covariances is

$$\text{avg } C_s(z) = \frac{1}{j_1} \sum_{j=0}^{j_1-1} c_{s,j,0} + \frac{z}{j_1} \sum_{j=0}^{j_1-1} c_{s,j,1} + \cdots \tag{10.38}$$

The averaging operation averages the coefficients of each power of z. This type of averaging is also known as *stacking*.

The covariance and average covariance are observable in the sense that we measure a set of $S_j(z)$ for the jth event, compute the $C_{s,j}$, and then average over j. Assuming the transmission function is the same for all events, we can write

$$\text{avg } C_s(z) = \frac{1}{j_1} \sum_{j=0}^{j_1-1} C_E(z) \, C_{A_j}(z) \tag{10.39}$$

The averaging operation averages the covariances of the source functions.

10.2.3 Subsurface Structure from Seismic Reverberations

Covariance methods of studying seismic reverberation in near-surface layers have several constraints: (a) A large data set is needed to average the covariances. (b) For simplicity arrivals that are near vertical incidence must be selected. Ideally an array of geophones is used. (c) The data are digitally recorded.

The small or microearthquakes that follow a large earthquake are good sources because they are frequent and usually nested or confined to a small region. To demonstrate the covariance method we have chosen data from microearthquakes or events near Kilauea volcano on the island of Hawaii.

The vertical and horizontal components of velocity were digitally recorded from a three-component geophone. A net of geophones was used to select events that were within a 2-km radius of the geophone and 8 to 10 km deep. Sample records from the vertical-component geophone are shown in Figure 10.6. The first arrivals were aligned, and the

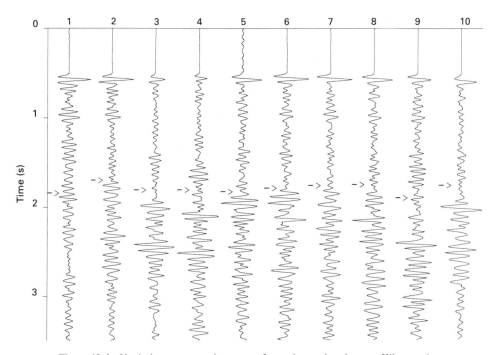

Figure 10.6 Vertical-component seismograms from microearthquakes near Kilauea volcano, island of Hawaii. The arrivals were aligned in time, and the phases and amplitudes were adjusted for display. The microearthquakes occurred within a 2-km radius around the geophone station and 8 to 10 km depth. Arrows indicate the arrival times of shear wave events. The times were gotten from horizontal-component geophones. (From M. R. Daneshvar, "Imaging of rough surfaces and planar boundaries using passive seismic signals," Ph.D. thesis, University of Wisconsin-Madison, 1987.)

phase and amplitudes were adjusted for display. Horizontal-component geophones recorded the shear waves. The arrival times of the shear waves are indicated by the arrows.

From the properties of elastic wave transmission in Section 2.2, vertically incident signals have vertical components of motion for the longitudinal waves (p-waves) and horizontal components for the shear waves (s-waves). Presumably, the vertical component of a three-component geophone gives the p-wave. The horizontal component of the geophone gives the s-wave part of the signal. In Figure 10.6 the initial part of the seismograms have the same character. After the arrival of the shear waves, indicated by the arrows, the character of the seismograms is different and variable. It appears that the change is associated with the arrival of shear waves.

At vertical incidence, s-waves are transmitted and reflected as s-waves, and p-waves are transmitted and reflected as p-waves. At vertical incidence, plane elastic waves do not convert from p-waves to s-waves or from s-waves to p-waves. Some explanations for the large-amplitude "wiggles" following the s-wave arrival times are needed. First, whatever disturbances cause the wiggles, they travel at s-wave velocities over most of their travel paths from the earthquake locations to the geophone. It seems reasonable to assume that the wiggles are due to s-waves from the earthquakes. The question is, How do vertically incident *s*-waves cause output signals from the vertical component of the geophone? Two explanations are: The geophone is imperfect and gives an output for horizontal motions. Inhomogeneities near the geophone scatter *p*- and *s*-waves, and some of the scattered waves have vertical components of motion at the geophone. Since we would expect the relative signs of the *p*-wave component and the *s*-wave components to vary for different earthquakes, alignment of the *p*-wave arrivals should cause confusion of the *s*-wave–related components. This confusion is apparent in Figure 10.6. We would expect a stacking operation to enhance the *p*-wave arrivals and to reduce *s*-wave–related components.

The autocorrelations or normalized covariances of the 10 signals are shown in Figure 10.7. An average correlation was computed for a total of 48 earthquakes. The trace on the right side of Figure 10.7 is the average correlation. The averaging procedure enhances compressional wave reflections in the layers and suppresses scattered components. The average autocorrelation was interpreted by assuming a trial structure-and-source correlation function and then computing a synthetic autocorrelation. Layer parameters were varied until the match of the experimental and synthetic autocorrelations was good enough. Data from the horizontal geophones were processed the same way. The results for vertical components are shown in Figure 10.8a. The shaded region of the experimental autocorrelation indicates the arrival times of the shear waves and scattered components. These

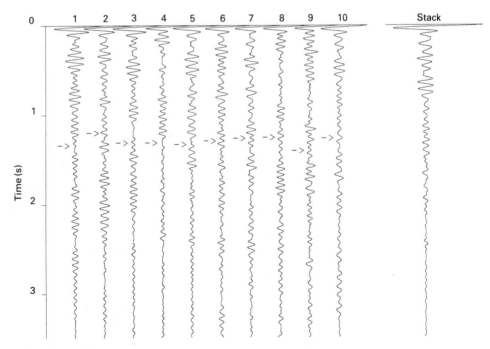

Figure 10.7   Autocorrelations of the 10 seismograms in Figure 10.6. The autocorrelations of each record are shown. These are typical examples of a set of 48 autocorrelations. The average of the 48 autocorrelations is the trace on the far right. The arrows indicate the arrival times for shear waves. (From M. R. Daneshvar, (1987); see caption to Figure 10.6.)

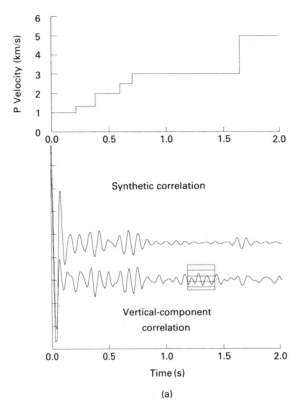

Synthetic correlation

Vertical-component
correlation

Time (s)

(a)

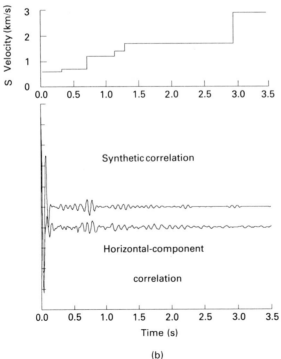

Synthetic correlation

Horizontal-component

correlation

Time (s)

(b)

Figure 10.8 Comparisons of experimental and synthetic average autocorrelations. The average autocorrelations were measured for the vertical components of motion and horizontal components of motion. Synthetic autocorrelations were computed for subsurface structures shown in (c). For the synthetic autocorrelation, the assumed autocorrelation of the sources was the autocorrelation of one cycle of a 13-Hz sine wave. (a) Average of vertical-component correlations and comparison with compressional (P) synthetic correlations. The shaded area indicates arrival time of shear waves. The seismic velocity structure is shown versus two-way travel time. (b) Average of horizontal component correlations and comparison with shear (S) synthetic correlations. The seismic velocity structure is versus two-way travel time. (c) Compressional (P) and shear (S) seismic velocity structures versus depth. (From M. R. Daneshvar, (1987); see Figure 10.6.)

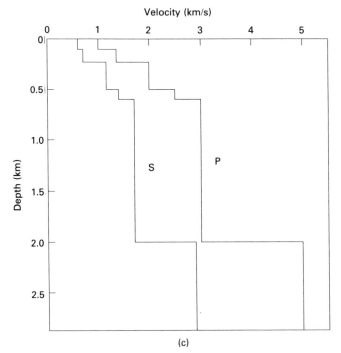

(c)   **Figure 10.8 (continued)**

interferences can be expected to disturb the match. The compressional wave velocity structure is shown versus time so that the interfaces and structure can be directly compared with the autocorrelations. Figure 10.8b shows similar results for shear waves. Seismic structure versus depth is shown in Figure 10.8c.

Geophysicists commonly regard reverberation in near-surface layers as a nuisance to be eliminated. They use all sorts of deconvolution and filtering schemes to reduce reverberations. Here we demonstrated the usefulness of reverberations to determine sub-surface structure, using data from microearthquakes beneath the geophone. This is an unusual situation. The technique requires only that the waves be nearly vertically incident at the geophone. The source can be nearby or very distant, such as on the other side of the earth.

The analysis of random seismic signals to determine shallow subsurface seismic structure has not been used very much. The somewhat analogous problem of using random electromagnetic waves from the ionosphere to measure subsurface structure has been extensively developed and used; it is known as the *magnetotelluric* method.

PROBLEMS

Purpose: to synthesize different examples of random signal and to compare their covariances. Let W(z) be a filter, transmission, or source function.

Step 1: Use the algorithm in Eq. (10.10) to compute a series of random numbers X(N). Here let X(N) = RND(1) − 0.5 to remove the mean value. A minimum number of terms is N1 = 256.

Step 2: Use the algorithm in Eq. (6.46) to convolve $X(z)$ and $W(z)$, $A(z) = W(z) * X(z)$.

Step 3: Recall Eq. (10.12), shift the summation limits

$$c_j = \frac{1}{n_1} \sum_{j}^{n_1+j} x_n x_{n-j}$$

write the covariance algorithm for $A(z)$ where $j_1 < n_1/5$. Let $j_1 = 32$. Use Eq. (10.28) as a model.

To aid comparisons, normalize the $C(J)$ by letting $C0 = C(0)$ and all $C(J) = C(J)/C0$. Plot the $C(J)$ from $J = 0$ to $J1$.

5. (a) Let $w_0 = 1$, $w_5 = 0.7$, and all other $w_m = 0$ where $m_1 = 5$. Compute the normalized covariance $C(J)$ of $A(z)$. Compare the $C(J)$ for $A(z)$ with $W(z)$ and $W(z^{-1}) * W(z)$ normalized.

  (b) Choose your own values of w_m and m_1 and try them.

6. (a) Let $m_1 = 8$ and $w_m = \sin(2\pi m/m_1)$.

  (b) Let $m_1 = 16$.

  Repeat Problem 5 for these functions.

7. Compare different lengths of time series for the same $W(z)$. Let $m_1 = 8$ and $w_m = \sin(2\pi m/m_1)$. Let $j_1 = 32$. Compute the normalized covariances of $A(z) = W(z) X(z)$ for $n_1 = 64, 256, 512$, and 1024. If your computer has enough memory try larger values of n_1. Compare the fluctuations of c_j.

8. Read F. K. Levin and D. J. Robinson (1969), "Scattering by a random field of surface scatterers," *Geophysics*, 34, 170–79, and H. B. Tatel (1954), "Note on the nature of a seismogram—II," *J. Geophys. Res.*, 59, 289–94. Do these laboratory experiments support the idea that scatterers near the detector can cause the seismogram to be noisy following an arrival?

11

Seismic Measurements from Controlled Sources

11.1 SEISMIC MEASUREMENTS FROM CONTROLLED SOURCES

Time-domain signal stacking is a technique for enhancement that works best when the noise is due to randomly located and randomly occurring noise sources. When the noise exceeds the signal transmission, the simplest method of enhancement is to increase the source level—that is, shoot more explosive or get a bigger hammer. If this is impractical, signals from many repeated transmissions can be added to achieve the equivalent of a larger source. In time-domain stacking we assume that all processes are linear and that all signal components can be added.

11.1.1 Signal Stacking in the Time Domain

Recalling the convolution expressions, Eqs. (6.38) and (6.39),

$$s_0 + s_1 z + s_2 z^2 + \cdots = (e_0 + e_1 z + \cdots)(a_0 + a_1 z + \cdots) \tag{11.1}$$

$$S(z) = E(z) A(z) \tag{11.2}$$

$S(z)$ is the z-transform of the signal for one transmission. Signal stacking (Figure 11.1) requires that the source and receiver be fixed on the earth. Also, the clocks or triggers at the source and receiver must be synchronized so that repeated transmissions add exactly in phase. For example, the time steps start with the synchronizing trigger, and the zeroth transmission gives

$$S_0(z) = E(z)A_0(z) \tag{11.3}$$

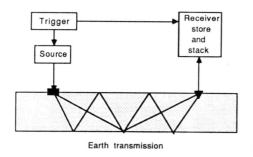

Figure 11.1 Stacking system. The receiver stores and stacks the repeated transmission initiated by the trigger.

Earth transmission

and subsequent transmissions give

$$S_1(z) = E(z)A_1(z) \tag{11.4}$$

$$S_2(z) = E(z)A_2(z) \tag{11.5}$$

The stack or sum of the transmissions is

$$S_S(z) = S_0(z) + S_1(z) + S_2(z) + \cdots \tag{11.6}$$

$$S_S(z) = E(z)\,(A_0(z) + A_1(z) + A_2(z) + \cdots) \tag{11.7}$$

where $E(z)$ is the same because the source and receiver are fixed on the earth. When the $A_n(z)$ are identical, the stack gives

$$S_S(z) = n_1 E(z)A(z) \tag{11.8}$$

for n_1 transmissions, where the subscript on $A(z)$ is dropped. This is the ideal situation. Actually, noise from other sources adds to the transmissions. Presumably the noises recorded for each transmission are uncorrelated with any other recording. The signal and noise, using $S + N$ as a subscript, for the nth transmission is

$$S_{S+N,n}(z) = s_0 + n_{0,n} + s_1 z + n_{1,n} z + \cdots \tag{11.9}$$

We expect the signal and noise to add linearly because the earth has a linear response for small seismic wave amplitudes. If we gather the signal in one polynomial and the noise in another, Eq. (11.9) becomes

$$S_{S+N,n}(z) = (s_0 + s_1 z + s_2 z^2 + \cdots) + (n_{0,n} + n_{1,n} z + n_{2,n} z^2 + \cdots) \tag{11.10}$$

The stack of n_1 transmissions is

$$S_{S+N}(z) = n_1(s_0 + s_1 z + s_2 z^2 + \cdots) + \sum_{n=0}^{n_1-1} n_{0,n}$$
$$+ z \sum_{n=0}^{n_1-1} n_{1,n} + z^2 \sum_{n=0}^{n_1-1} n_{2,n} + \cdots \tag{11.11}$$

Each of the summations of the noise terms is the sum of random numbers. The mean values of the sums of random numbers tend to zero as n_1 tends to infinity. Assuming that the noise powers are, on the average, the same for all terms, we can use the sum of

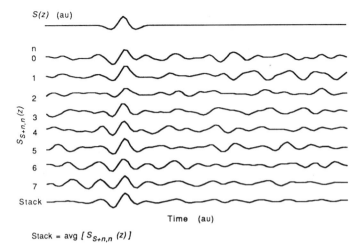

Figure 11.2   Individual transmissions and the stack for a Ricker wavelet. $S_{S+N}(z)$ are the individual transmissions. The signal is a Ricker wavelet. $w_b = 8$.

random numbers to estimate the fluctuations. Recall that the square of the sum of random numbers is

$$\sigma^2 \simeq \frac{1}{n_1} \left(\sum_{n=0}^{n_1-1} n_n \right)^2 \tag{11.12}$$

and the estimate of the rms fluctuations is σ. Solving for $(\cdot)^2$ in Eq. (11.12), we obtain an estimate of the fluctuations of the sum of n_1 random numbers,

$$(\text{fluctuations})^2 \simeq n_1 \sigma^2 \tag{11.13}$$

where σ is the rms noise. The stack of signals and noises gives

$$S_{S+N}(z) \simeq n_1 S(z) + (\text{fluctuations}) \tag{11.14}$$

The stack of n_1 terms causes the signal to increase with n_1. The estimate of the amplitude of noise fluctuations increases as $\sigma \, n_1^{1/2}$. Stated differently, the mean stacked signal is

$$\langle S_{S+N}(z) \rangle \simeq S(z) \pm \frac{\sigma}{n_1^{1/2}} \tag{11.15}$$

where $\pm \, \sigma/n_1^{1/2}$ is an estimate of the fluctuations due to noise. The individual transmissions of a wavelet type of signal and noises are shown in Figure 11.2. Comparisons of stacked signals for $n_1 = 8$, 64, and 128 are shown in Figure 11.3. It is apparent that a large improvement in the signal-to-noise ratio requires stacking a large number of transmissions.

11.2 CODED TRANSMISSIONS, MATCHED FILTERS, AND CORRELATION RECEIVERS

The concept of transmitting a coded signal having a long time duration to improve the signal-to-noise ratio probably originated in the 1940s. It also operates in the time domain. Van Vleck and Middleton described or specified a filter that gave the maximum output

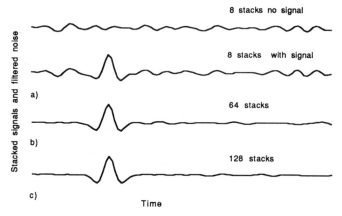

Figure 11.3 Comparison of different numbers of stacks. The transmission and stack are the same as Figure 11.2. (a) $n_1 = 8$. (b) $n_1 = 64$. (c) $n_1 = 128$.

signal-to-noise ratio for a known signal in 1946. The filter became known as the *matched filter* because it was matched to a particular coded transmission. The concept has been applied to radar, sonar, and seismic methods. It was also shown that the incoming signal could be correlated against a replica of the transmission to accomplish the matched filter operation. In geophysical exploration the coded signal transmission from a vibrating source is known as *Vibroseis*, which is a registered trade mark of the Continental Oil Company.

Systems

A typical system is shown in Figure 11.4. The source transmits a coded signal $A(z)$. The earth transmission $E(z)$ convolves with $A(z)$. The result is a complicated signal having a long duration. The geophone signal is

$$G(z) = A(z)E(z) \tag{11.15}$$

One input to the correlator is $A(z)$ and the other input is $G(z)$. The correlation operation gives

$$S(z) = A(z^{-1})A(z)E(z) \tag{11.16}$$

where we ignore a proportionality constant. We recognize $A(z^{-1})A(z)$ as being the correlation $C_A(z)$, and

$$S(z) = C_A(z)E(z) \tag{11.17}$$

We can replace the correlation operation by a matched filter having the response $A(z^{-1})$.

Let us do specific examples and choose the frequency-modulated signal, or *chirp*,

$$a(t) = \sin[2\pi(f_L + \alpha t)t] \tag{11.18}$$

At $t = 0$ it has the frequency f_L, and the frequency increases with time. The instantaneous frequency is the rate of phase change. Letting ϕ be the instantaneous phase, we have

$$\phi = 2\pi(f + \alpha t)t \tag{11.19}$$

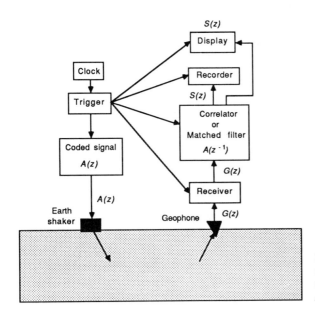

Figure 11.4 Coded signal transmission and signal processing system. The correlator is equivalent to a filter that has the response $A(z^{-1})$.

The instantaneous frequency is

$$f_i = \frac{1}{2\pi}\frac{d\phi}{dt} = (f_L + 2\alpha t) \tag{11.20}$$

We choose α by choosing the duration of the sweep, T_s, and the highest frequency, f_H

$$\alpha = \frac{f_H - f_L}{2T_s} \tag{11.21}$$

Examples of passing the signal $A(z)$ through its matched filter or equivalently the correlator are shown in Figure 11.5. The output is the covariance or correlation, depending on normalization. Three things are shown in the figure. First, the width of the central maximum of the match-filtered signal is proportional to $1/(f_H - f_L)$. Second, the wiggles have a duration of $2T_s$. And third, the filtered signal is symmetrical about its maximum.

The matched filter $A(z^{-1})$ has values for negative time, but this is not physcially realizable. That is, the filter would have an output before it had an input. A fixed time delay, T_s, is usually added to the filter to make the filter $A(z^{-1})\,z^s$, where s is the time step corresponding to the duration of the signal, and

$$A(z^{-1})\,z^s = a_s + a_{s-1}z + \cdots + a_0 z^s \tag{11.22}$$

This is the function used to calculate the curves in Figure 11.5. The time delay shifts all arrivals by the same amount at the output.

Figure 11.6 shows the convolution of a chirp-coded signal with a simple $E(z)$ having two reflectors. The matched-filter output has correlation maxima at the reflection times, where all times have been shifted by T_s.

An analysis that includes the effects of added noise is more complicated than the analysis of stacked transmissions in the previous section. It is sufficient to say that the

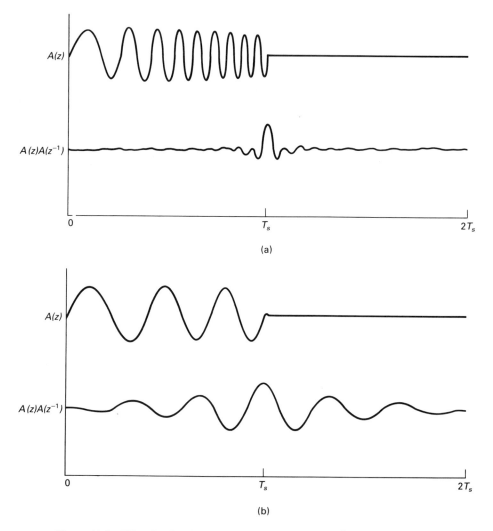

Figure 11.5   Chirp signals $A(z)$ and their correlations $A(z)A(z^{-1})$ $T_S = 1$ s, $T_0 = 1/64$ s. The peak of the correlation is delayed by T_s. (a) $f_H = 16$, $f_L = 2$. (b) $f_H = 4$, $f_L = 2$.

amplitude signal-to-noise improvement is proportional to $[(f_H - f_L)T_s]^{1/2}$, or the square root of the time frequency-bandwidth product.

PROBLEMS

1. Write and test an algorithm to make a chirp $A(z)$ for given f_L, f_H, and T_s. Display graphs.

2. Compute $A(-z)\,A(z)$ for your choices of f_L, f_H, and T_s. Display graphs.

3. Convolve $A(z)$ and $E(z)$. Let $E(z) = 1 + e_k z^k$. For a given choice of f_H, f_L, and T_s, how small can k be and still give resolved peaks in $S(z) = A(z^{-1})\,A(z)E(z)$?

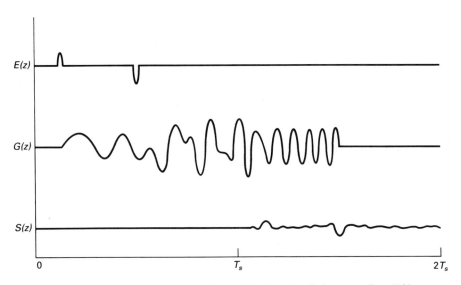

Figure 11.6   Chirp signal and two reflectors $E(z)$. $T_s = 1$ s, 64 time steps, $T_0 = 1/64$ s. (a) $E(z) = 0.5z^8 - 0.7z^{32}$. (b) $G(z) = A(z) E(z)$, received signal. (c) $S(z) = A(z^{-1}) A(z) E(z)$, the match-filtered or correlated signal.

12

Geophone Arrays

In the 1950s, exploration geophysicists performed many experiments to study the causes of noise in their reflection data. For reasonably quiet conditions and large enough sources, they found that the interfering signals were from the sources, because repeated transmissions gave the same records. If the interferences had been from random-noise sources, then repeated transmission would have given different records.

12.1 GROUND ROLL AND REFLECTIONS

Commonly, interferences are ground-roll arrivals that travel horizontally in near-surface layers. For example, a weathered top layer can have transmission velocities of a few hundred meters per second. As sketched in Figure 12.1, multiply reflected arrivals in the weathered layer can last a long time. Part of the radiation from the source is trapped in

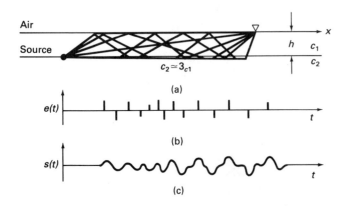

(a)

(b)

(c)

Figure 12.1 Weathered-layer arrival. (a) Multiple reflected arrivals in a weathered layer. The first arrival is the head wave along the $c_1 c_2$ interface. (b) Earth response $e(t)$. (c) Signal $s(t)$ for a Ricker wavelet transmission.

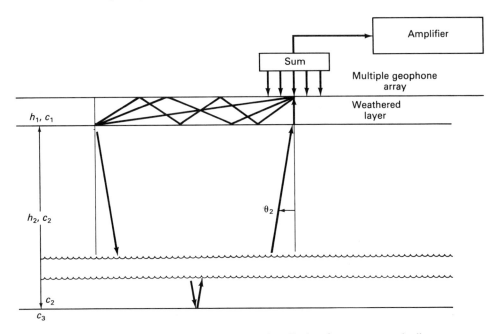

Figure 12.2   Weathered layer over a thick section. The interferences or ground roll are trapped in the shallow weathered layer. The reflection signal from a deep interface has nearly vertical incidence at the receivers.

the weathered and near-surface layers. From Section 2.1 the amplitudes of the trapped ground-roll arrivals decrease as $1/r^{1/2}$. Ground roll has vertical and horizontal components of motion. Ground-roll arrivals can mask reflection signals from deep subsurface interfaces. In some areas the properties and thickness of the weathered layer can cause the ground roll to have very low frequencies compared with the frequencies of reflection signals.

By carefully designing amplifiers and choosing band-pass filters, exploration geophysicists suppressed the ground-roll components of the signal and displayed the reflections. Figure 9.5 shows an example of filtering. But in many areas they could not eliminate the ground-roll interferences by electrical filters in the amplifiers. Since the interferences were traveling horizontally, and the reflection signals were nearly vertically traveling signals, it seemed reasonable to use arrays of geophones to directionally filter the signals (Figure 12.2).

We begin our discussion with a few fundamentals of the response of geophone arrays and give a discussion of arrays and sampling theory later.

12.2 ARRAY RESPONSE

Geophysicists often use continuous waves to describe array responses. We begin with a general areal array and then specialize to symmetrical line arrays. An example of an areal array, a five-leg star pattern, is shown in Figure 12.3a. Assuming the source is in the x-z plane, we project the geophone positions on the x-axis, Figure 12.3b. As in Figure

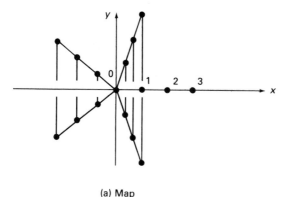

(a) Map

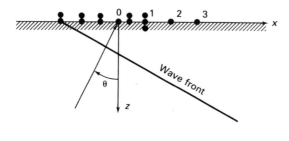

(b) Vertical

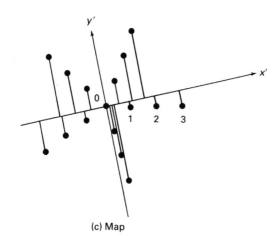

(c) Map

Figure 12.3 Array geometry. The example shows a five-leg star pattern. For simplicity, only the few geophones are numbered. (a) Map view of projections to the x-axis. (b) Vertical view of a wave front in the x-y plane. Some geophones in different legs are at the same distances from the origin. (c) Map view of projections to an x'-axis. This is used to compute response for a wave incident in the x'-z plane.

12.2, the signals from the individual geophones are added, and the source is "steered" to look vertically downward. The response to signals incident at the angle θ (Figure 12.3b) is one way to describe array response. For sources in other planes, for example, the x'-z plane, geophone positions are projected on that plane (Figure 12.3c).

If the source is in the x-z plane, the component of wave number along the x-axis is

$$k_x = \frac{2\pi \sin \theta}{\lambda} = \frac{2\pi f \sin \theta}{c} \tag{12.1}$$

For simplicity the incident wave has unit amplitude and frequency f.

The signal from the nth grophone has amplitude d_n and relative phase $k_x x_n$, where d_n is a sensitivity or weighting factor, and x_n is the projected position on the x-axis. The time-dependent signal is

$$s_n(t) = d_n e^{i(2\pi ft - k_x x_n)} \tag{12.2}$$

and the normalized sum of N geophones is

$$s(t) = e^{i2\pi ft} D(k_x)$$

$$D(k_x) \equiv \frac{1}{D_0} \sum_{n=0}^{N-1} d_n e^{-ik_x x_n} \tag{12.3}$$

$$D_0 \equiv \sum_{n=0}^{N-1} d_n$$

$D(k_x)$ is the array response as a function of k_x. For making comparisons of different basic designs of arrays it is convenient to change k_x to a "universal" variable u that contains f, c, $\sin \theta$, and L.

$$u \equiv \frac{Lf}{c} \sin \theta = \frac{L}{\lambda} \sin \theta \tag{12.4}$$

where L is defined for a specific array geometry and is an overall array dimension. The change of variable from k_x to u, using Eq. (12.1), gives us

$$\left. \begin{array}{c} k_x = \dfrac{2\pi u}{L} \\[2ex] k_x x_n = 2\pi u \dfrac{x_n}{L} \end{array} \right\} \tag{12.5}$$

It is possible to replace (x_n/L) with a dimensionless variable, but we think it better to keep L in the expressions. The array response becomes

$$D(u) = \frac{1}{D_0} \sum_{n=0}^{N-1} d_n e^{-i2\pi u x_n/L} \tag{12.6}$$

The absolute value and phase of $D(u)$ are computed by expanding the exponential, $\exp(ix) = \cos x + i \sin x$,

$$|D(u)| = \frac{1}{D_0} \left[\left(\sum_{n=0}^{N-1} d_n \cos\left(2\pi u \frac{x_n}{L}\right) \right)^2 + \left(\sum_{n=0}^{N-1} d_n \sin\left(2\pi u \frac{x_L}{L}\right) \right)^2 \right]^{1/2} \tag{12.7}$$

$$\Phi(u) = \arctan \left[\frac{-\sum\limits_{n=0}^{N-1} d_n \sin\left(2\pi u \dfrac{x_L}{L}\right)}{\sum\limits_{n=0}^{N-1} d_n \cos\left(2\pi u \dfrac{x_L}{L}\right)} \right] \tag{12.8}$$

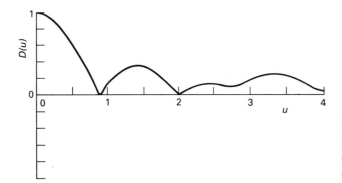

Figure 12.4 Response $|D(u)|$ of a five-leg star array. The 16-geophone array is shown in Figure 12.3a. The radius of the star is 2λ. The source is in the x-z plane, $u = L \sin \theta/\lambda$, and $L = 4\lambda$.

where $\Phi(u)$ is the relative phase. These are general expressions for an array. $D(u)$ is evaluated numerically for all but the very simplest arrays (see problems). An algorithm follows this section. The response of the five-leg star pattern is shown in Figure 12.4.

Line arrays are most often used because they are easy to place along roads and have good suppressions of ground-roll interferences from the source. They are usually symmetrical about the center. For an array center and origin at $x = 0$, d_n and x_n satisfy the symmetry relations

$$\left. \begin{array}{c} d_n = d_{-n} \\ x_n = -x_{-n} \end{array} \right\} \tag{12.9}$$

The overall length of the array is L.

Initially we let the number of geophones N_0 be odd and place the zeroth geophone at the origin. The summation from 0 to $N - 1$ can be changed to a pair of summations and a zeroth term as follows:

$$\begin{aligned} D(u) = \frac{1}{D_0} \Bigg[d_0 &+ \sum_{n=-1}^{-(N_0-1)/2} d_n \cos\left(2\pi u \frac{x_n}{L}\right) \\ &+ \sum_{1}^{(N_0-1)/2} d_n \cos\left(2\pi u \frac{x_n}{L}\right) - i \sum_{-1}^{-(N_0-1)/2} d_n \sin\left(2\pi u \frac{x_n}{L}\right) \\ &- i \sum_{1}^{(N_0-1)/2} d_n \sin\left(2\pi u \frac{x_n}{L}\right) \Bigg] \end{aligned} \tag{12.10}$$

If we replace n with $-n$ in the first and third summations and use Eq. (12.9), the sine terms cancel, and the summation reduces to

$$D(u) = \frac{1}{D_0} \left[d_0 + 2 \sum_{n=1}^{(N_0-1)/2} d_n \cos\left(2\pi u \frac{x_n}{L}\right) \right] \tag{12.11}$$

The same expression applies to an even number of geophones, where $d_0 = 0$ and $(N_0 - 1)/2$ becomes $N_e/2$. A symmetrical array algorithm follows this section.

In the 1950s and 1960s many acousticians and geophysicists studied the effects of non-uniform geophone spacings and unequal geophone sensitivities on array response.

They learned that rather simple adjustments of geophone sensitivities could significantly improve the response of geophone arrays. Concentrating the number of geophones toward the center of the array or increasing the response of the geophones near the center of the array improved the response.

Comparisons of array responses for three types of geophone weighting are shown in Figure 12.5. The array lengths are the same, and the odd number of geophones are equally spaced along the array. The insets on each graph show the type of weighting. An algorithm follows this section.

The uniform response, Figure 12.5a, has equal geophone weights and serves as a standard reference. The first null of the response is approximately $u = 1$. Our desire is to reduce the response in the range of u from 2 to 5. The triangular weighting, Figure 12.5b shows a large reduction of response for u in the range 2 to 5. The main lobe is broader, and the first null is at about 1.6. The $1 + \cos(x)$ type of taper has the best side-lobe reduction of the three. It has negligible response in the $u = 2$ to 5 range.

These examples demonstrate a rule of thumb. For a fixed number of geophones and array length, a geophone weighting choice that decreases the side lobes causes the width of the main lobe to increase.

12.3 ALGORITHMS

Algorithms for computing array responses are in this section. The algorithms assume that the geophone weights d_n, or D(N), and positions x_n, or X(N), have numerical values in the arrays. The D(N) are normalized, and the sum of the D(N) is 1. NG is the number of geophones. U is the ratio of array length LA to λ, $U = LA/\lambda$ for waves traveling along the surface or $\theta = 90°$. The array response is DS(M).

1. General array response
2. Symmetric array response
3. Position and geophone response weights for equally-spaced geophones.

12.3.1 General Array

```
REM  SEC 12.3.1 GENERAL ARRAY

REM  COMPUTE ARRAY RESPONSE FOR AN NG-ELEMENT ARRAY
REM  GEOPHONES ARE PROJECTED ON X-AXIS AT X(N)
REM  GEOPHONE SENSITIVITES ARE D(N)
REM  OUTPUT SUM IS DS(M)
REM  SOURCE IS IN X-Z PLANE
REM  ARRAY LENGTH = L.
REM  U = (L / LAMBDA) * SIN(THETA). THETA = 90 DEG
REM  NS = NUMBER STEPS OF U

     DIM  D(30), DS(30), X(30)
     NS=20
```

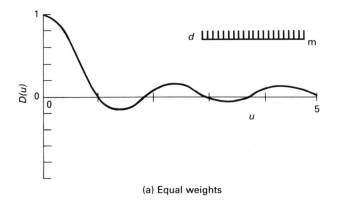

(a) Equal weights

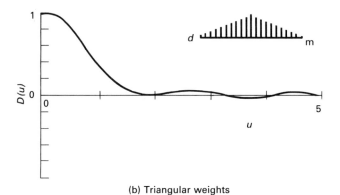

(b) Triangular weights

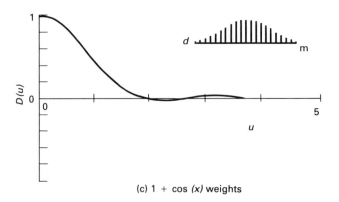

(c) 1 + cos *(x)* weights

Figure 12.5   Array responses for shaded or weighted arrays. All arrays have $L/\lambda = 5$. The 21 geophones are equally spaced. The continuous wave signal is traveling horizontally along the array. The insets show the relative geophone sensitivities. (a) Equal weights. (b) Triangular weights. (c) $1 + \cos(x)$ weights.

```
10    PRINT" INPUT ARRAY LENGTH=";:INPUT L
      PRINT" INPUT NUMBER OF GEOPHONES";:INPUT NG
      PRINT" INPUT LAMBDA=";:INPUT LAMBDA

      U = L / LAMBDA
      FOR N = 0 TO NG - 1
         X(N) =N* L / (NG-1)
         D(N) = 1
      NEXT N

      PI = 4 * ATN(1)
      DU = 2 * PI * U / NS

      FOR M = 0 TO NS
         R = 0
         S = 0
         MU = M * DU

           FOR N = 0 TO NG - 1
              R = R + D(N) * COS(MU * X(N) / L)
              S = S + D(N) * SIN(-MU * X(N) / L)
           NEXT N

           DS(M) = SQR(R^2 + S^2)
      NEXT M

      PRINT" U","ABS RESPONSE"

      FOR M = 0 TO NS
         PRINT M*U/NS, DS(M)/DS(0)
      NEXT M

      PRINT "NEW RUN Y OR N ";:INPUT Q$
      IF Q$ = "Y" GOTO 10

      END
```

12.3.2 Symmetrical Array

```
REM  SEC 12.3.2 SYMMETRIC ARRAY

REM  COMPUTE ARRAY RESPONSE FOR AN NG-ELEMENT ARRAY
REM  GEOPHONES ARE PROJECTED ON X-AXIS AT X(N)
REM  GEOPHONE SENSITIVITES ARE D(N)
REM  OUTPUT SUM IS DS(M)
REM  SOURCE IS IN X-Z PLANE
REM  ARRAY LENGTH = L.
REM  U = (L / LAMBDA) * SIN(THETA). THETA = 90 DEG
REM  NS = NUMBER STEPS OF U

     DIM  D(30), DS(30), X(30)
     NS=20

10   PRINT" INPUT ARRAY LENGTH=";:INPUT L
     PRINT" INPUT ODD NUMBER OF GEOPHONES";:INPUT NG
     PRINT" INPUT LAMBDA=";:INPUT LAMBDA

     U = L / LAMBDA
     FOR N = 0 TO (NG - 1) / 2
        X(N) =N* L / (NG-1)
        D(N) = 1
     NEXT N

     PI = 4 * ATN(1)
     DU = 2 * PI * U / NS

     FOR M = 0 TO NS
        R = 0.5
        MU = M * DU

           FOR N = 1 TO (NG - 1) / 2
             R = R + D(N) * COS(MU * X(N) / L)
           NEXT N
           DS(M) = 2 * R

     NEXT M

     PRINT"  U ","RESPONSE"

     FOR M = 0 TO NS
         PRINT M*U/NS, DS(M)/DS(0)
     NEXT M

     PRINT "NEW RUN Y OR N ";:INPUT Q$
     IF Q$ = "Y" GOTO 10

     END
```

12.3.3 x_n, d_n, and D_0 for Triangular and $1 + cos$ (x) *Weighted Equally Spaced Arrays*

```
REM  SECTION 12.3.3 GEOPHONE WEIGHTS

REM  COMPUTE GEOPHONE WEIGHTS FOR AN ODD NG-ELEMENT ARRAY
REM  TRIANGULAR WGT GEOPHONE SENSITIVITES ARE DTRI(N)
REM  (1 + COS(X)) WEIGHTS ARE DCSN(N)
REM  GE0PHONES ARE EQUALLY SPACED AND SYMMETRIC ABOUT 0
     DIM DTRI(30), DCSN(30)

10   PRINT" INPUT ODD NUMBER OF GEOPHONES";:INPUT NG
     NH = (NG - 1) / 2
     PI = 4 * ATN(1)

     FOR N = 0 TO NH
        DTRI(N) = 1 + NH - N
        DCSN(N) = 1+COS(N*PI/(NH+1))
     NEXT N
     STRI = DTRI(0)
     SCSN = DCSN(0)

     FOR N = 1 TO NH
        STRI = STRI + 2 * DTRI(N)
        SCSN = SCSN + 2 * DCSN(N)
     NEXT N

     FOR N = 0 TO NH
        PRINT N, DTRI(N),DCSN(N)
     NEXT N

     PRINT "SUM TRI WGT = "; STRI;"   SUM COS WGT = ";SCSN

     PRINT "NEW RUN Y OR N ";:INPUT Q$
     IF Q$ = "Y" GOTO 10

     END
```

12.4 SPATIAL SAMPLING AND MULTIPLE GEOPHONE ARRAYS

A spread of geophone stations and their multiple geophone arrays are sketched in Figure 12.6. The objective is to record signals from each station. Some of the seismic events have horizontal components of wavelength λ_x that are less than the station separations. We need to consider this problem because proper sampling in space is as important as proper sampling in the time domain. Later in Chapter 16, we give cross correlation processing methods that can relax the spatial sampling rules.

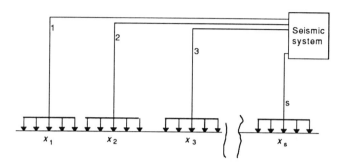

Figure 12.6 Local geophone arrays and a set of receiving stations. The geophones in the array are connected to give the sum. The sums go to the inputs of seismic amplifiers.

12.4.1 Spatial Sampling Theorem

Spatial samples of signals must be taken at separations that are less than half the shortest wavelength of any signal, noise, or interference along the coordinate axis. For example, the separations L_0 along the x-direction must be less than $\lambda_x/2$, where λ_x is the shortest wavelength. If the source and receiver stations are along the x- and y-directions, then the spatial sampling theorem applies to both directions. As in the time domain, we can state the sampling theorem in terms of the wave numbers k_x and k_y along x and y.

There are roughly three ways to satisfy the sampling theorem, and geophysicists use a combination of all three. (1) Choose band-pass filters in the frequency domain to remove as many components of ground motion having short wavelengths as possible. The band-pass filter must pass the signals. (2) Place the stations at the separations L_0. A long profile may require an enormous number of stations. (3) Use multiple geophone arrays as low-pass spatial filters to remove components of ground motion that have λ_x less than some value of interest.

12.4.2 Interferences and Noise

Most of the short-wavelength events are seismic disturbances that travel along the surface and in near-surface layers. The sources of these disturbances are the source, cultural noises (trains, trucks, animals, etc.), and environmental noises (wind, rain, earthquakes, etc.). Cultural and environmental noises are extremely variable and can come from many directions. Surface-wave and ground-roll disturbances from the source usually travel along the paths between the source and receiving stations.

Interference studies are commonly made by firing shallow shots and recording ground-roll signals at closely spaced single geophones. The separations of the geophone stations must be small enough that the same phase of a cycle of ground roll can be followed from one station to the next, as shown in Figure 12.7a. If we let the separation be ΔL, the phase velocity along x is

$$v_x = \frac{\Delta L}{\Delta t} \qquad (12.12)$$

where Δt is the travel time for the peak. From the figure, ΔL is 3 m, and the distance across six stations is 15 m. The time delay Δt of the blackened peaks is about 0.04 s. These values give $v_x = 370$ m/s. The period of the event, the time between the pairs of blackened peaks, is $T = 0.04$ s. The corresponding frequency is about 25 Hz. The component of wave number k_x is

$$\left. \begin{aligned} k_x &= \frac{2\pi f}{v_x} \\ k_x &\simeq 0.42 \text{ m}^{-1} \end{aligned} \right\} \tag{12.13}$$

There are many arrivals and a way of displaying ground-roll data is shown in Figure 12.7b. Both the frequencies f and amplitudes A are indicated on the time/distance plot. The phase velocities are the reciprocals of the slopes of the lines.

Another way to display the analysis of ground-roll data is the wave number or spectrum shown in Figure 12.8a. It is derived from Figure 12.7b by using f and v_x to

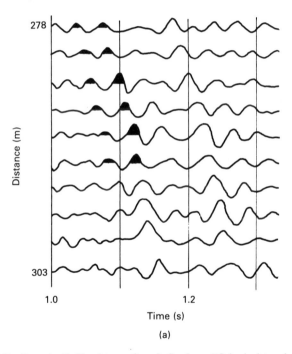

(a)

Figure 12.7 Ground roll. The data are from J. Graebner, "Seismic data enhancement: a case history," (*Geophysics*, 25 (1960), 283–311, Figs. 4 and 11.) The original data were traced and figures were redrawn. (a) The sample of a ground-roll or interference seismogram is redrawn from a larger seismogram. The shot was at 12 m (40 ft), and the distance to the first geophone was 278 m (910 ft). The spacing between geophones was 3 m (10 ft). (b) The simplified travel-time distance graph is from a large set of seismograms. The phase velocities are the reciprocal of the slopes. Events having phase velocities less than 400 m/s are interferences and ground-roll arrivals. The first arrival is the head wave. The frequency and relative amplitudes of the arrivals are indicated as (f, A). Events having nearly horizontal lines are signals or reflections. The frequency range of signals is 35 to 45 Hz. The frequency range of ground-roll events is 22 to 40 Hz.

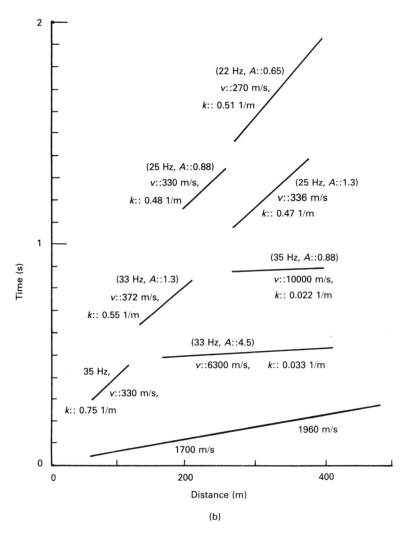

Figure 12.7 (continued)

calculate k_x. The graph shows ground-roll arrival amplitudes $A(k)$ versus k_x. Two regions are identified as being the range of k_x for interferences and the range of k_x for reflection signals.

12.4.3 Array Design

Our purpose is to reject seismic disturbances having large k_x and to pass the seismic signals having small k_x. Since the interferences and noise can come from any direction, we have to decide whether to use an areal pattern or a line array of geophones. For simplicity, we assume that all interferences come from the source and design a line array of multiple geophones.

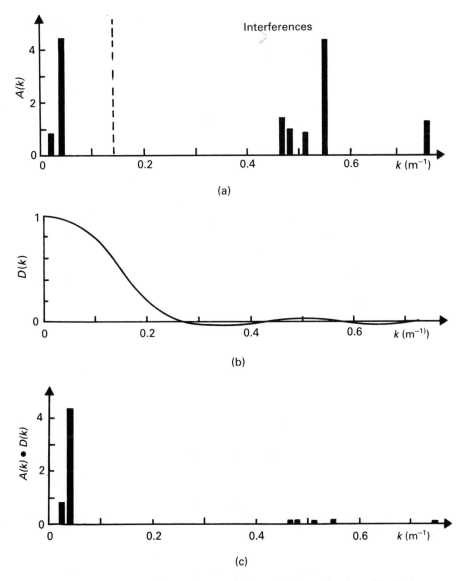

Figure 12.8   Ground roll, array response, and array output as functions of k_x. (a) Approximate spectra of events shown in Figure 12.7. The $A(k)$ are relative amplitudes. Ground-roll events have k_x greater than roughly 0.4 m^{-1}. Signals or reflections have k_x less than 0.05 m^{-1}. (b) Response of the $1 + \cos(x)$ array $D(k_x)$. The array has 21 geophones spaced 2 m apart. The array length is 40 m. (c) The output of the multiple geophone array is the product or convolution $A(k_x) \cdot D(k_x)$.

For a numerical example, assume that the bedrock seismic velocities of layers beneath the weathered layer are 1900 m/s, and the reflection signals can arrive at a 30° incident angle. A 50-Hz seismic signal has k_x

$$\left.\begin{aligned} k_x &= \frac{2\pi f \sin\theta}{c} \\ k_x &= 0.083 \text{ m}^{-1} \end{aligned}\right\} \tag{12.14}$$

The array should pass signals having $k_x < 0.1$ m^{-1}. If the signal is increased to 100 Hz, then the array should pass signals having $k_x < 0.2$ m^{-1}. Ideally the array will reject components having $k_x > 0.1$ m^{-1} for the 50-Hz signal or $k_x > 0.2$ m^{-1} for the 100-Hz signal.

The sampling theorem and the event having the largest k_x give the maximum spacing between geophones. In Figure 12.8a the largest k_x is 0.75 m^{-1}, or λ_x is about 8.4 m. Correspondingly, the spatial sampling theorem requires that the signal be sampled at separations less than 4.2 m, that is, $\lambda_x/2$.

From Figure 12.8c the $1 + \cos(x)$ tapered array is nearly zero for u more than 2. The main lobe is greater than 0.4 for u less than 1. The choice to reject all components having $k_x > 0.3$ is reasonable. From the definition of u (Eq. (12.4)), the expression for u using k_x is

$$\left.\begin{aligned} u &\equiv \frac{L \sin\theta}{\lambda} \\ 2\pi u &= k_x L \end{aligned}\right\} \tag{12.15}$$

The substitution of $k_x = 0.3$ m^{-1} and $u = 2$ gives $L \simeq 42$ m. For a spacing of 4.2 m between geophones, the minimum number of geophones is 11. A conservative design uses $N_0 = 21$ geophones and $L = 40$ m. The geophone array response is shown in Figure 12.8b. The signal components passed by the array are $A(k)D(k)$ and are shown in Figure 12.8c.

The spatial sampling theorem gives the maximum spacing between stations. The multiple geophone arrays pass signals having k_x less than about 0.2 m^{-1}. The corresponding λ_x is about 31.4 m. For a $\lambda_x/2$ sampling criterion, the maximum spacing between stations is 15 m. This choice is consistent with the choice of 50 Hz for the maximum signal frequency.

PROBLEMS

The problems are designed to help you explore many important aspects of array design and response. They are only a beginning. The problems should be done on a computer with a graphics output. For simplicity we use equally-spaced symmetrical line arrays for the problems in this set (Eq. (12.11)).

1. Equally-weighted arrays. Write a program to compute and graph array responses as a function of u. (An algorithm is at the end of this section.) Test your program for the example in Figure 12.5a.

2. Equally-weighted arrays, $D(u)$. Use the same length of array, $L/\lambda = 9$, for all parts of this problem. Use the range $0 \leq u \leq 9$. Compute and graph $D(u)$ for the following numbers of geophones: (a) $N = 9$; (b) $N = 13$; (c) $N = 15$; (d) $N = 21$; (e) $N = 37$. A maximum step in u of 0.2 will display the shapes of the curves. Notice that the shapes of the main lobes, $D(u)$ for $0 \leq u \leq 1$, are about the same for all numbers of geophones. Notice that the $D(u)$ are about the same for $N > = 13$. Why does $D(u)$ have a maximum of 1 for 9 geophones at $u = 8$? *Hint*: What are the geophone separations in wavelengths?

3. Equally-weighted arrays. Write a modified program to compute and graph the array response as a function of the incident angle θ. (Recall the definition of u.) The choices of $c = 1000$ m/s, $L = 120$ m, and $N = 13$ geophones give nice results. Vary θ from 0 to 90°. Explore the response as a function of frequency. (a) $f = 10$ Hz; (b) $f = 25$ Hz; (c) $f = 50$ Hz; (d) $f = 100$ Hz. Notice the dependence of the first null on frequency. Why is there a maximum response at $\theta = 90$ for $f = 100$ Hz?

4. Weighted arrays, $D(u)$. Modify your program for $D(u)$ to run for symmetrical weighted arrays. Test your program for the examples shown in Figures 12.5b and 12.5c. (An algorithm for computing triangular and $1 + \cos(x)$ geophone weights appears at the end of this section.)

5. Weighted arrays, $D(u)$. Repeat the parts of Problem 2 for the triangular and $1 + \cos(x)$ geophone weights. In addition, compare these results with the uniform weighted results.

6. $D(u)$ for different array weights. The purpose is to compare center maximum weighting and end-of-the-array maximum weighting designs. Let $L/\lambda = 5$; use the range $0 \leq u \leq 5$; $N = 9$ geophones.
   (a) Center maximum. $d_0 = 5$, $d_1 = 4$, $d_2 = 3$, $d_3 = 2$, $d_4 = 1$. $D_0 = 25$. Remember that $d_n = d_{-n}$.
   (b) Ends are maximum, reverse triangular. $d_0 = 1$, $d_1 = 2$, $d_2 = 3$, $d_3 = 4$, $d_4 = 5$. $D_0 = 29$.

Notice the dependence of the main lobe width on the type of weighting. Notice the sizes of the side lobes. When would it be desirable to use the reverse triangular weighting?

7. Algebraic reduction of $D(u)$ for uniform equally-weighted arrays. Evaluate the summation in Eqs. (12.10) and (12.11). *Hint*: Using Eq. (12.10), express the sum of the cosine and i sine terms as the sum of exponentials. What are the x_n for an equally-spaced array? Does the result look like a familiar series?

13

Stacking of Multichannel Data

The art of enhancing multiple channel seismic reflection data has been extensively developed by exploration geophysicists in their search for petroleum. Enhancement of reflection data by signal stacking is based on decades of geophysical experience. What geophysicists did by hand and what they saw in the old wiggly line records has become the basis of computational algorithms.

13.1 CORRECTIONS TO DATUM ELEVATION

Some of the seismic data processing was and still is tedious. An important first step is to move all the seismic time traces to a common datum elevation. Seismic measurements are made over variable terrane. The geophones are on the surface, and the sources are either on the surface or at some convenient depth. Commonly, geophone and source elevation vary across the spread, as shown in Figure 13.1. The records must be corrected for the effect of shot depth, weathered layer thickness, and geophone elevations. Normally, the first arrivals are the head waves that travel along the "basement" at the base of the weathered layer. Seismic interpreters use head waves or refraction techniques to calculate the weathered layer thickness along the spread. Weathered layer and elevation corrections are made for each geophone station. The weathered layer usually has a low velocity, and small elevation changes give large time shifts. The arrival times of seismic events or reflections are transferred to a datum elevation by finding the difference between the travel times in the weathered layer and the rock between the weathered layer and a lower datum elevation. Figure 13.1 shows the projected shot and geophone positions on the datum.

The corrected seismograms for each spread are then referred to a common datum elevation for the area. These corrections are called *static corrections* because they give constant time shifts for the individual seismic traces.

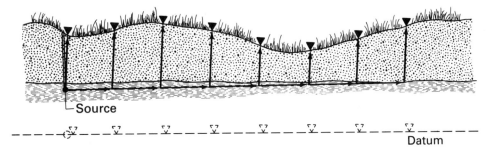

Figure 13.1 The source is s, and the geophones are the ▼. The ray paths are head waves along the top of the layer beneath the weathered layer. The seismic velocity in the weathered layer is so low that the ray paths are nearly vertical. The datum elevation is the dashed line. Projections of the source and geophones are shown on the datum. The arrival at the geophone above the shot gives the *uphole time* and can be used to estimate the seismic velocity in the weathered layer.

Static corrections are often larger than subsurface structures. Mistakes or jumps in the static corrections can break the alignment of reflection signals from a deep planar interface and incorrectly indicate a fault.

13.2 REFLECTION STACKING

We develop the concept of *normal moveout* and normal moveout correction in this section. The stacking of corrected records follows easily. Here the seismic data are from a simple single-ended profile. The source is at $x = 0$, and the geophones are along the x-axis. The next section contains a description of a modified procedure, *common depth point stacking*. The results for the method in this section and the common depth point stacking are the same for data taken over a horizontally stratified medium. Recall Section 3.4 and Problems 10 and 11 of Chapter 3.

For simplicity, we use the term *geophone* to represent a station having a single geophone or an array of geophones. Similarly *source* means a single source or an array of sources. For the most part we treat the multilayer reflection problem as being a single composite layer to the reflecting interface. The composite layer has the root mean square velocity $c_{n,\text{RMS}}$. A derivation of $c_{n,\text{RMS}}$ is in Section A1.2 of Appendix A. The derivation is based on MacLaurin's expansion of the ray-tracing expressions in a time squared–distance squared expression. The first two terms of the expression for a horizontally layered half-space are (recall Section 3.4)

$$t_n^2(x) \simeq t_n^2(0) + \left(\frac{x}{c_{n,\text{RMS}}}\right)^2 \tag{13.1}$$

$$\tau_i \equiv \frac{2h_i}{c_i} \tag{13.2}$$

$$c_{n,\text{RMS}}^2 = \frac{\sum\limits_i c_i^2 \tau_i}{\sum\limits_i \tau_i} \tag{13.3}$$

$$t_n(0) = \sum_i \tau_i$$

where the coefficient of x is zero because the dip is zero.

Equation (13.1) is accurate to more than three significant figures for maximum x less than half the depth to the interface and modest changes of c_i. One can use the multilayer reflection ray trace (Section A1.4.3) to test the applicability of Eq. (13.1) to a given structure.

13.2.1 Normal Moveout

Geophysical interpreters look for events that behave like reflections, namely, the times should fit Eq. (13.1). The increase of $t_n(x) - t_n(0)$ as x increases is called the *normal moveout* (NMO):

$$\text{NMO} \equiv t_n(x) - t_n(0) \tag{13.4}$$

The numerical seismogram in Figure 13.2 shows the normal moveout of reflections. For simplicity, direct arrivals, surface waves, and multiple reflections are omitted. Interpreters use the dependence of NMO curves on $t_n(0)$ and $c_{n,\text{RMS}}$ to estimate $c_{n,\text{RMS}}$ or the *stacking velocity*. The stacking velocity is a function of depth or reflection time. Since the stacking velocity is a function of depth for many different reflection times, we label it c_s.

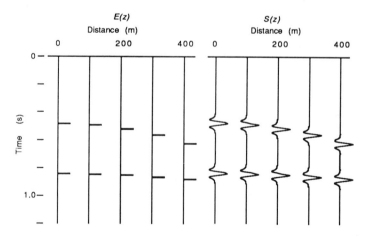

Figure 13.2 Numerical or synthetic reflection seismogram. Horizontally traveling waves and multiple reflected waves are omitted. $E(z)$ is the impulse response. $A(z)$ is the source, a Ricker wavelet. $S(z)$ is the convolution of $E(z)$ and $A(z)$. The peak of the Ricker wavelet is set on the arrival time. Parameters: $c_1 = 1000$ m/s, $h_1 = 240$ m, $c_2 = 2000$ m/s, $h_2 = 360$ m, and $c_{2,\text{rms}} = 1512$ m/s. Vertical incidence reflection times are 0.48 and 0.84 s.

Normal moveout correction is a remapping operation, and we use z-transforms to describe it. The z-transformation expressions for the signal at position x_i and the ith geophone are

$$S_i(z) = A(z) \, E_i(z) \tag{13.5}$$

$$S_i(z) = s_{i0} + s_{i1} \, z + s_{i2} \, z^2 + \cdots \tag{13.6}$$

where $E_i(z)$ is the impulse response, and $A(z)$ is the source function. The remapping operation takes information s_{iq} at time step q and moves it to time step j, that is,

$$s_{iq} \rightarrow s'_{ij} \tag{13.7}$$

$$S_i(z) \rightarrow S'_i(z) \tag{13.8}$$

The remapping of information at i and q to i and j uses an assumed stacking velocity c_s. Ideally, c_s is $c_{n,\mathrm{RMS}}$ for the nth reflection interface. In time steps, q and i are

$$q = \mathrm{INT} \, [(j^2 + i^2)^{1/2} + 0.5] \tag{13.9}$$

$$i = \mathrm{INT} \left[\frac{x_i}{c_s t_0} + 0.5 \right] \tag{13.10}$$

where c_s is the stacking velocity, t_0 is the sampling interval, and $+0.5$ is for round off. We let c_s be CS(J) and a function of time step J in the following basic remapping algorithm:

```
REM  REMAPPING OF S%(I,Q) TO SP%(I,J) FOR NMO CORRECTION
REM  THE SHOT IS AT XSHOT. THE STACKING
REM  VELOCITY CS(J) IS A FUNCTION OF TRAVEL TIME.
REM  USE ARRAYS OF INTEGERS S%(I,Q) AND SP%(I,J)
REM  TO REDUCE MEMORY NEEDS.
REM  JL AND JH ARE THE MIN AND MAX VALUES OF J.
REM  THE LIMIT OF QMAX = 280 IS BASED
REM   ON THE VERTICAL PIXEL RANGE OF 0 TO 300.

     QMAX = 280 : REM QMAX SETS A MAXIMUM TIME
```

$$\text{FOR J =JL TO JH} \tag{13.11}$$
$$\text{FOR I= 0 TO IM - 1} \tag{13.12}$$
$$\text{XI = (X(I)-XSHOT)/(CS(J)*T0)} \tag{13.13}$$
$$\text{Q = INT(SQR(J\char`^2 + XI\char`^2)+0.5)} \tag{13.14}$$
$$\text{IF Q< QMAX THEN SP\%(I,J) = S\%(I,Q)} \tag{13.15}$$
$$\text{NEXT I} \tag{13.16}$$
$$\text{NEXT J} \tag{13.17}$$

Where IM is the number of geophones, and JL and JH set the lower and higher times for the remapping operation. Q is not a linear function of J, and the INT operation in Eq. (13.14) may give the same values of Q for several values of J. Duplications of Q are more likely to happen at large I or X(I). Duplications can stretch the mapping of a single value of S(I,Q) into several SP(I,J) locations.

Presumably the signals in each of the channels are the same after correction for normal moveout. If so, a stacking operation like that in Section 11.1 can enhance the reflections. A signal-stacking algorithm follows:

```
REM  STACK SIGNALS

    FOR J=JL TO JH
      S = 0
        FOR I = 0 TO IM - 1
          S = S + S%(I,J)
        NEXT I
      SS(J) = S / (IM+1) : REM AVERAGE
    NEXT J
```
(13.18)

Where SS(J) is the average stacked signal. Successful enhancement of stacked signals requires good estimates of c_s or CS(J) as functions of t or J.

Examples of the normal moveout correction and stacking of signals are shown in Figure 13.3. The stacking velocities are constant. The stacking velocity of 1000 m/s corrects the normal moveout for the first reflection and does not correct the normal moveout of the second reflection. The stack enhances the first reflection and not the second. The stacking velocity of 1500 m/s corrects the normal moveout of the second reflection and does not correct the normal moveout of the first reflection. Correspondingly, the stacked trace of the first reflection is large for a stacking velocity of 10000 m/s, and the stacked trace of the second reflection is large for a stacking velocity of 1500 m/s.

We can use the dependence of the stacked signal on stacking velocity to estimate the dependence of the stacking velocity or $c_{n,\mathrm{RMS}}$ on reflection time. Figure 13.4 shows the dependence of the stacked signal on stacking velocity. The stacked signal for the first

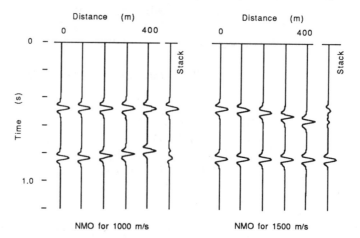

Figure 13.3 Normal moveout corrections. The seismogram in Figure 13.2 is given normal moveout corrections for constant stacking velocities. The trace marked "stack" is a stack of the other five traces. (a) Stacking velocity c_s = 1000 m/s. (b) Stacking velocity c_s = 1500 m/s.

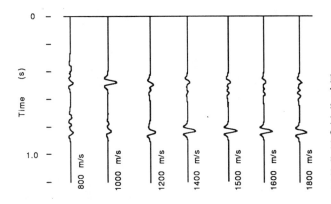

Time (s)

0

1.0

800 m/s
1000 m/s
1200 m/s
1400 m/s
1500 m/s
1600 m/s
1800 m/s

Figure 13.4 Stacking velocity analysis. The structure and synthetic seismogram is shown in Figure 13.2. Stacking velocity analysis compares stacked signals for a set of constant stacking velocities. The stacking velocity of 1000 m/s enhances reflection 1. The stacking velocity of 1512 m/s, or here 1500 m/s, enhances reflection 2 a little more than 1400 m/s or 1600 m/s.

reflection is clearly a maximum for $c_{1,\text{RMS}} = 1000$ m/s. The second reflection has a broad maximum that ranges from 1400 m/s to 1600 m/s. We choose a stacking velocity of 1500 m/s for the second reflection.

13.2.2 Depth Dependence of c_s

Full normal moveout correction requires the inclusion of the time or depth dependence of the stacking velocity. If we knew, as in our numerical examples, that the structure was two uniform layers, then we could use Eq. (13.3) to compute the stacking velocity. Assume as a numerical example that layer 1 has $c_1 = 1000$ m/s and $t_1 = 0.48$ s. Assume that layer 2 has a $c_2 = 2000$ m/s and variable thickness. From Eq. (13.3) the two-layer expression is

$$c_{2,\text{RMS}}^2 = \frac{c_1^2 t_1 + c_2^2 t_2}{t_1 + t_2}$$

$$t = (t_1 + t_2) \tag{13.19}$$

A graph of a numerical evaluation is shown in Figure 13.5. The stacking velocity c_s is constant by assumption for $t < t_1$. The stacking velocity varies smoothly from 1000 m/s to 1500 m/s for $t >$ greater t_1. Since the seismic velocity in a real layer usually changes from top to bottom, and we do not know what it is, we use a simple estimation of the stacking velocity. A linear interpolation between $c_{1,\text{RMS}}$ at t_1 and $c_{2,\text{RMS}}$ at t_2 is reasonable

$$\left.\begin{array}{ll} c_s = c_1 & \text{for } t \leq t_1 \\ c_s = c_1 + b(t - t_1) & \text{for } t > t_1 \end{array}\right\} \tag{13.20}$$

The corrected seismogram and its stack are shown in Figure 13.6.

The normal moveout correction is not a linear transformation. The nonlinearity causes the obvious distortions of the reflections at the 300- and 400-m stations. Other stations and events have less distortion.

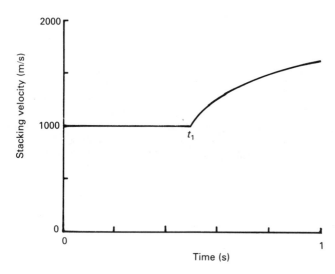

Figure 13.5 Stacking velocity versus reflection time. The stacking of RMS velocity depends on the thickness and seismic velocities in the layers. The thickness of layer 2 is varied. Except for the thickness of layer 2, the parameters are the same as in Figure 13.2.

13.2.3 Interval Velocity

Given estimates of the reflection times and $c_{n,\text{RMS}}$ to the reflection, we can estimate the interval velocity. As shown in Section A1.2 of Appendix A, rearrangement and manipulations of Eqs. (13.1) to (13.3) gives the Dix-Dürbaum equation,

$$c_2^2 = \frac{c_{2,\text{RMS}}^2\, t_2(0) - c_{1,\text{RMS}}^2\, t_1(0)}{t_2(0) - t_1(0)} \tag{13.21}$$

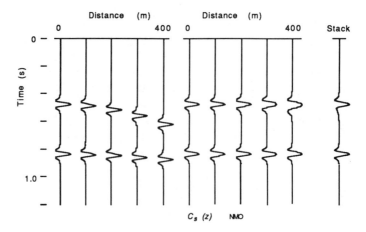

Figure 13.6 Normal moveout-corrected section. The synthetic seismogram in Figure 13.2 has been corrected for normal moveout. The stacking velocities are

$$c_s = 1000 \text{ for } 0 \leq t \leq 0.48 \text{ s}$$
$$c_s = 1000 + 1388\,(t - 0.48) \text{ for } t > 0.48 \text{ s}$$

The substitution of

$$\left.\begin{array}{r} c_{1,RMS} = 1000 \text{ m/s} \\ t_1(0) = 0.48 \text{ s} \\ c_{2,RMS} = 1500 \text{ m/s} \\ t_2(0) = 0.84 \text{ s} \end{array}\right\} \tag{13.22}$$

gives $c_2 = 1979$ m/s. The error is less than 2% of the true value, 2000 m/s.

As an optical analogy, the combination of normal moveout correction and stacking focuses the geophone array on an image of the source. The time shifts in the normal moveout correction serve the same function as the lens in an optical system. The image at the focus of a lens is the coherent sum of many rays that pass through different parts of the lens. It is the same for ray paths to different geophones. The focusing lens enhances the image of the object and degrades contributions from other sources. Reflection stacking enhances the reflected image and degrades the seismic waves from other sources.

The numerical examples demonstrate the focusing character of reflection stacking. For example, in Figure 13.3 the stacking velocity of 1000 m/s focuses the array on image sources along the z-axis in a medium having a seismic velocity of 1000 m/s. The image source at 0.84 s in a medium having a composite velocity of 1512 m/s is out of focus. Sources in other directions are out of focus. The choice of $c_s = 1512$ m/s brings the second image reflection at 0.84 s into focus and defocuses the first source. The small array of five geophones shows an impressive difference between being focused on a reflected image or not.

13.3 COMMON MIDPOINT STACKING

The normal moveout removal and stacking procedure in Section 13.2 assumes that the reflection interface is a planar surface. As shown in Figure 13.7a, the seismic waves interact in the region AA', and here the interface should be a plane. After adding extensions to remove edge effects, AA' is greater than half the spread length. When the structure is smaller than AA', the stacking procedure in Section 13.2 does not work.

A minor change of the geometry and stacking procedure eliminates most of the trouble. As shown in Figure 13.7b, let the source and receiver move symmetrically away from the midpoint. The seismic waves reflect at the same patch of interface for each transmission. This technique is known as common depth point (CDP) or common midpoint (CMP). CMP techniques require a large number of source positions, geophone stations, and data processing steps. Figure 13.8 shows the geometry and nomenclature for a CMP stack. The algebraic relations are

$$x_m = \frac{(r - s)}{2} \tag{13.23}$$

$$m = \frac{(r + s)}{2} \tag{13.24}$$

where x_m is the half-offset distance.

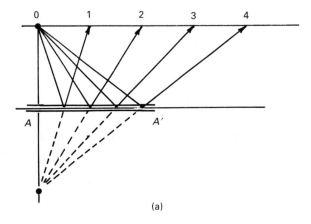

(a)

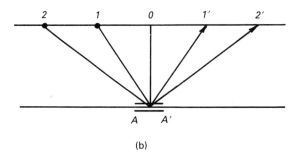

(b)

Figure 13.7 Geometry for measurements. (a) Common or fixed source. The interaction region AA' extends beyond the ray-path intersections to eliminate edge effects. (b) Common or fixed midpoint. Source at 0 transmits to a geophone at 0. Source at 1 transmits to a geophone at $1'$, and so forth. The distance AA' depends on interface depth, geometry, and seismic wavelength.

For common midpoint stacking, we hold m constant and chose a set of x_m that satisfies Eqs. (13.23) and (13.24). Figure 13.9 shows the positions of geophones and sources on orthogonal axes. The lines of constant m help in determining the set of source and receiver positions for a stack at m. In Figure 13.9a, for example, the combinations of source at 1 and geophone at 6, source at 2 and geophone at 5, source at 3 and geophone

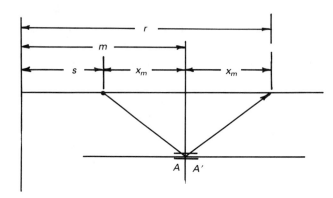

Figure 13.8 Common mid-point geometry. The source and receiver are located relative to an arbitrary coordinate system. x_m is the half-offset distance. The reflecting interface is AA'.

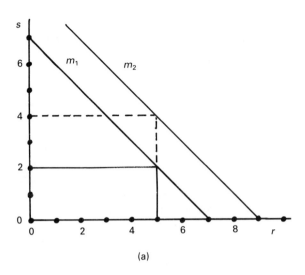

(a)

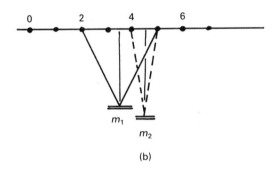

(b)

Figure 13.9 Common midpoint. (a) s is the coordinate of source positions. r is the coordinate of receiver positions. The common midpoint lines m_1 and m_2 are shown. A geophone at 5 and source at 2 are on the m_1 midpoint. A geophone at 5 and source at 4 are on the m_2 midpoint. (b) Ray paths for midpoints m_1 and m_2. For illustration m_1 and m_2 are shown as being at different depths.

at 4, and so forth, all stack on the position m_1. The stacks on m_2 use different combinations of the source and geophone positions. At least in a simple way, the stack at m_1 and the stack at m_2 could locate interfaces at different depths (Figure 13.9b).

The common midpoint technique was introduced by W. H. Mayne in 1962. The technique is so powerful that it has become a standard processing technique, but it has limitations. For example, the interfaces are assumed to be horizontal, and ray paths are drawn as if the rays reflect at a point. Seismic signals are waves and reflect at an area. The quality of the reflection depends on the roughness and dimensions of the area.

PROBLEMS

The normal moveout correction uses the remapping algorithm (13.11)–(13.17). It moves information from an array S(I, Q) to an array SP(I, J). Use the data shown in Figure 13.2 and compute an $E_i(z)$ for first reflection for receiver positions from 0 to 600 m. These $E(z)$ will be used in the problems. Convolve $E_i(z)$ with a Ricker wavelet to get $A_i(z)$.

1. Give and state your reasons for choosing an ideal impulsive waveform for a signal in normal movement correction and stacking operations. (*Hint*: Compare normal movement corrections on $E_i(z)$ and $S_i(z)$.)

2. Compare the waveforms of the reflected signal $S_i(z)$ for $x = 0$ and $x = 600$ m. Do the normal moveout operation and compare $SP_i(z)$ for $x = 0$ and $x = 600$ m.

3. Suppose there is a reverberating signal such as in the examples in Section 7.1. Show graphically what the normal moveout correction does to such a signal.

4. Consider the following sequence of operations: deconvolution, normal moveout correction, and stacking. Compare the results of doing deconvolution and normal movement correction with normal moveout correction and deconvolution. Does the order of operations matter? Why or why not?

5. Read J. W. Dunkin and F. K. Levin, "Effect of normal moveout on a seismic pulse," *Geophysics*, 38 (1973), 635–42. Do your answers and reasons agree with Dunkin and Levin?

14

Reflections
at Rough Interfaces

We begin an exploration of an earth that is more realistic than a stack of plane parallel layers. We want to include reflections at finite patches of an interface, roughness of an interface, and faults. Essentially the medium can be inhomogeneous. The analysis of the reflection and scattering of seismic waves in such media requires more powerful expressions than we have developed so far. Introductory physics texts show the usefulness of the Huygens-Fresnel principle and its constructions in developing solutions in many optical diffraction problems. The same wavelet constructions work in geophysics. Mathematical formulations of the Huygens-Fresnel principle are known as the Kirchhoff diffraction formula and alternatively as the Helmholtz-Kirchhoff theory.

14.1 HUYGENS-FRESNEL PRINCIPLE AND KIRCHHOFF'S DIFFRACTION FORMULA

The story of a theory for the diffraction and scattering of waves by a randomly rough surface spans three centuries. Skipping many actors, including Newton (1642–1727), we give a brief history of wave theory. Christian Huygens (1629–95) gave a conceptual basis for describing traveling waves in a complicated medium. His hypothesis was that *each point on an advancing wave front can be considered as a source of secondary waves that move forward as spherical wavelets in an isotropic medium. The envelope of the wavelets is a new wave front.* The geometric construction is sketched in Figure 14.1 Augustin Jean Fresnel (1788–1827) added two very important postulates to Huygens' hypothesis: (1) *The secondary wavelets mutually interfere.* This postulate states that superposition applies to the fields of the wavelets. (2) *Wavelet strengths are strongest in the forward direction.* The dependence of wavelet strength on direction is known as the *Fresnel obliquity factor*

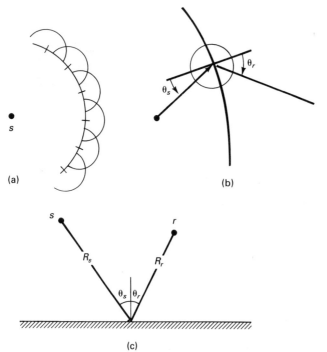

(a)

(b)

(c)

Figure 14.1 Huygens-Fresnel constructions. (a) Huygens construction of wavelets: (b) Geometry for the Fresnel obliquity factor in transmission. (c) Geometry for the obliquity factor in reflection.

$f(\theta_s, \theta_r)$. The combination is known as the Huygens-Fresnel principle. George G. Stokes (1819–1903) and later Gustav Kirchhoff (1832–87) derived the obliquity factor

$$f(\theta_s, \theta_r) = \frac{ik(\cos \theta_s + \cos \theta_r)}{(4\pi)} \tag{14.1}$$

Kirchhoff put the theory of diffraction on a sound mathematical basis.

Rather than derive the Kirchhoff diffraction formula, we give a heuristic justification. The geometry is shown in Figure 14.2. We use (signal) pressures in the development because the pressure contributions are scalar and add algebraically. The source is at s and has the angular frequency ω. A pressure wave from the source spreads spherically, and the pressure of the incident wave at dS is

$$P_s \sim \frac{e^{i(\omega t - kR_s)}}{R_s}$$

$$k \equiv \frac{\omega}{c} \tag{14.2}$$

We use Huygens' hypothesis and let dS be the source of an outgoing wavelet. The strength of the wavelet is proportional to the incident pressure, the reflection coefficient R_{12}^p, Fresnel's obliquity factor, and dS,

$$\text{Wavelet source} \sim \frac{ik}{4\pi} (\cos \theta_s + \cos \theta_r) R_{12}^p \, p_s \, dS \tag{14.3}$$

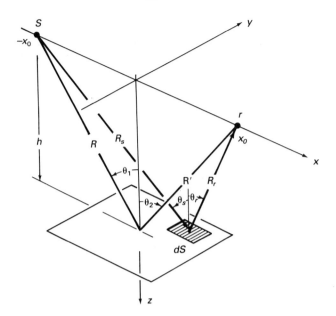

Figure 14.2 Geometry for Huygens scattering. In specular direction $R = R'$ and $\theta_1 = \theta_2$.

The wavelet spreads spherically, and the differential pressure at the receiver is

$$dp_r \sim \frac{ik}{4\pi R_s R_r} (\cos \theta_s + \cos \theta_r) \, e^{i[\omega t - k(R_s + R_r)]} \, dS \qquad (14.4)$$

from Fresnel's superposition postulate, the pressure is the sum or integral over all wavelets,

$$p_r \simeq \frac{ikB}{4\pi} \int_S R_{12}^p \frac{(\cos \theta_s + \cos \theta_r)}{R_s R_r} \, e^{i[\omega t - k(R_s + R_r)]} \, dS \qquad (14.5)$$

where B replaces the proportionality constant. Equation (14.5) is our version of the Kirchhoff diffraction formula for reflection geometry. We obtain the standard version of the diffraction integral by letting $R_{12}^p = 1$. Section C1.1 of Appendix C gives a derivation of Eq. (14.5) from the Helmholtz-Kirchhoff integral.

The diffraction of sound waves by a periodic corrugated surface was given by John Strutt (Lord Rayleigh, 1842–1919) in his *Theory of Sound* (1896). Rayleigh used an incident plane wave and an expansion of outgoing waves to satisfy the boundary conditions at the interface. Surprisingly there was little activity in rough-surface theory until the early 1950s when L. M. Brekhovskikh and Carl Eckart (1902–73) applied Kirchhoff theory to rough-surface problems.

Eckart showed that one could combine a statistical description of the sea surface and a mathematical form of the Huygens-Fresnel principle to derive statistical descriptions of the scattered signals. Since he wrote his paper, many people have used and extended his theory. Eckart's formulation and its extensions give good estimates of scattered signals for the following conditions: The smallest radii of curvature of the rough surface should be much larger than the wavelength. The estimates are good from normal incidence to perhaps 45°. Estimates may become poor at larger angles of incidence.

During the past few years some of us have been using Biot and Tolstoy's exact solution for an impulsive point source near an ideal wedge as the basis of a rough-surface theory. Section C1.2 of Appendix C gives the final equations of the Biot-Tolstoy solution and an example.

14.1.1 Reflections at a Patch of Interface and Fresnel Integrals

The Kirchhoff diffraction equation is a nice place to start a discussion of reflections at a patch of interface. The geometry is shown in Figure 14.2. The seismic properties are ρ_1 and c_1 above the interface and ρ_2 and c_2 below the interface. The source is at coordinates $(-x_0, 0, 0)$. The interface is a plane at $z = h$. The distances to the scatterer dS are

$$\left.\begin{aligned}R_s^2 &= (x + x_0)^2 + h^2 + y^2 \\ R_r^2 &= (x - x_0)^2 + h^2 + y^2\end{aligned}\right\} \tag{14.6}$$

A sequence of algebraic manipulations and expansions follows to get a convenient approximate form for $R_s + R_r$. We define R as follows:

$$R^2 \equiv x_0^2 + h^2 \tag{14.7}$$

and square $(x + x_0)$ and $(x - x_0)$ in Eq. 14.1.6. We substitute R^2 for $x_0^2 + h^2$ and take the square roots,

$$\left.\begin{aligned}R_s &= R\left(1 + \frac{x^2}{R^2} + \frac{y^2}{R^2} + \frac{2x_0x}{R^2}\right)^{1/2} \\ R_r &= R\left(1 + \frac{x^2}{R^2} + \frac{y^2}{R^2} - \frac{2x_0x}{R^2}\right)^{1/2}\end{aligned}\right\} \tag{14.8}$$

The binomial expansion needs the second-order term

$$(1 + a)^{1/2} = 1 + \frac{a}{2} - \frac{a^2}{8} + \cdots \tag{14.9}$$

The expansions of Eq. (14.8) are

$$\left.\begin{aligned}R_s &\simeq R + \left(1 - \frac{x_0^2}{R^2}\right)\frac{x^2}{2R} + \frac{y^2}{2R} + \frac{x_0x}{R} + \cdots \\ R_r &\simeq R + \left(1 - \frac{x_0^2}{R^2}\right)\frac{x^2}{2R} + \frac{y^2}{2R} - \frac{x_0x}{R} + \cdots\end{aligned}\right\} \tag{14.10}$$

where we keep terms of the order x, x^2, and y^2 and drop the higher terms. The sum $(R_s + R_r)$ is, using Eq. (14.7),

$$(R_s + R_r) \simeq 2R + \cos^2\theta_1 \frac{x^2}{R} + \frac{y^2}{R} \tag{14.11}$$

$$\left.\begin{aligned}\cos\theta_1 &= \frac{h}{R} \\ \sin\theta_1 &= \frac{x_0}{R}\end{aligned}\right\} \tag{14.12}$$

The substitution of Eq. (14.11) in the Kirchhoff diffraction formula, Eq. (14.5), gives

$$p_r \simeq \frac{ikB}{4\pi} e^{i(\omega t - 2kR)} \iint_S R_{12}^p \frac{(\cos\theta_s + \cos\theta_r)}{R_s R_r}$$

$$\times \exp\left[-ik \left(\cos^2\theta_1 \frac{x^2}{R} + \frac{y^2}{R} \right) \right] dy\, dx \qquad (14.13)$$

This expression is difficult to evaluate, so people use a number of approximations and simplifications. A common simplification is to remove the slowly varying terms from under the integral sign. For many source and receiver geometries, the slowly varying terms are R_{12}^p, $(\cos\theta_s + \cos\theta_r)$, R_s, and R_r. The complex exponential is a rapidly varying term. The integrals of functions such as $\exp(i\, a\, g^2)$ are known as Fresnel integrals, and mathematical handbooks give their properties and numerical tables. The following development uses the slowly varying approximation, that is, it moves the terms out of the integral and evaluates the Fresnel integral.

14.1.2 Finite Scattering Areas

We know from our own experience of looking in mirrors that a finite and even small interface gives a fully reflected image. The same is true for reflected seismic signals. Our purpose in this section is to determine the dimensions of a patch that give the same reflected amplitude as an infinite interface. We call this area a *Fresnel* patch.

The first task in evaluating the diffraction formula (Eq. (14.13)) is to remove the slowly varying terms from under the integral sign. For patch dimensions that are small compared with R_s and R_r, the approximations are

$$\left.\begin{array}{c} R \simeq R_s \simeq R_r \\[4pt] 2\cos\theta_1 \simeq \cos\theta_s + \cos\theta_r \\[4pt] R_{12}^p \text{ evaluated at } \theta = \theta_1 \end{array}\right\} \qquad (14.14)$$

The double integral over dx and dy is equivalent to the product of a pair of integrals. After these changes, p_r becomes

$$p_r \simeq \frac{ikB \cos\theta_1}{2\pi R^2} R_{12}^p\, e^{i(\omega t - 2kR)}\, I_x I_y \qquad (14.15)$$

$$I(x_s) \equiv \int_{-x_s}^{x_s} \exp\left(-\frac{ik}{R} \cos^2\theta_1\, x^2 \right) dx \qquad (14.16)$$

$$I(y_s) \equiv \int_{-y_s}^{y_s} \exp\left(-\frac{ik}{R} y^2 \right) dy \qquad (14.17)$$

where x_s and y_s are the limits of integration on the surface S. Since the exponential functions in both Eqs. (14.16) and (14.17) are even functions, the limits of integration can be changed as follows:

$$I(x_s) = \int_{-x_s}^{x_s} = \int_{-x_s}^{0} + \int_{0}^{x_s} = 2\int_{0}^{x_s} \qquad (14.18)$$

and the pair of integrals become

$$I(x_s) = 2 \int_0^{x_s} \exp\left(-\frac{ik}{R} \cos^2 \theta_1 \, x^2\right) dx$$

$$I(y_s) = 2 \int_0^{y_s} \exp\left(-\frac{ik}{R} y^2\right) dy \qquad\qquad (14.19)$$

Expansion of the exponentials as cosine $+ i$ sine gives

$$I(x_s) = 2 \int_0^{x_s} \cos\left(\frac{k}{R}\cos^2 \theta_1 \, x^2\right) dx - i2 \int_0^{x_s} \sin\left(\frac{k}{R}\cos^2 \theta_1 \, x^2\right) dx$$

$$I(y_s) = 2 \int_0^{y_s} \cos\left(\frac{k}{R} y^2\right) dy - i2 \int_0^{y_s} \sin\left(\frac{k}{R} y^2\right) dy \qquad (14.20)$$

The integrals in Eq. (14.20) have the form of the standard Fresnel integrals $C(w)$ and $S(w)$

$$C(w) \equiv \int_0^w \cos\left(\frac{\pi}{2} u^2\right) du$$

$$S(w) = \int_0^w \sin\left(\frac{\pi}{2} u^2\right) du$$

$$F(w) \equiv [C(w)^2 + S(w)^2]^{1/2} \qquad\qquad (14.21)$$

$$C(\infty) = 0.5$$

$$S(\infty) = 0.5$$

$$F(\infty) = 0.707$$

$C(w)$ and $S(w)$ are evaluated numerically in the *Handbook of Mathematical Functions* (M. Abramowitz and I. A. Stegun, Nat. Bur. Stand. Washington, D.C.: Govt. Printing Office, 1964). Graphs of $C(w)$, $S(w)$, and $F(w)$ are shown in Figure 14.3.

The changes of variables from u to x and y or vice versa are

$$x = \left(\frac{\pi R}{2 k \cos^2 \theta_1}\right)^{1/2} u$$

$$y = \left(\frac{\pi R}{2 k}\right)^{1/2} u \qquad\qquad (14.22)$$

and the corresponding limits are

$$x_s = \left(\frac{\pi R}{2 k \cos^2 \theta_1}\right)^{1/2} w_x$$

$$y_s = \left(\frac{\pi R}{2 k}\right)^{1/2} w_y \qquad\qquad (14.23)$$

With these changes of variables, the product $I(x_s)\, I(y_s)$ becomes

$$I(x_s)\, I(y_s) = \frac{2\pi R}{k \cos \theta_1} [C(w_x) - iS(w_x)] [C(w_y) - iS(w_y)] \qquad (14.24)$$

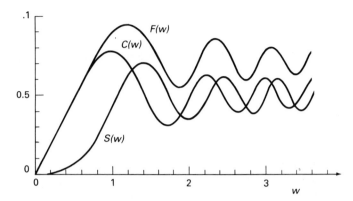

Figure 14.3 Fresnel integrals.

$$C(w) = \int_0^w \cos(\pi u^2/2)\, du$$

$$S(w) = \int_0^w \sin(\pi u^2/2)\, du$$

$$F(w) = [C^2(w) + S^2(w)]^{1/2}$$

The limiting values as w tends to ∞ are

$$C(w) = S(w) = 0.5 \qquad F(w) = 2^{-1/2}$$

and the reflected signal (Eq. (14.15)) is

$$p_r \simeq \frac{iB}{R} R_{12}^p\, e^{i(\omega t - 2kR)}\, [C(w_x) - iS(w_x)] \times [C(w_y) - iS(w_y)] \tag{14.25}$$

p_r is the signal reflected by a patch of surface having the dimensions $2x_s$ by $2y_s$. From Figure 14.3 the Fresnel functions increase and then have damped oscillations about their asymptotic limits as w tends to infinity. At infinite limits the Fresnel functions are

$$\left.\begin{aligned}
C(\infty) - iS(\infty) &= \frac{(1 - i)}{2} \\[2mm]
[C(\infty) - iS(\infty)]^2 &= \frac{-i}{2} \\[2mm]
F^2(\infty) &= 0.5
\end{aligned}\right\} \tag{14.26}$$

and the corresponding reflected signal is

$$p_r \simeq \frac{B}{2R} R_{12}^p\, e^{i(\omega t - 2kR)} \tag{14.27}$$

and this is the image or ray-path reflection. Integration of the Kirchhoff diffraction formula gives the "right answer."

The dimensions of a finite patch that give the same amplitude of reflection as the asymptotic limits is defined as a *Fresnel patch*. From Figure 14.3 $F(w)$ crosses the asymptotic limiting value 0.707 at

$$w_1 \simeq \frac{3}{4} \tag{14.28}$$

from Eqs. (14.23) and (14.28) the corresponding half-width dimensions are

$$\left. \begin{aligned} x_1 &= \frac{3}{8} \frac{(\lambda R)^{1/2}}{\cos \theta_1} \\ y_1 &= \frac{3}{8} (\lambda R)^{1/2} \end{aligned} \right\} \tag{14.29}$$

We call a patch having the dimensions $2x_1$ by $2y_1$ a Fresnel patch. The Fresnel patch is analogous to the first Fresnel zone in the transmission of light through a circular aperture.

In Chapter 5 and particularly in Section 5.2, we introduced the concept of a ray tube and the patch of a plane-wave front that travels with the ray. Without going into a long argument, it is evident by analogy that the Fresnel patch corresponds to a patch of wave front traveling with the ray. (Remember that x_1 is the projection on a horizontal interface.) The dimensions grow with the square root of distance along the ray path.

PROBLEMS

Purpose: to explore the dependence of a reflected signal on the dimensions of a Fresnel patch, geometry, and wavelength. For all problems assume $c_1 = 2000$ m/s, $\rho_1 = 2000$ kg/m^3, $h = 1000$ m, $c_2 = 3000$ m/s, and $\rho_2 = 2800$ kg/m^3.

1. At vertical incidence for a colocated source and receiver, compute the Fresnel patch dimensions, $2x_1$ *and* $2y_1$, for 10-, 20-, . . . , and 100-Hz signals. (*Answer*: 10 Hz; width = 335 m)

2. A plane-reflecting facet or patch has the dimensions $2x_1$, or width = 100 m, and $2y_1$, or length = 400 m. For a colocated source and receiver over the patch compute and graph the relative amplitude of reflected signals for signal frequencies of 10, 20, . . . , and 200 Hz. Let B = 1 m. Use the reflection of the signals from an infinite plane interface as the reference. *Hint*: Consider Figure 14.3. (a) Graph the results versus frequency. (b) Explain why the patch reflections differ from 1. (c) How good is the approximation that the *patch reflection coefficient* is 1 for patch widths greater than $2x_1$. (d) Give an approximate expression for patch reflection coefficients for patch widths less than $2x_1$.

14.2 SIGNAL STACKING AND THE REFLECTION AND SCATTERING OF SIGNALS AT ROUGH SURFACES

It is well known that the reflection signals from an interface can vary from strong to weak along a seismic profile. In Chapter 7 we found that a change of the properties of a layer— for example, the layer thins—can cause the reflection signal to weaken. This situation describes sedimentary layers. Unconformities, eroded surfaces, and sea floors at spreading centers are rough. The roughness of the interface affects the quality of a reflection. If the roughness of an interface changes along a seismic profile, then we can expect the quality of the reflected signal to change.

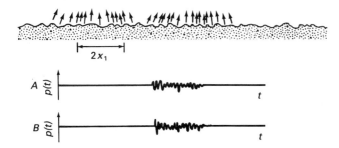

Figure 14.4 Signals scattered at a rough interface. An impulsive source (black dot) transmits, and the outgoing wave front interacts at the rough interface. The up-going arrows indicate sourcelets of Huygens wavelets. The receiver is ∇. The width of a Fresnel patch $2x_1$ is indicated. The same measurements are made at positions A and B. Sketches of signals $p(t)$ for these positions are shown.

Figure 14.4 illustrates the problem. The lateral dimensions of the rough features are much less than the dimensions of a Fresnel patch, so no one feature reflects like a mirror. If we imagine that each element of roughness scatters Huygens wavelets, then the sum of the wavelets at the receiver gives a complicated signal. Over a randomly rough interface the sequences of contributions add to give a random signal. The signals at source and receiver positions A and B are different. A stack of signals over a rough bottom must include the effects of roughness.

14.2.1 Roughness in the Helmholtz-Kirchhoff Integral

The scattering of waves at a rough interface is one of the more intractable problems in geophysics. Accordingly, "solutions" have numerous simplifications and limitations. Even so, the analysis is interesting because it shows how signal-stacking procedures enhance one component of the signals and suppress other components.

To insert roughness in the Kirchhoff formula (Eq. (14.13)), let ζ be the depression from a horizontal plane at depth h, Figure 14.5. The values of R_s and R_r are

$$\left. \begin{aligned} R_s^2 &= (x + x_0)^2 + y^2 + (h + \zeta)^2 \\ R_r^2 &= (x - x_0)^2 + y^2 + (h + \zeta)^2 \\ \zeta &\ll h \end{aligned} \right\} \tag{14.30}$$

where ζ is a function of x and y, and the depth to dS is

$$z = h + \zeta \tag{14.31}$$

The algebraic manipulations are the same as in Eqs. (14.6) to (14.12). Keeping the terms in x^2, y^2, and ζ, we have

$$R_s + R_r \simeq 2R + \cos^2\theta_1 \frac{x^2}{R} + \frac{y^2}{R} + 2\zeta \cos \theta_1 \tag{14.32}$$

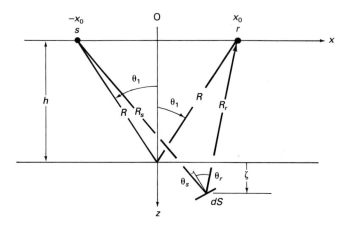

Figure 14.5   Scattering at a rough surface. The y-dependence is not shown for simplicity. The scattering area dS is at (x, y, z), $z = h + \zeta$.

This modification gives a new Kirchhoff diffraction formula,

$$p_r \simeq \frac{ikB}{4\pi} e^{i(\omega t - 2kR)} \iint_S R_{12}^p \frac{(\cos\theta_s + \cos\theta_r)}{R_s R_r}$$

$$\times \exp\left[-ik\left(\cos^2\theta_1 \frac{x^2}{R} + \frac{y^2}{R}\right) - i2k\zeta\cos\theta_1 \right] dy \, dx \tag{14.33}$$

We want to simplify the integral by moving the slowly varying terms out. Again, since the patch is small,

$$R \simeq R_s \simeq R_r$$

and $R_s R_r$ becomes R^2. The values of $\cos\theta_s + \cos\theta_r$ depend on the local slope of dS. The approximation

$$2\cos\theta_1 \simeq \cos\theta_s + \cos\theta_r \tag{14.34}$$

is acceptable for small slopes. For example, a local slope of $20°$ causes the right side of Eq. (14.34) to be about 6% less than the left side at vertical incidence, $\theta_1 = 0$. R_{12}^p also depends on the local angle of incidence. Except near critical angles, R_{12}^p is a slowly varying function of θ_1. The simplified diffraction integral, including roughness, is

$$p_r \simeq \frac{ikB R_{12}^p \cos\theta_1}{2\pi R^2} e^{i(\omega t - 2kR)}$$

$$\times \iint_S \exp\left[-\frac{ik\cos^2\theta_1}{R}x^2 - \frac{iky^2}{R} - i2k\zeta\cos\theta_1 \right] dy \, dx \tag{14.35}$$

Evaluation of Eq. (14.35) requires specific assumptions about the nature of ζ. It is difficult to evaluate the integral theoretically for an arbitrary roughness ζ as a function of x. Although the algebra and expansions are tedious, a specific example such as $\zeta = b\cos kx$ can be evaluated in approximation. Perhaps surprisingly, the computation of the statistical properties of p_r for simple randomly rough surfaces is relatively easy. The next section gives the coherently reflected component of the signal reflected and scattered at a rough surface.

14.2.2 Eckart's Coherent Reflection from a Randomly Rough Interface

Eckart's short paper (see Problem 3) introduced a technique for estimating the mean signal and mean squared signal reflected at a randomly rough sea surface. We use his technique here.

In the signal-scattering literature, terms such as mean signal, coherent component, first moment of the signal, and stacked signal generally mean the same thing. Equation (14.35) is a good starting place. The stack or average of p_r over an ensemble of rough surfaces is

$$\langle p_r \rangle \simeq \frac{ikB\, R^p_{12}}{2\pi R^2} e^{i(\omega t - 2kR)}$$
$$\times \left\langle \iint_S \exp\left(-\frac{ik\cos^2\theta_1}{R}x^2 - \frac{iky^2}{R} - 2ik\zeta\cos\theta_1\right) dy\, dx \right\rangle_\zeta \tag{14.36}$$

where $\langle\ \rangle_\zeta$ indicates average over ζ, and ζ is a function of x and y. Eckart evaluated the integral by averaging it over the probability density function (PDF) of ζ. The PDF of ζ, $w(\zeta)$, is defined as follows: The probability of observing ζ between ζ_1 and $\zeta_1 + d\zeta$ is $w(\zeta_1)\, d\zeta$. If we follow Eckart and ignore the dependence of ζ on x and y, the average operates only on $\exp(-i2k\zeta\cos\theta_1)$ in Eq. (14.36). The rest of the integral is the same as in Eq. (14.15). If we let the surface S have infinite limits, the double integral reduces to Eq. (14.27) or p_0.

$$\left.\begin{array}{l} p_0 \equiv \dfrac{B\, R^p_{12}}{2R} e^{i(\omega t - 2kR)} \\[3mm] |p_0| \equiv \dfrac{B\, R^p_{12}}{2R} \end{array}\right\} \tag{14.37}$$

Thus, Eq. (14.36) takes a simpler form,

$$\langle p_r \rangle_\zeta \simeq p_0 \langle\exp(-i2k\zeta\cos\theta_1)\rangle_\zeta \tag{14.38}$$

If we use the PDF, Eq. (14.38) becomes

$$\langle p_r \rangle_\zeta = p_0 \int_{-\infty}^{\infty} w(\zeta) \exp(-i2k\zeta\cos\theta_1)\, d\zeta \tag{14.39}$$

The infinite limits are convenient. Some readers will recognize Eq. (14.39) as having the form of a Fourier integral transformation, Section B1 of Appendix B. In the absence of data for a specific surface, many people assume that the surface has a normal (Gaussian) distribution

$$w(\zeta) \equiv (2\pi\sigma^2)^{-1/2} \exp\left(-\frac{\zeta^2}{2\sigma^2}\right) \tag{14.40}$$

where σ is the RMS roughness. The integration for the normal PDF follows directly by using the integral

$$\int_{-\infty}^{\infty} e^{-au^2 + ibu} \, du = \left(\frac{\pi}{a}\right)^{1/2} e^{-b^2/4a} \tag{14.41}$$

where

$$\left. \begin{array}{l} a = \dfrac{1}{2\sigma^2} \\[2em] b = -2k \cos \theta_1 \end{array} \right\} \tag{14.42}$$

and

$$\langle p_r \rangle_\zeta \simeq p_0 \, e^{-2k^2\sigma^2 \cos^2 \theta_1} \tag{14.43}$$

Many measurements of underwater sound signals scattered at windblown sea surfaces have demonstrated the usefulness of Eq. (14.43). A set of laboratory data is shown in Figure 14.6. Even though the PDF of the surface is not normal, Eq. (14.43) shows less than 10% error over the range $0 < k\sigma \cos \theta_1 < 1$.

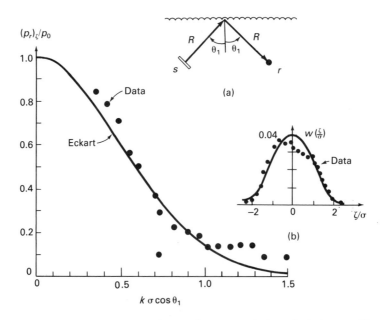

Figure 14.6 Stacked or coherent reflection. The geometry is shown in the inset (a). The angle of incidence was 45°. The frequency of the signal was varied. The rough surface was a laboratory wind-driven sea $\sigma = 0.45$ cm and $\Lambda = 16$ cm. Inset (b) shows the measured $w(\zeta)$ or $w(\zeta/\sigma)$ of the surface. The solid curve is the Gaussian fit to the data, Eq. (14.40). The coherent components are $\langle p_r \rangle / p_0$. Eckart is Eq. (14.43). $\theta_1 = \theta_2 = 45°$ and $R = 3.8$ m. The data are replotted from N. Mayo, H. Medwin, and W. M. Wright, *J. Acoust. Soc. Am.*, 45 (1970), 112(a), and also C. S. Clay and H. Medwin, *Acoustical Oceanography*, Fig. (10.4.4). New York: John Wiley, 1976.

The averaging operations in Eqs. (14.36) and (14.38) are equivalent to signal stacking. Signal stacking enhances the reflections from a plane interface and reduces information from rough interfaces. In sedimentary basins, depositional process gives nearly plane interfaces. These plane interfaces may be altered by later tilting, faulting, and erosional processes. Even so, $k\sigma \cos \theta_1$ may be small, and signal stacking gives excellent results in the search for petroleum in sedimentary basins. Cratonic and volcanic regions are different. Here the structures may be very irregular, rough, and inhomogeneous. The roughness $k\sigma \cos \theta_1$ may be large, and we would expect Eq. (14.43) to give very small values of the mean signal. A direct transfer of seismic exploration methods from sedimentary basins to cratonic and volcanic regions may give poor results.

PROBLEMS

3. Read C. Eckart; "The scattering of sound from the sea surface," *J. Acoust. Soc. Am.*, 25 (1953), 566–70, and E. O. LaCase and P. Tamarkin; "Underwater sound reflection from a corrugated surface," *J. Appl. Phys.*, 27 (1956), 138–48. LaCase and Tamarkin give comparisons of Rayleigh, Brekhovskikh, and Eckart theories. They help in explaining Eckart's paper. D. F. McCammon and S. T. McDaniel; "Application of a new theoretical treatment to an old problem; sinusoidal pressure release boundary scattering," *J. Acoust. Soc. Am.*, 78 (1985) 149–56, show that satisfaction of the correct boundary conditions gives a remarkably good fit of theory and the LaCase and Tamarkin data. The theory is advanced and you can read about the difficulty of the Kirchhoff type of solution. They show that the Kirchhoff is pretty good for specular reflections and surfaces having $h/\lambda < 0.1$.

4. The sound velocity is 200 m/s, and the interface is 2000 m below the surface. For collocated source and receivers and vertical incidence, compute the relative mean or stacked reflection $\langle p_r \rangle_\zeta / p_0$ for an rms roughness of $\sigma = 1$ m and $\sigma = 3$ m. Compare the results for $f = 10$, 20, . . . , 100 Hz.

5. The stacked or mean reflection can be used to estimate the roughness of an interface. At vertical incidence the relative mean reflections are $f = 10$ Hz, $\sim 1 \, p_0$; $f = 50$ Hz, $0.96 \, p_0$; and $f = 100$ Hz, $0.54 \, p_0$. Use these data to estimate σ.

6. A common midpoint stacking operation adds the signals from many source and receiver separations after normal moveout correction. Consider how a rough interface would affect the quality of the stack. Suppose that the interface is 1000 m deep, and the maximum source-receiver separation is 2000 m. For a roughness $\sigma = 3$ and signal frequency of 60 Hz estimate a relative magnitude of $\langle p_r \rangle_\zeta / p_0$, that is, 90%, 80%, and so forth. After normal moveout correction, explain why you might expect the pair of signals from source-receiver separations of 100 m and 0 m to be more alike than the pair of signals from 2000-m and 0-m separations.

15

Huygens' Construction, Migration, and Seismic Holography

Migration and seismic holography have an interdisciplinary history. In 1936 Frank Rieber used seismic models to make pictures of wave fronts reflected and diffracted by simple structures. His pictures are textbook examples of the application of Huygens' hypothesis and constructions. Rieber applied these ideas to subsurface seismic exploration. He needed electrically reproducible seismic records for his research, so calling on his experience as a motion picture engineer, Rieber built a special recording system that used motion picture–type sound tracks to record the signals from a spread or array of geophones. These records displayed the signals as variable-density ribbons instead of wiggly lines. He steered the array to locate sources of reflected and diffracted energy. Rieber's idea was to use the records to locate the positions of reflectors and diffractors without using ruler-and-compass constructions.

Rieber's data display had much higher visual impact than the wiggly-line seismograms. About two decades later geophysicists placed Rieber-type variable-density records next to each other and got images that looked like geologic structure. We call these *seismic pseudosections*. They are also graphic pictures of seismic holograms.

By the geometry of reflections from dipping reflectors, geophysicists knew that the reflection interfaces in pseudosections were not in the right places. By ruler-and-compass constructions they moved or "migrated" the reflections to the correct positions. Various elaborations of ruler-and-compass migration techniques were used from roughly 1950 until 1970 and the start of digital data processing.

By 1970 the advances in digital computers made seismic imaging of the subsurface possible. As geophysicists started doing migrations digitally, they realized that they had to be much more sophicticated in using solutions of the wave equation for subsurface structures. Claerbout introduced a finite-difference formulation of the wave equation to the geophysical literature. Since solutions of the wave equation can be run forward or backward, the problem of migration took a new look. The new look has its roots in optical holography.

Optical holography and holographic images were demonstrated by Dennis Gabor in 1948. The invention of laser single-frequency light sources made holography practical. Gabor won the 1971 Nobel prize in physics. He used a photographic plate to record the amplitudes and phases of the wave fronts scattered by an object. The positive transparency of the plate is called a *hologram*. The hologram can be illuminated and an image of the object can be viewed. The analogy is obvious: The seismic data set becomes the hologram, and the digital computer constructs the image.

By tradition, the definition of the term *migration* has been extended past its original meaning of ruler-and-compass methods to include sophisticated theoretical and numerical methods. We prefer the names *seismic holography* or *impulse holography* because they combine the concepts of time-dependent signals and solutions of the wave equation.

15.1 SEISMIC PSEUDOSECTIONS, DIFFRACTIONS, AND HUYGENS' CONSTRUCTIONS

Raw seismic data are displayed as profiles that show the horizontal distance x along the horizontal axis and plot the reflection times of events vertically downward. Sometimes the reflection times are converted to depth by using a velocity versus depth function. Common mid- (or depth) point processing gathers the reflection data from a spread of geophones into a single trace at the midpoint. The CMP trace effectively corresponds to a source in the middle of a receiving array, where the array is focused on reflection points beneath the source. In a simple analytic approximation we can trace CMP data as if the source and receiver are coincident. Because the profile displays events *as if* they are vertically beneath the source, the profile is a seismic pseudosection. (Here, "pseudo" means a close or a deceptive resemblance.)

Two examples of marine seismic profiles or pseudosections are shown in Figure 15.1. The upper profile was taken over the continental shelf and slope east of Cape Hatteras. The lower profile was taken over an abyssal hill area in the mid-Atlantic Ocean. The continental shelf profile shows many nearly parallel events (interfaces). Beneath these interfaces the pseudosection is chaotic and has many over-lapping crescent-shaped features. The abyssal hill profile has a few short sections of parallel interfaces at and beneath the water-sediment interface. Elsewhere, the pseudosection shows many overlapping crescent features. On both profiles the creasent features are concave downward. We need to explain the origin of the crescents and to develop techniques to interpret them.

Huygens' constructions

Suppose the crescents are due to diffractions from abrupt changes of interfaces, such as the edge of a wedge. There are two steps in demonstrating the supposition. The first step is to make a Huygens wave-front construction for a wedge, and the second step is to use the construction to make the corresponding seismic pseudosection. If the construction gives crescents like the data, then it is most likely that we have an explanation for the crescents.

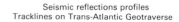

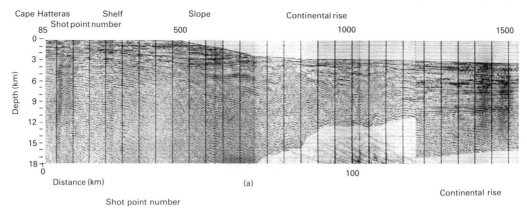

Figure 15.1 Multichannel marine seismic profiles. The source and receiver were closely spaced: (a) Continental shelf east of Cape Hatteras; (b) abyssal hills. The times are two-way travel times. The distance is in kilometers. The display from 0 to 6 s is not shown. The data are from P. A. Rona, (NOAA Atlas 3, The Central North Atlantic Ocean Basin and Continental Margins: Geology, Geophysics, Geochemistry and Resources, including Trans-Atlantic Geotraverse (TAG). U.S. Department of Commerce, National Oceanic and Atmospheric Administration, Environmental Research Laboratories, February 1980).

Huygens' constructions give the direct, reflected, and diffracted wave fronts. The sketches in Figure 15.2 show the wave fronts at a sequence of travel times. The conversion of the wave-front construction to a seismic pseudosection follows with the aid of Figure 15.3. For simplicity, the source and receiver are coincident. Assuming that the direction of the arrival of an event is unknown, we plot the time of arrival vertically downward. We move the source and receiver along the profile and make seismic measurements at a set of positions. The diffraction arrivals $d_1, d_2, d_3, \ldots$ form crescent-shaped features. Recalling Figure 15.1, we guess that each crescent originates at some kind of a wedge. It is easy to use travel times of wave fronts to construct a pseudosection of a profile. Conversely, we can use the locations of crescent peaks to guess the locations of wedges. The profile sketched in Figure 15.4a was constructed to match a short section of the profile in Figure 15.1 and here Figure 15.4b. A pair of synthetic pseudosections is shown in

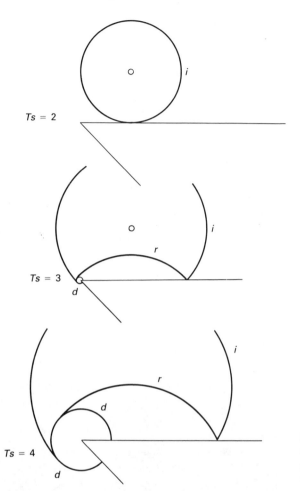

$Ts = 2$

$Ts = 3$

$Ts = 4$

Figure 15.2 Huygens wave-front constructions for a diffraction at time steps 2, 3, and 4. *i* is the incident wave front. *r* is the reflected wave front. *d* is the diffracted wave front.

Figures 15.4c and 15.4d. The pseudosection shown in Figure 15.4c includes the reflections from the facets between wedge apexes, and Figure 15.4d does not include the reflected components. We think the original data (Figure 15.4b) look more like Figure 15.4c. This means that the reflected components are small and that the sea-floor segments between wedge apexes are rough. The computations were made with the Biot-Tolstoy impulsive solution for diffractions, Section C1.2 of Appendix C.

15.2 MIGRATION CONSTRUCTION

Originally, *migration* meant a graphic procedure in seismic data interpretation. As sketched in Figure 15.5a, the seismic pseudosection shows a reflection from a dipping interface. Contrary to the appearance on the pseudosection, the reflections for coincident source

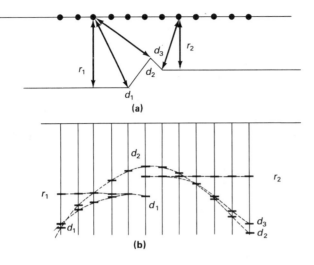

Figure 15.3 Construction of a seismic profile over a ridge crest. The source and receiver are effectively coincident. (a) The position of the ship carrying the source and receiver. (b) Seismic profile.

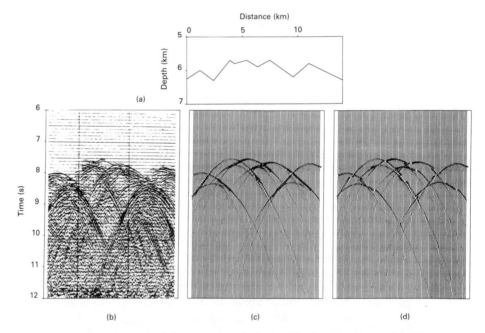

Figure 15.4 Comparison of a real and synthetic pseudosection. The data are from 97 to 115 km along the seismic profile in Figure 15.1b. The seismic signals appear to have been low-pass filtered with a 25-Hz filter. The computations were made by using the exact Biot-Tolstoy solution for the diffraction and reflections at an ideal wedge. The Biot-Tolstoy solution gives the response to an impulsive point source. The numerical results were (digitally) low-pass filtered to match the data. Reza Daneshvar furnished the computations. (a) Assumed profile. (b) Sections of data from Figure 15.1b. (c) Synthetic pseudosection having only diffracted components. (d) Synthetic pseudosection having both reflected and diffracted components. The figure is from M. R. Daneshvar, and C. S. Clay, "Imaging of rough surfaces for impulsive and continuously radiating sources," *J. Acoust. Soc. Am.*, 82 (1987), 360–69.

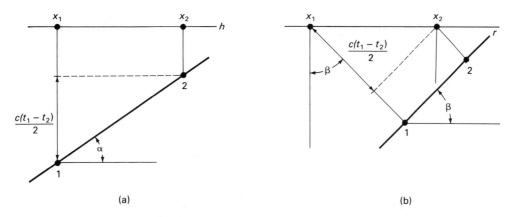

(a)   (b)

Figure 15.5 Migration of a dipping interface. The seismic pseudosection is plotted versus depth without vertical exaggeration. The construction assumes that the medium has a constant seismic velocity c. Dashed lines are for construction. (a) Seismic pseudosection. (b) Migrated section.

and receivers are normal to the interface (Figure 15.5b). For a constant seismic velocity c, the geometric construction gives

$$\frac{c(t_1 - t_2)}{2} = (x_2 - x_1)\tan \alpha$$

$$\frac{c(t_1 - t_2)}{2} = (x_2 - x_1)\sin \beta \tag{15.1}$$

$$\tan \alpha = \sin \beta$$

The seismic interpreter moves or migrates the reflector up-dip and steepens the dip of the reflector. Equation (15.1) is known as the *migration equation*. For sources and receivers on the surface, the limiting dip is $\beta = 90°$. Correspondingly, the maximum dip on the pseudosection is $\alpha = 45°$.

The migration of a crescent follows the same procedure. Each short section of a crescent is treated as if it is a facet or short segment of an interface (Figure 15.6). Migrations of the facets move them to the same place. Migration of the dipping interfaces completes the wedge structure.

As geophysicists started doing automatic or machine migration, they realized that it was difficult to determine the dip α in complicated pseudosections. Recalling Huygens-Fresnel constructions, they assumed that an event could have been scattered by a scatterer at any position on a constant-time surface. As sketched in Figure 15.7, the construction consists of drawing an arc of a circle to represent the locus of a possible scatterer. In a medium having constant seismic velocity, the constant-time surface is a circle having the radius $ct/2$ for an event at t. The constructive interference of many circles locates, as in the sketch, the apex of a wedge. Also, the envelope of the circles indicates the reflecting plane. The construction in Figure 15.7 does not use the slope information. Geophysicists usually limit the arcs of the circles to less than $45°$. The sketches in Figure 15.7 show that a wave-front migration construction is a mapping and stacking operation.

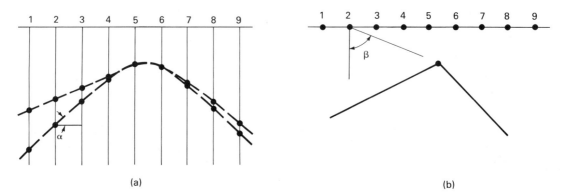

(a)   (b)

Figure 15.6 Migration of a crescent and dipping interface. (a) The crescent is treated as if it is constructed of many reflecting facets of a structure. The dipping interfaces are shown. (b) Each facet is migrated according to the migration equation, tan α = sin β.

15.3 RECONSTRUCTED WAVE FRONTS AND SEISMIC HOLOGRAPHY

From the beginnings of seismic exploration, geophysicists dreamed of having direct methods of imaging subsurface structures from seismic data sets. A great deal of theoretical effort, huge computers, and digital graphic displays have brought the dream close to reality. The dream has its roots in Huygens' principle, solutions of the wave equation, and holography.

Sketches of Gabor's optical holographic technique are shown in Figure 15.8. Light from a monochromatic point source passes through an object. Normally a lens is used to focus the wave fronts and form an image as shown in Figure 15.8c. However, Gabor placed a photographic plate at *H* and recorded the relative phases and amplitudes of the scattered, diffracted, and direct waves. The means of getting the phases in optics is not important here because we can record the phases in seismology. The photographic plate is developed, and a positive plate, the hologram, is made. The optical hologram is a two-

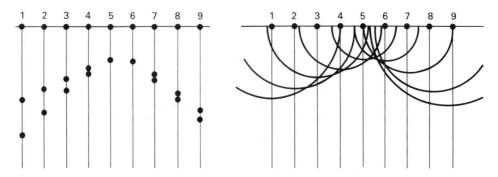

Figure 15.7 Wave-front migration. (a) The pseudosection is the same as in Figure 15.6. (b) The wave fronts are surfaces of constant travel time. Here the wave fronts are circles because the medium has constant seismic velocity.

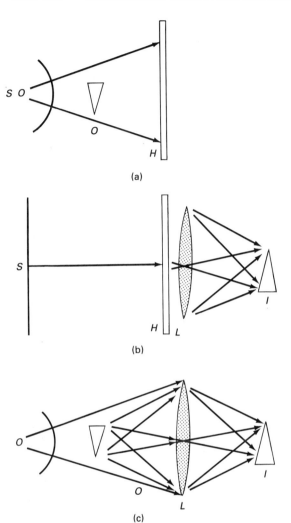

(a)

(b)

(c)

Figure 15.8 Holograms and reconstructed wave fronts. The sketches show a holographic reconstruction for transmission. The ray paths indicate where wave fronts go. The light source S radiates a single-frequency light. (a) Direct radiation and the light scattered and transmitted by the object O are intercepted by a photographic plate H. Although not shown, the plate is also illuminated by a plane wave reference light from the same source. The photographic plate records the amplitude and relative phases of the wave fronts. The reference wave is normally incident on H. (b) A positive of the plate is placed at H. This is the hologram, and it is illuminated by the source. The outgoing waves have the same relative phases and amplitudes as the wave incident on the plate H in (a). The lens focuses the waves (indicated by ray paths) to an image. (c) For comparison with a simple optical system, a lens focuses waves or rays from the object into an image.

dimensional record of the relative phase and amplitude of the outward-traveling wave front at each point on the plate. It is placed at H (Figure 15.8b) and is illuminated. From Huygens' principle for outgoing waves, each point on the hologram becomes a sourcelet of Huygens wavelets. The Huygens wavelets have the phases and amplitudes of intercepted waves. Thus, the hologram changes the simple wave that travels outward from the source into an outgoing replica of the intercepted waves. The replica or reconstructed wave fronts continue and can be viewed or focused to form images.

Let us compare seismic data sets and signal processing with optical holography. The data from a large multichannel areal array of geophones are a seismic equivalent of a hologram. Each channel contains the time history of the upward-traveling waves at each point on the surface. In comparison with the optical hologram, the seismic hologram is three-dimensional in x, y, and t. The following are the major differences between optical holography and seismic holography:

1. Relative to their wavelengths, the dimensions of optical holograms are huge compared with the usual dimensions of geophone arrays.

2. Whether it is an advantage or disadvantage, optical holography uses single-frequency signals, while seismic data sets use transient signals having wide frequency ranges. This is the reason that impulse or seismic holograms include a third dimension, time.

3. Optical holography is done in air and is used to image objects in a transparent medium. The earth is a complicated medium, and our purpose is to construct images of the medium.

4. In optics the rods and cones of our retinas sense images by measuring the patterns of the relative intensities of colored light. In seismic imaging the pixels in a graphics display play the role of rods and cones. Each pixel in the image field has a brightness that corresponds to our choice of how to represent the wave field. We can stack the wave fields linearly or algebraically or, following an eye model, let pixel brightness be proportional to wave-field intensity.

In seismic holography the data are digital, and all imaging operations are also digital computations. We can devise algorithms to make the reconstructed wave fronts travel upward or backward and downward or any other way that pleases us. We can use graphic displays to display the waves as they travel and interfere and then freeze the action at interesting times.

15.3.1 Seismic Applications Using Impulsive Signals

Figure 15.9a shows a simple set of seismic measurements. The source and received signals are brief impulsive signals—here, delta functions. The coincident source and receiver measure the waves scattered by the object O. The signals are recorded as two-way travel times in a medium having seismic velocity c. The records start at T_s and end at T_e. These signals are changed into a seismic hologram for the numerical and graphic imaging by time reversal. The time-reversed signals are shown in Figure 15.9b. For numerical wave-front reconstruction the seismic velocity is reduced to $c/2$ so that we can use two-way travel times and compute the correct one-way distance.

The time-reversed signals are transmitted through sources located at the receiver positions. The transmissions start at T_e and end at T_s. At time T_m the wave fronts have moved short distances into the medium (Figure 15.9c). The display or computations can be stopped or frozen at time T_s. The sum of all the wave fronts gives the image I. Presumably, the wave fronts away from the image region destructively interfere as in Huygens' constructions. The quality of the image depends on the aperture, number of signal channels, and signal bandwidth.

The construction of backward-propagation algorithms and the inclusion of time require care. Figure 15.10 gives a sequence of data processing steps for one of many channels. The sketch in Figure 15.10a shows an impulsive source and closely spaced receiver on a weathered layer and over a half-space. A signal $p(t)$ is recorded for digital processing. As sketched in Figure 15.10b, individual signals are oscillatory. The oscillatory character is due to some combination of shallow-layer reverberation, source transmission,

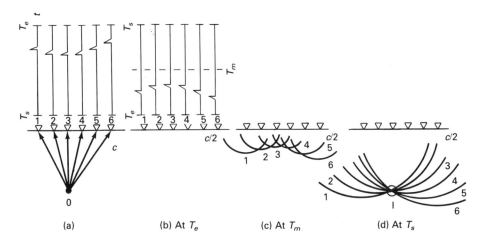

(a)   (b) At T_e (c) At T_m (d) At T_s

Figure 15.9 (a) Coincident source and receiver on a medium having seismic velocity c. The signals are from an object O. The positions are 1 through 6. T_s is the beginning, and T_e is the end of the record. (b) Time reversal of the signals. The time "at T_e" means that we have just started to transmit the time-reversed signals through sources at the geophone position. The seismic wave velocity is $c/2$. (c) Positions of wave fronts at time T_m. (d) The wave fronts converge and form an image I at time T_s.

and amplifier filters. The signal needs to be changed to a brief impulsive signal by passing it through a deconvolving filter. The Wiener optimum prediction filter, Section B1.5 of Appendix B, is routinely used. Section B1.5, Figures B.3b and B.3e display numerical examples of reverberation reduction and signal simplification. The oscillatory signal in Figure B.3b becomes two spikes, one for each reflector.

In Figure 15.10 ideal deconvolution of $p(t)$ in Figure 15.10b gives $d(t)$. "Perfect deconvolution" causes an arrival to become an impulsive *delta function*, $\delta(t)$, where

$$\delta(t) = \frac{1}{t_0} \qquad \text{for } t = 0$$
$$\delta(t) = 0 \qquad \text{for } t \neq 0 \tag{15.2}$$

and t_0 is the time step. In numerical computations we can replace $t = 0$ with $-t_0/2 < t < t_0/2$, and so forth. The ideal deconvolution of $p(t)$ gives

$$d(t) = a_1\delta(t - t_1) + a_2\delta(t - t_2) \tag{15.3}$$

where a_1 and a_2 are amplitude factors, and t_1 and t_2 are the two-way travel times from the source-receiver station to diffractors 1 and 2.

15.3.2 Time-Reversed Signal

The time-reversal operation replaces t with $-t'$, and $d(t)$ becomes $d(-t')$. To partially compensate for spreading losses, we arbitrarily multiply $d(-t')$ by t^2. The compensation is cosmetic. The compensated and time-reversed signal, Figure 15.10d, is

$$b(t') = t^2 d(-t') \tag{15.4}$$

from Eq. (15.3)

$$b(t') = a_1 t_1^2 \,\delta(-t' - t_1) + a_2 t_2^2 \,\delta(-t' - t_2) \tag{15.5}$$

where t_1^2 and t_2^2 are cosmetic amplitude compensations.

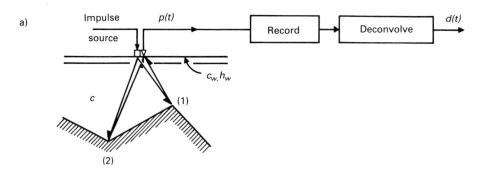

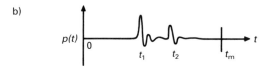

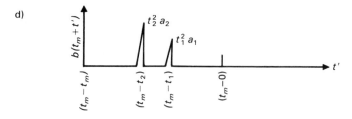

Figure 15.10 Wave-front construction by backward transmission. (a) The physical measurement shows a coincident source and receiver on the surface of a weathered layer (c_w, h_w) over a half-space. The seismic velocity of the half-space is c. A deconvolution (or optimum prediction filter) operation is included to reduce the reverberations caused by the weathered layer. (b) $p(t)$ is the recorded signal. The oscillatory signals are due to multiple reflections or reverberation. (c) $d(t)$ is the deconvolved signal. (d and e) The signal is conditioned for cosmetic reasons reversed in time to give $b(t)$. $b(t)$ is transmitted as $Db(t)$ in a half-space having seismic velocity $c/2$. D is a directional transmission function. The wave fronts are frozen at the instant $t = 0$.

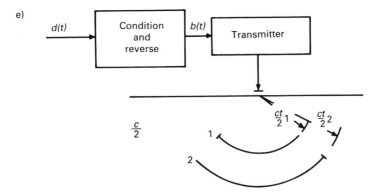

Figure 15.10 (continued).

The holographic construction consists of transmitting $b(t')$ from the source-receiver station. The transmission of $b(t')$ gives a down-going wave field $g(t - 2R/c)$, where the seismic velocity is $c/2$, and R is the radius of the wave front. If we assume an amplitude and directional transmission function D, the wave field is, with t' in $b(t')$ becoming $(t - 2R/c)$,

$$g\left(t - \frac{2R}{c}\right) = D\left[a_1 t_1^2 \delta\left(\frac{2R}{c} - t - t_1\right) + a_2 t_2^2 \delta\left(\frac{2R}{c} - t - t_2\right)\right] \quad (15.6)$$

The wave field in the half-space is frozen at $t = 0$ and becomes

$$g(R) = D\left[t_1^2 a_1 \delta\left(\frac{2R}{c} - t_1\right) + t_2^2 a_2 \delta\left(\frac{2R}{c} - t_2\right)\right] \quad (15.7)$$

The field $g(R)$ is plotted as a function of position. These are the wave fronts marked 1 and 2 in Figure 15.10e. The sketch shows $g(R)$ as being the map of wave fronts for a single station. The superposition of wave fronts from many stations gives an image of subsurface structure. This procedure is also known as reverse-time migration.

The example and analysis are for a brief impulsive wave front. After we remove the reverberation effects of the weathered layer by deconvolution, we ignore the weathered layer for the backward propagation. Clearly, we can use signals having longer duration and propagate them backward after time reversal. Correspondingly, the wave fronts become broad, and the images become fuzzy.

15.3.3 Algorithms and a Numerical Example

We make a minor generalization of Eq. (15.7) by letting $g(R)$ become $g_m(R)$, t_i become $t_{m,i}$, and so on, for the mth station, and

$$\left.\begin{aligned} g_m(R) &= \sum_i D_m\, t_{m,i}^2\, a_{m,i}\, \delta\left(\frac{2R}{c} - t_{m,i}\right) \\ t_{m,i} &= it_0 \end{aligned}\right\} \quad (15.8)$$

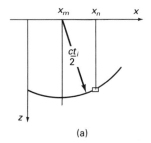 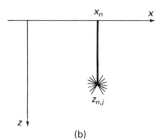

Figure 15.11 Wave-field superposition. (a) Geometry for a station at m and the wave field at $(x_n, z_{n,j})$; (b) Superposition of several wave fields at $(x_n, z_{n,j})$.

where the delta function is zero except at $R = c\, t_{m,i}/2$. The sketches in Figure 15.11 show the geometry. The superposition of waves at $(x_n, z_{n,j})$ is

$$
\left.
\begin{aligned}
s_{n,j} &= \sum_{m=n-m_1}^{n+m_1} \sum_i t_{m,i}^2\, a_{m,i}\, \delta\!\left(\frac{2R}{c} - t_{mi}\right) \\[4pt]
&\text{where } D_m = 1 \quad \text{for } m = n - m_1 \text{ to } m = n + m_1 \\[4pt]
&\;\; D_m = 0 \quad \text{otherwise}
\end{aligned}
\right\}
\tag{15.9}
$$

$$
\left.
\begin{aligned}
R^2 &= \left(\frac{ct_i}{2}\right)^2 = z_{nj}^2 + (x_n - x_m)^2 \\[6pt]
z_{n,j} &= \left[\left(\frac{ct_i}{2}\right)^2 - (x_n - x_m)^2\right]^{1/2} \\[6pt]
j &= \text{INT}\left(\frac{2\,z_{nj}}{ct_0} + 0.5\right)
\end{aligned}
\right\}
\tag{15.10}
$$

where the 0.5 is round-off. The algorithm steps through the values of i and sums in array elements $s_{n,j}$, where j is given by Eq. (15.10). Programs to make sample signals or pseudosections and to process the pseudosections are in Appendix C.

A numerical example is shown in Figure 15.12. The geometry was chosen to give reflection and diffraction arrivals. The source is an impulsive function having the duration of two time steps. The reflection arrivals are the large positive events in traces 4 to 8. The diffraction arrival is positive on traces 0 to 3 and is negative on traces 4 to 8. The change of sign is characteristic of diffraction arrivals on either side of the vertical incidence reflection. Since the reflection arrivals are a set of impulsive functions having a width of $2t_0$, and the diffraction arrivals are small, deconvolution is not necessary. The wave fronts were stacked from $n - 2$ to $n + 2$ in Eq. (15.9). The sum of the frozen wave fronts is shown in Figure 15.12b. The procedure broadens the reflection arrival. The apex of the wedge is at -100 m or between stations 2 and 3. The method gives a strong image beneath station 3 and a weaker image at station 2. As shown in Figure 15.12b, stations 0, 1, 7, and 8 have incomplete stacks.

The reconstruction of the pseudosection in Figure 15.12a gives a fuzzy image of the actual structure. The quality of the images can be improved by using more sophisticated algorithms and many more source-receiver stations. For example, the array-weighting

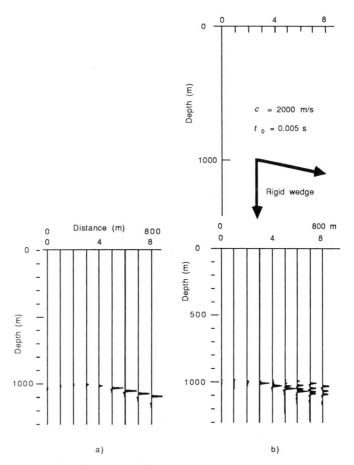

Figure 15.12 Seismic pseudosection and wave-front imaging. (a) Seismic pseudosection. (b) Image from the backward or downward propagation of reconstructed wave fronts. The amplitudes of the "frozen wave fronts" are added algebraically at each point in the x-z space.

functions such as shown in Figure 12.5 decrease the side-lobe response. By regarding the Huygens sourcelets for backward propagation as elements of an array, we can apply a weighting function to the seismic hologram and reduce side lobes or extraneous images.

The concept of transmitting a time-reversed signal back into the medium (seismic velocity of $c/2$) changes the geometric construction into a wave-propagation problem; namely, we transmit a set of $b_m(t)$ into a best guess of the subsurface structure and then freeze the wave fronts. The numerical formation of images requires a huge number of calculations. Much research has been focused on the creation of fast and accurate algorithms to make wave-field extrapolations and migrations. References to Claerbout's wave-field extrapolation technique and other wave-front imaging or migration techniques are given in the selected readings.

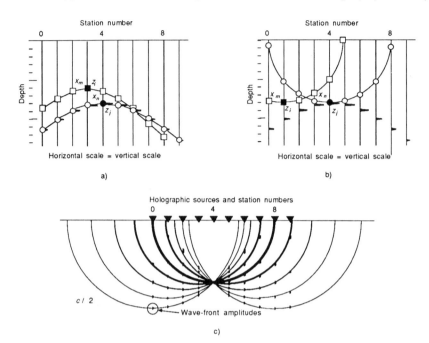

Figure 15.13   Diffraction stack, wave-front stacking, and seismic hologram. The signals for a point diffractor were computed using the algorithms in Appendix C. For this example, the signal amplitudes include a $(1/t^2)$ factor that represents the geometrical spreading from the source and diffractor. The cosmetic factor t^2 in Eqs. (15.5) to (15.19) was omitted. The diffractor locations are the solid circles at (x_n, z_j). Trial or assumed diffractor locations are indicated by the solid squares at (x_m, z_i). (a) The diffraction stack sums the signal amplitudes at the grid points on the diffraction arrival lines. The diffraction stack for the amplitude at (x_n, z_j) is the sum of the signal amplitudes at the open circles. The stack at (x_m, z_i) sums amplitudes at the open squares. (b) The wave-front stack has a different order. Wave-front migration spreads the amplitude of an event along the equal-time wave front. The signal amplitude at (x_n, z_j), the solid circle, is put at each grid point (the open circles) along the wave front. Similarly, the signal amplitude at (x_m, z_i), the solid square, is put at the open squares. The wave-front stack at a grid point is the sum of all contributions. (c) The holographic image is made by transmitting the set of time-reversed signals. Here, the wave fronts were drawn by connecting the samples of the wave front beneath each station. To show the relative signal amplitudes, the widths of the wave fronts are drawn as being approximately proportional to the amplitudes.

PROBLEMS

1. Two common methods of imaging or migration are known as *diffraction summation* and *wave-front migration*. Examples are sketched in Figure 15.13. For both technqiues the source and receiver are coincident. Assume that a diffractor exists at $(x_n, z_j = ct_j/2)$. Derive a summation algorithm for a diffraction stack. In this type of processing we *assume* that a diffractor exists at $(x_m$ and $z_i = ct_i/2)$ and then stack along the corresponding hyperbola. The diffractor stack matches at the actual diffractor when $m = n$ and $j = i$.

2. Using the same assumptions as in Problem 1, derive a wave-front migration algorithm. Here, for example, the amplitude at $(x_n, ct_j/2)$ is the superposition of all wave fronts passing through the point.

3. (a) Compare the results of Problems 1 and 2 and show that the operations are equivalent.

   (b) Compare these results with the backward-transmission algorithm (Eqs. (15.8) to (15.10)). You may wish to delete the time-varying gain t_m^2, i in Eqs. (15.8) and (15.9) for the comparison.

   (c) Derive a time-reversed imaging algorithm. To improve the images of the wave fronts, calculate wave-front amplitudes on several vertical lines between stations.

4. Test your algorithm on a computer by computing a synthetic seismogram, six stations, for a single diffractor and then imaging the result.

5. Suppose that the source is at x_1 and the receiver is at x_2. What is the locus of a scatterer for an event or signal from a diffractor? Derive a stacking algorithm for a source and a set of receivers along the x-axis. *Hint*: Recall the properties of conic functions from analytical geometry.

The next problems combine concepts from wave propagation, sound scattering at rough surfaces, and stacking.

6. Suppose that an interface has an rms roughness σ of about $\lambda/2$. If you were to do a common midpoint stack, how would you expect the stack of these signals to compare with the stack of

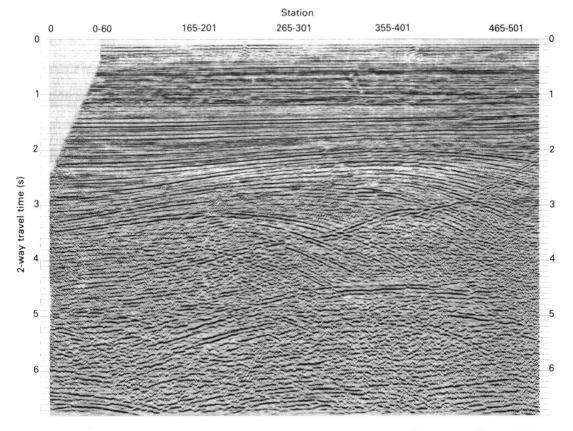

Figure 15.14 Seismic pseudosection. The section is a common midpoint stack. Figures 15.14 through 15.16 were furnished by Dr. J. D. Robertson, ARCO.

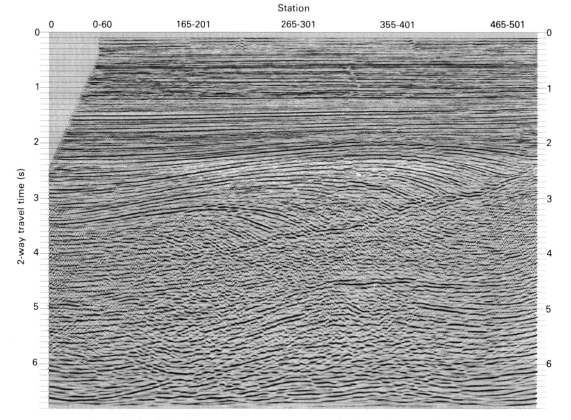

Figure 15.15 Holographic or migrated seismic section. The pseudosection in Figure 15.14 was imaged by using a finite-difference downward-continuation program. The figure was furnished by Dr. J. D. Robertson, ARCO.

reflections from a smooth plane interface? Assume near-vertical incidence reflection geometry, and give an estimate of relative amplitude.

7. The common midpoint stacking operation enhances reflections from smooth interfaces. The wavefront imaging operation appears to stack on sources of diffractions. Discuss the implications of using these apparently contradictory assumptions about the nature of the subsurface interface.

15.4 REAL DATA

Petroleum reserves are found in sedimentary basins where the sediments form layered sequences. Of course, the layers may be distorted or broken by faults. Erosion may have caused unconformable contacts between upper and lower layers. Salt domes may have forced their way through tremendous piles of sediments. The enhancement and imaging of these structures has been the task of exploration geophysicists.

In the beginning we mentioned the fictional detective's use of a few clues and reason to deduce the structure. Now it seems as if all the sleuth's observational power, ability

165-201   365-401

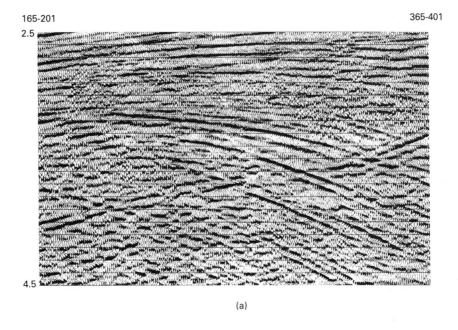

(a)

165-201   365-401

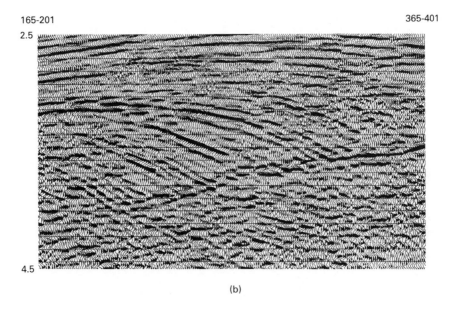

(b)

Figure 15.16 Comparisons of enlargements from Figures 15.14 and 15.15. The figures were furnished by Dr. J. D. Robertson, ARCO. (a) Seismic pseudosection. Shots 200 to 400 and 2.5 to 4.5 s. (b) Wave-front imaged section.

to see past the irrelevant details, and thought processes have become part of an expert computer system. The workhorses of the expert system are deconvolution, common midpoint stacking, and wave-front imaging or migration. Our choice of the word "seems" is deliberate. Geophysicists must know what information is enhanced and what information is suppressed by the expert computer systems. After all, an expert system is like a rigid bureaucracy. Some of the knowledge is in the exploration geophysics literature, but we suspect that most of the knowledge is passed privately among the users.

The string of enhancing operations, deconvolution, common midpoint stacking, and migration give better images. An example of a seismic pseudosection is shown in Figure 15.14 on page 187. The section shows a downthrown roll over anticline. It is bounded by a major growth fault at about 2.5 s on the right side of the section. The holographic or migrated section is shown in Figure 15.15 on page 188. The reduction of crescents and simplifications is apparent. Figure 15.16 compares enlargements of the sections between shots 200 and 400 and times 2.5 to 4.5 s. The images in these figures were made by using Claerbout's finite-difference and downward continuation technique.

Geophysicists learned fairly early that stacking and migration of seismic reflection sections gave better images. Stacking and migration algorithms are the result of hundreds of people-years of labor. The computational effort is huge, and the exploration geophysical industry uses the largest and fastest computers in the world.

16

Reconstructed Wave Fronts and Image Processing

In optical holography the reconstructed wave fronts continue propagating away from the hologram as if they were coming from the object. Images are formed by focusing the waves on a sensitive surface—for example, the lens in our eye focuses light on our retina. The components of optical waves add to form the image; the retina senses the intensity of the image at each rod and cone. We use the intensities, the colors, and the shapes of objects and process the images in our brains. Since our optical systems work so well in complex visual environments, it is reasonable to explore similar image-processing systems in seismic environments.

16.1 FOCUSED IMAGING

We use the analogy between an optical hologram and a seismic data set. The sketches in Figure 16.1 show the use of a lens to form a real image. The space beneath the hologram *H* is object space, and the space above it is image space. For simplicity the object is either illuminated or is a source.

For digital processing of a seismic hologram, *SH* in Figure 16.1c, the time delays D_0, D_1, and so forth, replace the function of the optical lens. The time delays are chosen to bring the wave front from a grid cell or point in object space to focus in a grid cell or point in image space. To carry out the optical analogy, we choose the image space coordinates to follow a standard ray path construction for a lens. For the reconstruction of seismic holograms, we usually put the image space beneath *SH* and let the coordinates in image space be the same as in object space (Figure 16.1d). In digital imaging we focus the waves on each cell or grid in image space and then display cell contents.

Compared with optical lenses, digital focusing and imaging gives more freedom in combining the signals or information in a channel with signals from other channels. Instead

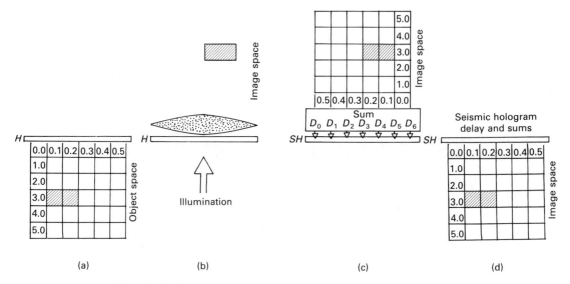

Figure 16.1 Optical and seismic focusing operations. The object occupies two cells in object space. Scattered and diffracted waves from the object are recorded for the hologram H. The lens and time-delay summing operations are equivalent. Both add components of the wave front. (a) Object in object space. (b) Lens focuses the reconstructed wave fronts in image space. (c) Digital or time-delay focusing of reconstructed waves from the seismic hologram SH. The time delays are D_0, D_1, and so forth. For illustration, the image space has the same reversal as given by an optical lens. The "Sum" operation gives the equivalent operation on seismic data to optical image formation. (d) Conventional seismic image for image space having the same coordinates as object space.

of having to form an image by adding the components of wave fronts as in optics, we can combine the components in various ways. For example, we can add them, add and square them, or compute the covariance of pairs of channels.

Our choice of what we want to see or what we wish to image controls our data processing. For example, the signals in adjacent cells or grid points are likely to have the same phase for the reflection from a plane surface. The focused components of signals scattered at a rough surface are likely to have different phases in adjacent cells or grid points. A simple amplitude display gives an image of the source for reflection from a plane interface. Over a rough surface, a simple amplitude display is noisy. To image rough surfaces we may wish to measure intensity (summed and squared signals) in each cell or grid point. There is optical precedence for summing and squaring operations. Our eyes use the lens to focus and sum wave fronts, and the rods and cones measure the relatively intensities of the waves.

16.1.1 Seismic Imaging

The seismic data have been recorded and can be reproduced. Digital focusing operations start with the data or a seismic hologram. To keep the algebra simple, we use a two-receiver array. We can extend the analysis to N channels by replacing the sum on two channels with a sum on N channels.

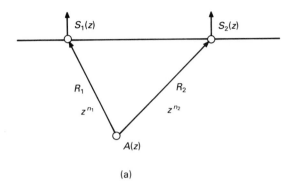

(a)

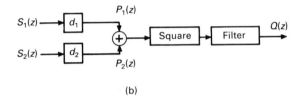

(b)

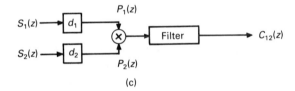

(c)

Figure 16.2 Nonlinear array processing. (a) Geometry. (b) Delay, sum, and square. (c) Delay and cross correlate d_1 and d_2 are time delays in time steps. A low-pass filter following squaring or cross correlation may be necessary.

Signals

In Figure 16.2a the source transmits $A(z)$, and if we ignore the spherical spreading amplitude change, $S_1(z)$ and $S_2(z)$ are time-delayed versions of $A(z)$:

$$\left.\begin{array}{l} S_1(z) = A(z)\, z^{n_1} \\ S_2(z) = A(z)\, z^{n_2} \end{array}\right\} \tag{16.1}$$

$$\left.\begin{array}{l} n_1 = \text{INT}\left(\dfrac{R_1}{t_0 c} + 0.5\right) \\[2mm] n_2 = \text{INT}\left(\dfrac{R_2}{t0c} + 0.5\right) \end{array}\right\} \tag{16.2}$$

where t_0 is the time step. As shown in Figures 16.2b and c, d_1 and d_2 are time delays in time steps. It is evident from the figure that a two-element array can determine direction. More elements are needed to focus or determine distance.

The signal-processing operations are primarily sensitive to the difference of the time delays $(d_2 - d_1)$. The time-delayed signals are

$$\left.\begin{array}{l} P_1(z) = S_1(z) z^{d_1} \\ P_2(z) = S_2(z) z^{d_2} \end{array}\right\} \tag{16.3}$$

and the substitution of Eq. (16.1) gives

$$P_1(z) = A(z) \, z^{n_1 + d_1} \atop P_2(z) = A(z) \, z^{n_2 + d_2}$$

$$(16.4)$$

In conventional beam steering, we choose d_1 and d_2 to make $P_1(z)$ and $P_2(z)$ identical. The next operations use summing, absolute squaring, and covariance operations. Readers can refresh their memories by reviewing Section 10.2.1 for covariance functions.

Summation or linear processing

The wave or signal components in each cell or grid point are summed. For two channels the sum is the sum of $P_1(z)$ and $P_2(z)$,

$$L(z) = P_1(z) + P_2(z) \atop L(z) = A(z) \, [z^{n_1 + d_1} + z^{n_2 + d_2}]$$

$$(16.5)$$

The choice of delays that make

$$n_1 + d_1 = n_2 + d_2 \qquad (16.6)$$

gives a general matching, steering, or focusing condition. For a two-element array,

$$L(z) = 2A(z) \, z^{n_1 + d_1} \qquad (16.7)$$

The difference between being steered and not steered is roughly a factor of 2 in the expression for $L(z)$. Linear summation needs many more channels of receivers for effective array steering. If one has N channels, then $L(z)$ becomes $NA(z) \, z^{n_1 + d_1}$ at the correct steering direction. Of course, one would choose the delays to focus a large multielement array on the source.

Delay, sum, and square

Wave or signal components in a cell or grid point are added to give $L(z)$; then $L(z)$ is squared to give intensity. The absolute square of the sum $L(z)$ is

$$Q(z) = L(z^{-1})L(z) \qquad (16.8)$$

The terms in Eq. (16.8) are

$$Q(z) = P_1(z^{-1})P_1(z) + P_2(z^{-1})P_2(z) \atop + P_2(z^{-1})P_1(z) + P_1(z^{-1})P_2(z)$$

$$(16.9)$$

The first two terms are the absolute squares of $P_1(z)$ and $P_2(z)$. The last terms or cross terms contain the comparison of $P_1(z)$ and $P_2(z)$. The sensitivity of $Q(z)$ to steering is in these terms. The substitution of Eq. (16.4) and factoring gives

$$Q(z) = A(z^{-1})A(z) \, [2 + z^{-(n_2 + d_2) + (n_1 + d_1)} + z^{(n_2 + d_2) - (n_1 + d_1)}] \qquad (16.10)$$

This equation shows the dependence of $Q(z)$ on the delays d_1 and d_2. At the matched condition Eq. (16.6), $Q(z)$ becomes $4A(z^{-1})A(z)$ at the time delay $n_1 + d_1$. The sensitivity of $Q(z)$ to steering is small for a two-element array because of the presence of the sum of squares, that is, the first two terms in Eq. (16.9). As with linear processing, delay, sum, and square processing needs many more channels of receivers for good array steering. An N-channel array has N-squared terms and $(N^2 - N)$ cross terms. At large N, the cross terms dominate $Q(z)$. The cross terms are the cross covariances of pairs of channels and give the steering or directional information.

16.2 CROSS-COVARIANCE SIGNAL PROCESSING AND EXAMPLES

Cross-covariance or correlation signal processing has roots in statistical signal theory, radio astronomy, and acoustics. In the 1950s receiver channels were very expensive, so radio astronomers and underwater acousticians used statistical signal theory to reduce the number of signal channels for directional systems. From statistical signal theory the covariance of a complicated signal (having a very wide frequency bandwidth) has a sharp peak and small side lobes. The astronomers and acousticians reasoned that they could measure the signals at a pair of widely separated receivers and then use the cross correlation of time-delayed signals to determine the direction of a source. Figure 16.2c shows a cross-covariance processing system.

In common usage, covariance, cross covariance, and cross correlation usually refer to the same operation. Cross correlation is a normalized cross covariance. The cross covariance of $P_1(z)$ and $P_2(z)$ is

$$\left. \begin{aligned} C_{12}(z) &= P_1(z^{-1})\, P_2(z) \\ C_{12}(z) &= A(z^{-1})\, A(z)\, z^{[-(n_1 + d_1) + (n_2 + d_2)]} \\ C_{21}(z) &= C_{12}(z^{-1}) = P_2(z^{-1}) P_1(z) \end{aligned} \right\} \tag{16.11}$$

The products $P_1(z^{-1})\, P_2(z)$ and $P_2(z^{-1})P_1(z)$ are terms in Eq. (16.9). $C_{12}(z)$ is the covariance (or autocovariance) of the signal at the time delay $[-(n_1 + d_1) + (n_2 + d_2)]$. The maximum appears at 0 when the steering condition (Eq. (16.6)) is satisfied.

Examples of the covariances of a sequence of 32 random numbers are shown in Figure 10.3. If the original source $A(z)$ was a sequence of random numbers, it would be easy to determine the focused condition. Also, if the source was an impulse function, with $a_0 = 1$ and all other $a_m = 0$, the covariance $c(j)$ would have the value 1 at $j = 0$ and $c(j) = 0$ for all other j. Both of these signals have narrowly peaked covariance functions.

Signal-to-noise gain

Cross-covariance signal processing has a very important attribute, namely, a lot of signal-to-noise gain. For a very simple discussion of the signal-to-noise gain, we use the following assumptions or conditions:

1. The source transmits a continuous random signal $A(z)$. The peak of the autocovariance of $A(z)$ decreases to zero or the side-lobe fluctuation level in M_0 time steps. The duration of $A(z)$ is M_A. In the absence of noise, the RMS amplitudes of the signals at receivers 1 and 2 are the same and σ_A.

2. The receivers include filters that are adjusted to pass the signals. The noises pass through the same filters.

3. The noises at the receivers are $N_1(z)$ and $N_2(z)$. The noises $N_1(z)$ and $N_2(z)$ are uncorrelated. $N_1(z)$ and $N_2(z)$ are uncorrelated with $A(z)$. In the absence of signals, the RMS amplitudes of the noises at receivers 1 and 2 are the same and are equal to σ_N.

The signal has a duration of M_A time steps; however, each of these time steps is not an independent (uncorrelated) sample of the signal. For a signal-to-noise calculation we need the number of independent observations. M_0 is an estimate of the number of time steps for observations to be independent. The number of independent observations of $A(z)$ is approximately

$$N_I \simeq \frac{M_A}{M_0} \tag{16.12}$$

The addition of noise to the signals Eq. (16.1) gives

$$\left.\begin{array}{l} S_1(z) = A(z)\,z^{n_1} + N_1(z) \\ S_2(z) = A(z)\,z^{n_2} + N_2(z) \end{array}\right\} \tag{16.13}$$

The delay and cross-covariance operation gives

$$\begin{aligned} C_{12}(z) = \ & A(z^{-1})A(z)z^{[-(n_1+d_1)+(n_2+d_2)]} \\ & + N_1(z^{-1})N_2(z)z^{-d_1+d_2} + N_1(z^{-1})A(z)z^{-d_1+n_2+d_2} \\ & + A(z^{-1})N_2(z)z^{-n_1-d_1+d_2} \end{aligned} \tag{16.14}$$

The first term is signal, and the rest are fluctuating noise terms. A measure of the peak value of $A(z^{-1})A(z)$ that includes the width of the peak is $N_I\sigma_A^2$. For an estimate of the fluctuating terms we use

$$\sigma_F^2 = \sigma_n^2 + 2\sigma_N\sigma_A \tag{16.15}$$

and then write

$$\text{Peak } C_{12}(z) \simeq N_I\sigma_A^2 \pm \sigma_F^2 \tag{16.16}$$

where $\pm\sigma_F^2$ is a fluctuating term. The output signal-to-noise ratio, $(S/N)_{\text{out}}$, is

$$(S/N)_{\text{out}} \simeq \frac{N_I\sigma_A^2}{\sigma_F^2} \tag{16.17}$$

The output signal-to-noise is proportional to the number of independent samples of $A(z)$. Even though the input signal-to-noise power ratios may be poor, that is, $\sigma_A^2/\sigma_F^2 < 1$, we can use a large N_I to give a strong output signal.

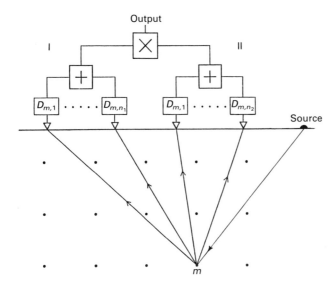

Figure 16.3 Cross-covariance array processor. The inhomogeneity at m scatters waves from the source. The signals from the Huygens sourcelet m are delayed and summed in arrays I and II; then the cross covariance of the array output is computed. The time delays are $D_{m,1}$, $D_{m,2}$, and so forth.

16.3 SPLIT-ARRAY CROSS-COVARIANCE SIGNAL PROCESSING AND IMAGING

Cross-covariance signal processing can be used with controlled sources, natural events, and random signals. Experience in ocean acoustics has shown that two-receiver systems are inadequate in complicated environments; more than two data channels are necessary. In the 1950s acousticians and geophysicists tried many nonlinear schemes to extend two-channel cross-covariance processing to more channels. Techniques such as multiple correlation and multiple polarity coincidence methods usually made the output signal-to-noise ratio worse than two-channel cross correlations. Of the various ideas, Jacobson's split-array cross-covariance technique emerged as a method of choice because it was simple and robust (see Selected Readings).

A split-array cross-covariance system is shown in Figure 16.3. The arrays of receivers can be area or volume arrays. The sources can be anywhere. The sketch is a projection on the x-z plane. Arrays I and II are focused or beam formed on signals from a source at m. The outputs of the arrays are treated as the receivers in a two-channel cross-covariance processor. The output signal-to-noise ratio gain is a combination of array gain and cross-covariance processing gain.

16.3.1 Covariance of Signals Due to Continuously Radiating Random Sources

One can make an image of sourcelet locations by focusing the arrays on a grid of points and plotting the magnitude of the covariance as intensities on the grid points or pixels. The resolution of sourcelet location depends on array aperture and the width of the autocovariance function of the signal. The example in Figure 16.4a is a simulation that

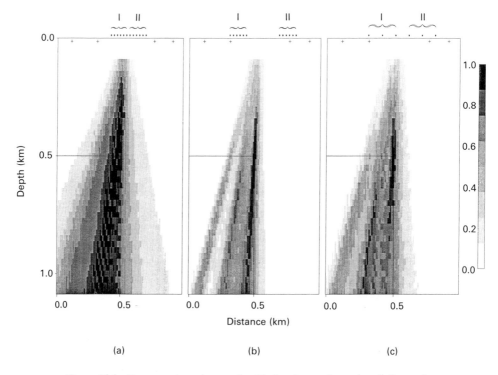

Figure 16.4 Cross-covariance images of an ideal wedge: continuously radiating random sources. The sources are driven by independent random function generators. The cross covariance is normalized, and the intensity scale gives the peak values. Source positions (+), receiver positions (*), the two arrays (I and II), and the model are shown. (a) Continuous-array layout; (b) split-array layout; and (c) sparse-array layout. The figures are from M. R. Daneshvar, and C. S. Clay, "Imaging of rough surfaces for impulsive and continuously radiating sources," *J. Acoust. Soc. Am.*, 82 (1987), 360–69.

has continuously radiating independent random sources on the surface. Each source gives a reflected and diffracted wave. The diffraction signal from the wedge appears as an intensity maximum slightly below the apex of the wedge. Here, the cross covariances are normalized. Intensity values of 0.6 and greater have an image that is smeared from about 500 to 700 m depth. The horizontal smear is less than 100 m.

The other two intensity maxima are at about 1000 m depth. These corresponded to the image reflection positions for two of the sources. If there were many more radiating sources on the surface of the ground, their images would appear as a plane of images at 1000 m depth. The diffracted components would strengthen the image near the wedge apex. An example of cross-covariance processing for a pair of sparse arrays is shown in Figure 16.4b. The images are about the same as in Figure 16.4a.

Pure geometric focusing gives a correct image at the wedge apex or diffractor and an "incorrect" surface at twice the interface depth. This result demonstrates the difference between seeing a rough surface and reflections in a mirror. Without other information, mirrors give indirect indications of their presence.

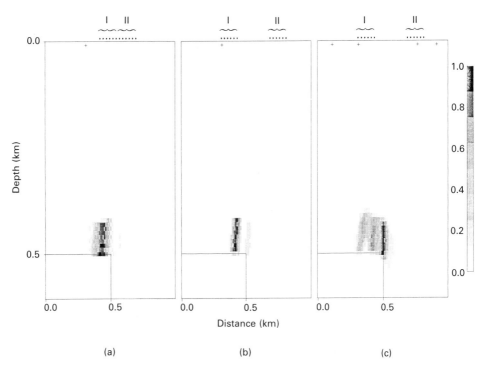

Depth (km)

Distance (km)

(a)   (b)   (c)

Figure 16.5   Cross-covariance image of an ideal wedge: impulsive source. The conditions are the same as in Figure 16.4. Source positions (+), receiver positions (*), the two arrays (I and II), and the model are shown. The signal is the impulse response low-pass filtered below 25 Hz. The apparent shallow image is an artifact of the computation using the impulse response. (a) Single-source correlation image for continuous-array layout; (b) single-source correlation image for split-array layout; and (c) stacked correlation image for split-array layout. The figure is from Daneshvar and Clay (1987); see caption to Figure 16.4.

16.3.2 Covariance of Signals Due to Known Impulsive Sources

When the location and transmission time of an impulsive source transmission are known, the images can be greatly improved. For cross-covariance calculations, the travel time is from the source to a grid point and then to the receivers. An example of the image for an impulsive source is shown in Figure 16.5. The structure and arrays are the same as shown in Figure 16.4a. Here, image maxima are on the interface. The horizontal plane of the wedge is identified. Even though the arrays are focused on the grid points at 500 m depth, the depth of focus is large enough to give an adequate response to the actual wave front coming from "image sourcelets" at 1000 m depth. This example shows that travel times reduce the location ambiguities of reflections and diffractions.

The last examples show images for a more complicated interface. The interface is the wedge model of sea floor shown in Figure 15.4. The set of images for impulsive sources are in Figure 16.6. In both figures the images show intensity maxima at wedge apexes.

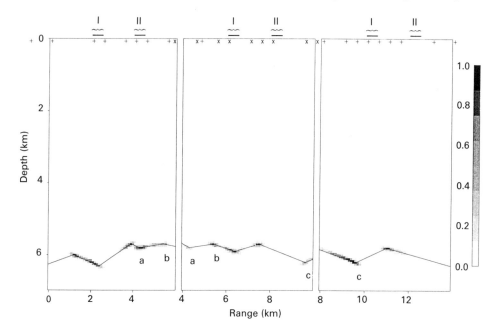

Figure 16.6 Cross-covariance images of a model sea floor: impulsive sources. The sea-floor model is the same as in Figure 15.4. The figure shows reflections and diffraction arrivals. Each panel is for a short section. The arrays and sources are moved relative to the bottom for each panel; there is much overlap. The figure is from Daneshvar and Clay (1987); see caption to Figure 16.4.

PROBLEMS

1. Section 12.3 includes a discussion of spatial sampling and multiple geophone arrays. Here, one can compute array responses and show that overspaced arrays have poor directional response patterns. The cross-covariance methods in Section 16.2 suggest that one can use a pair of widely spaced receivers and get good directional patterns.
   (a) What are the basic assumptions in Section 12.3?
   (b) What are the assumptions in Section 16.2?
   (c) Explain why both techniques are correct within their operating conditions.
   (d) If you find contradictions, resolve the differences.

2. Make a table of common types of measurements (reflection, wide-angle, or refraction) and typical subsurface structures. For each technique and type of structure, indicate applicability of standard linear geophysical methods and the nonlinear cross-covariance methods to measurements of the structures.

16.4 FIELD MEASUREMENTS: SPLIT-ARRAY CROSS-COVARIANCE PROCESSING

Cross-covariance or correlation processing of signals from a split geophone array was used to image internal structure in the Puritan batholith, northern Wisconsin. The Puritan batholith is part of a greenstone-granite terrane of late Archean age. Figure 16.7 is a map

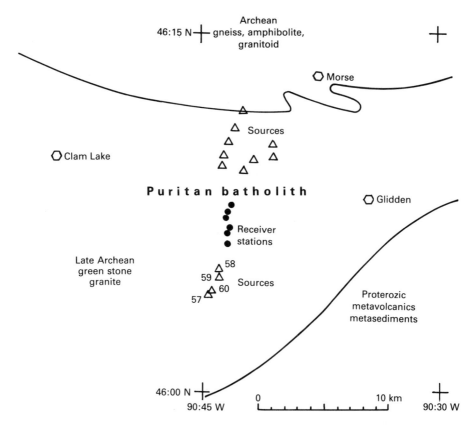

Figure 16.7 Map of receiver stations. The receiver and source locations are in the Puritan batholith in northern Wisconsin. Images are projected on a vertical plane through the line of receiver stations. Figures 16.7–16.10 are from W. D. Doll, "Seismic diffraction processing applied to data from Ashland County, Wisconsin," Ph.D. thesis, University of Wisconsin-Madison, 1983, and W. E. Doll, and C. S. Clay, "Seismic imaging of the Puritan batholith, Wisconsin, using split-array cross-correlation processing," *J. Geophys. Res.*, Vol. 93, No. B7 (1988), 8023–34. Copyright by the American Geophysical Union.

of receiver stations and source locations. The Vibroseis source was driven by a downswept chirp of 30 s duration and sweep from 28 Hz to 12 Hz. To improve the signal-to-noise ratio each receiver station had a five-leg star array of six geophones, and the legs were 30 m long. The signals were match-filtered or correlated in a multichannel real-time correlator. A discussion of coded signal transmissions and matched filters is given in Section 11.2. Refraction in the area gave 5.75 km/s at the surface and a gradient of 0.1 (km/s)/km to 0.3 km depth. Beneath 0.3 km the gradient decreased to 0.1 (km/s)/km. Ray traces showed that the curved ray paths due to the seismic velocity gradient caused small displacements of the ray paths relative to straight ray paths. At 30° incident angle, the shift was 0.2 km at 10 km depth. An average constant seismic velocity of 5.8 km/s was used for signal processing.

An example of the seismograms is shown in Figure 16.8. Since the seismic velocity is 5.8 km/s, signals arriving at about 4 s are from diffractors or reflectors about 10 to 12

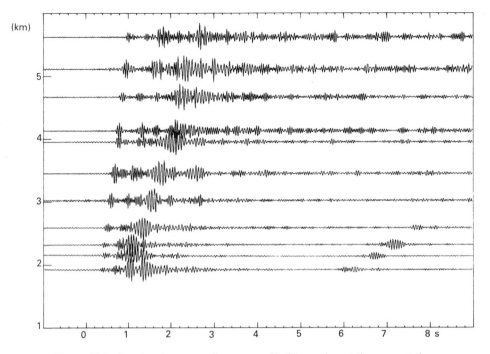

Figure 16.8 Sample seismograms from source 59. The receiver stations nearest the source (two bottom traces) show very late arrivals. These are surface waves. From Doll (1983); see caption to Figure 16.7.

km deep. There is plenty of energy, but obvious reflection signals are absent. Delay and sum linear array processing also gave poor (uninterpretable) results. A running average of the absolute values of these records was used to make an empirical time-varying gain for covariance processing.

The array of stations was split into two groups of three stations for cross-covariance processing. The only change from Figure 16.3 was the use of polar coordinates to display images. Polar coordinates were chosen because an array has nearly constant angular resolution. Since the transmission times and source positions were known, travel times from source to diffractor to arrays were included in the signal processing, as in Figures 16.5 and 16.6.

Images from four source positions are shown in Figure 16.9. The letter A marks a break in the character of the steeply dipping images above A. At an average velocity of 5.8 km/s, A is at 3.5 km depth. Images from a little below a line from B to C and 1.2 s or ~ 7 km depth to a depth of A are strong. Beneath 7 km the character of the images changes and the images are weaker. The letters C, D, and E indicate dipping images. Using the assumption that the image of an inhomogeneity should stay in the same place for different source positions, we can construct a composite profile of the images. Figure 16.10 shows a composite sketch of inhomogeneities within the batholith. The images are drawn as if they were in a vertical sheet that is along the line of receivers.

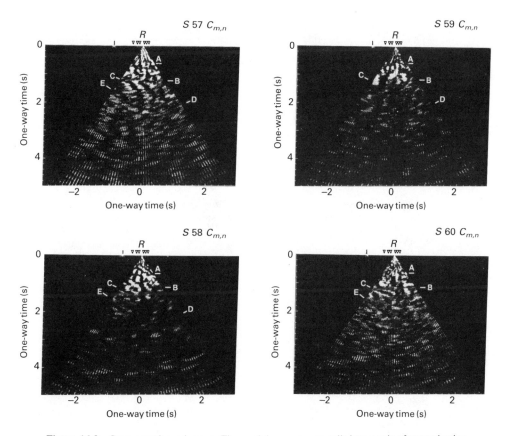

Figure 16.9 Cross-covariance images. The receiving array was split into a pair of arrays having three receiver stations in each subarray. The images are for four different source positions. The average seismic velocity is 5.8 km/s. From Doll (1983); see caption to Figure 16.7.

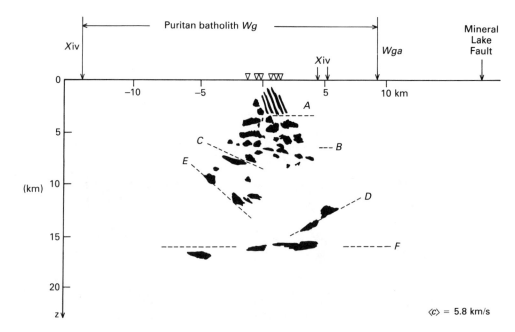

Figure 16.10 Composite sketch of cross-covariance processing. The sketch is based on an overlay of images from many different sources, including sources both north and south of the array. From Doll and Clay (1988); see caption to Figure 16.7.

Unit III—Suggested Readings

Signal enhancement uses many techniques. The readings are separated into groups for a primary classification. To give a sense of the intellectual development of the fields, each section is in chronological order. Some of the readings, and particularly the books, include many topics and could be in any group. Our placements are based on particular qualities. We have chosen representative books and papers from a very rich literature.

Signal Theory

The analysis of random signals is a standard undergraduate electrical engineering course. Thus, the market for texts is very competitive, and there are many good books. Usually, these books are slanted toward communications. Geophysical applications are different. We list several books by Robinson because they are useful and consistent in style and notation.

VAN VLECK, J. H., AND D. MIDDLETON, "A theoretical comparison of the visual, aural, and meter reception of pulsed signal in the presence of noise," *J. Appl. Phys.* 17 (1946), 940–71. The authors developed the matched filter. This is a fundamental paper in signal theory. Although the experimental examples are obsolete, the physical and mathematical development is particularly lucid. They show that the matched filter is an optimum filter.

NEITZEL, E. B., "Seismic reflection records obtained by dropping a weight," *Geophysics*, 23 (1958), 48–80. Neitzel introduced a signal stacking and dropping weight source. Although the scale is different, hammer seismic stacking systems are direct descendants.

CRAWFORD, J. M., W. E. N. DOTY, AND M. R. LEE, "Continuous signal seismography," *Geophysics*, 25 (1960), 95–105. The authors used the transmission of a chirp signal from a vibration on the surface for reflection measurements. Their analyzer was a matched filter or correlator. This paper described a major innovation, the Vibroseis of Continental Oil Company.

ROBINSON, E. A., *Multichannel Time Series Analysis with Digital Computer Programs*. San Francisco: Holden-Day, 1967. An excellent treatment of seismic data processing for digital data. It is one of the first texts to demonstrate the importance and power of linear algebra and z-transforms in signal- and wave-propagation problems.

WATERS, K. H., *Reflection Seismology*. New York: John Wiley, 1978. A well-written description of exploration geophysical techniques.

ROBINSON, E. A., AND S. TREITEL, *Geophysical Signal Analysis*. Englewood Cliffs, N.J.: Prentice-Hall, 1980. Robinson and Treitel put their journal articles together in coherent treatment.

KANASEWICH, E. R., *Time Sequence Analysis in Geophysics* (3d ed.). Edmonton, Alberta, Canada: University of Alberta Press, 1981. A comprehensive treatment of the subject, including convolutions, z-transforms, correlation and covariance, spectrum analysis, deconvolution, velocity (array) filters, and maximum entropy and likelihood methods of spectral analysis.

ROBINSON, E. A., AND M. T. SILVIA, *Digital Foundations of Time Series Analysis. Vol. 1: The Box-Jenkins Approach. Vol. 2: Wave-Equation Space-Time Processing.* San Francisco: Holden-Day, 1981. Volume 1 takes people who have a background in elementary statistics through convolution and *z*-transforms, correlation and autocorrelation, digital filters, deconvolution and Box-Jenkins modeling. Volume 2 applies digital methods to multidimensional problems, namely, the wave equation. Topics include a two-dimensional sampling theorem, spatial filtering, downward continuation of potential field data, and array theory and spectral methods.

CLARK, R. D., "Enders Robinson," *Geophysics: The Leading Edge*, 4 (1985), 16–20.

PROUBASTA, D., "Sven Treitel," *Geophysics: The Leading Edge*, 4 (1985), 24–28. Robinson and Treitel were members of the legendary Geophysical Analysis Group at MIT. Separately, together, and with other coauthors, Robinson and Trietal wrote many of the fundamental papers in geophysical signal theory.

Arrays and Multichannel Processing

Several of the books in the signal theory section extend one-dimensional treatments to multidimensional filters and arrays. The exploration geophysical literature has many papers on multiple geophone arrays. A few books and papers have been selected because of author preference.

KLIPSCH, P. W., "Some aspects of multiple recording in seismic prospecting," *Geophysics*, 1 (1936), 365–77. Klipsch considered the problem of improving the signal-to-noise ratio by using multiple geophone arrays. In many respects this is a fundamental paper in array theory. He computed the signal-to-noise gain as being $n^{1/2}$, where *n* is the number of geophones. In 1936 he wrote that real improvement would require over 100 geophones per trace, an absurdity.

RIEBER, F., "A new reflection system with controlled directional sensitivity," *Geophysics*, 1 (1936), 97–105. Rieber made variable-density seismic records on film. He used photoelectric cells in a summing playback to steer the array. His insight and technology were decades ahead of standard practice in the exploration industry.

JACOBSON, M. J., "Analysis of a multiple receiver correlation system," *J. Acoust. Soc. Am.*, 29 (1957), 1342–47. Jacobson gives a cross-covariance system of processing for multiple receivers. It is a nonlinear technique. His purpose is to detect and determine the direction to a source that radiates a random signal in a noisy ocean.

GRAEBNER, R. J., "Seismic data enhancement: a case history," *Geophysics*, 25 (1960), 283–311. This is one of the papers on the use of multiple geophones to improve record quality.

MAYNE, W. H., "Common reflection point horizontal data stacking techniques," *Geophysics*, 27, p. II (1962), 927–38. Field examples of a technique that became an industry standard.

SCHOENBERGER, M., "Optimization and implementation of marine seismic arrays," *Geophysics*, 35 (1970), 1038–53. A comparison of several optimum methods of array design and the sensitivity of array responses and side lobes to small changes of hydrophone sensitivity.

SENGBUSH, R. L., *Seismic Exploration Methods.* Boston: International Human Resources Development Corp., 1983. An introduction to seismic data processing. The sections on seismic noise and noise reduction by seismic arrays are good. Sengbush uses effective illustrations to develop the material with a minimal amount of mathematical formalism.

PROUBASTA, D., "Harry Mayne," *Geophysics: The Leading Edge*, 4 (1985), 18–24. A delightful biographical sketch of Mayne and his contributions. I like the description of his early efforts to do common reflection (or mid-) point processing.

Scattering and Diffraction of Sound (Scalar) Waves

Scattering and diffraction computations are difficult. Except for special cases, approximations of one kind or another are used. Some solutions have larger errors than others. In the last few years people have been doing the calculations numerically for specific cases. An important question has been the magnitude of the errors introduced by the Kirchhoff approximation. The suggested readings start with Rayleigh (he did not use the Kirchhoff approximation). Born and Wolf follow Rayleigh because much of the discussion refers to developments in the same era. There is a mixture of theoretical and experimental readings.

RAYLEIGH, J. W. S. (J. W. STRUTT), *The Theory of Sound*, Vol. 2, Sec. 272a, 89–96 (2d ed., 1896). Reprint. New York: Dover, 1945. Rayleigh gives a solution for an incident plane wave on a cosinusoidally corrugated surface. The sections just preceding this one contain derivations of the reflection at an interface and a thin layer. The discussions are concise.

RIEBER, F., "Visual presentation of elastic wave patterns under various structural conditions," *Geophysics*, 1 (1936), 196–218. Rieber did a set of laboratory model experiments in air that displayed the wave fronts of diffracted waves. The pictures are very impressive and clearly show the diffracted wave fronts for simple and complicated diffracting interfaces. Since the experiments were in air, elastic waves in the title are not accurate.

ECKART, C., "The scattering of sound from the sea surface," *J. Acoust. Soc. Am.*, 25 (1953), 566–70. Eckart wrote one of the fundamental papers in scattering theory. He showed how to put a randomly rough surface in the Helmholtz-Kirchhoff integral and to calculate useful results.

LACASE, E. O., AND P. TAMARKIN, "Underwater sound reflection from a corrugated surface," *J. Appl. Phys.*, 27 (1956), 138–48. A good comparison of the Rayleigh, Brekhovskikh, and Eckart theories with a set of experimental data. It includes an English version of Brekhovskikh's theory.

BIOT, M. A., AND I. TOLSTOY, "Formulation of wave propagation in infinite media by normal coordinates with an application to diffraction," *J. Acoust. Soc. Am.*, 29 (1957), 381–91. Biot and Tolstoy use their normal coordinate expressions to solve the diffraction problem in closed form. The solution is for a point impulsive source and point receiver in a wedge bounded by infinitely rigid reflectors. The solution can be transformed to perfect pressure release by a minor change in the algebra. I suggest reading Jebsen and Medwin (1982) for a better explanation of source functions. See also I. Tolstoy and C. S. Clay *Ocean Acoustics*, New York: American Institute of Physics, 1987, Appendix 5.

BORN, M., AND E. WOLF, *Principles of Optics* (3d ed.). Oxford: Pergamon 1965. Chapter 8, Elements of the Theory of Diffraction, gives a careful and complete discussion of the theory of diffraction of scalar waves, i.e., sound waves. It starts with the Huygens-Fresnel principle, derives the intergal theorem of Holmholtz and Kirchhoff, and gives many examples. The Kirchhoff assumption is given. Section 8.9 gives the so-called boundary diffraction wave that leads to the Rubinowicz representation of the Kirchhoff diffraction integral. I often use this book as a reference.

HILTERMAN, F. J., "Three-dimensional seismic modeling," *Geophysics*, 35 (1970), 1020–37. Hilterman used a laboratory model to measure diffracted and reflected waves. The model was an air medium over hard boundaries. The source was a spark. The receiver was a small microphone near the surface. He showed good agreement between his numerical calculations and experiment.

TROREY, A. W., "A simple theory for seismic diffractions," *Geophysics*, 35 (1970), 762–84. Trorey uses the Laplace transform to evaluate the Helmholtz-Kirchhoff integral for an impulsive point source. Jebsen and Medwin (1982) showed that his Kirchhoff solution has large errors at grazing angles. A Fourier transform version is in Chapter 10 of Clay and Medwin (1977).

TOLSTOY, I., *Wave Propagation*. New York: McGraw-Hill, 1973. Tolstoy develops wave-propagation theory for a wide range of media and situations. Here, the development of the wedge diffraction problem is more complete than the Biot-Tolstoy paper.

CLAY, C. S., AND H. MEDWIN, *Acoustical Oceanography*. New York: Wiley Interscience, 1977. Chapters 10 and A10 give derivations of the Helmholtz-Kirchhoff diffraction theory for underwater sound.

MEDWIN, H. AND J. C. NOVARINI, "Backscattered strength and the range dependence of sound scattered from the ocean surface," *J. Acoust. Soc. Am.*, 69 (1981), 108–11. An introduction to the wedge-assemblage or facet-ensemble approximation.

JEBSEN, G. M., AND H. MEDWIN, "On the failure of Kirchhoff assumption in backscatter," *J. Acoust. Soc. Am.*, 72 (1982), 1607–11. The Kirchhoff assumption that the reflected component of a signal is approximately the local plane-wave assumption is used in many wave-scattering papers. The authors compare Trorey's Kirchhoff and boundary-wave solution, the exact Biot-Tolstoy solution, and experiment. The experiment and Biot-Tolstoy solution agree. The Trorey solution has large errors.

KINNEY, W. A., C. S. CLAY, AND G. A. SANDNESS, "Scattering from a corrugated surface: comparison between experiment, Helmholtz-Kirchhoff theory, and the facet-ensemble method," *J. Acoust. Soc. Am.*, 73 (1983), 183–94. The authors represent an almost sinusoidal corrugated surface by a set of plane facets. The intersection of a pair of facets forms a wedge. They compute the scattering from the surface by using the Biot-Tolstoy wedge diffraction solution at each wedge. They give a Fourier transformation of the Biot-Tolstoy solution. They call this the *facet-ensemble method*, which is the same as the wedge-assembly method. The paper compares the two numerical methods and experiment. Both agree with experiments for angles of incidence out to 40°.

MACDONALD, J. A., G. H. F. GARDNER, AND F. J. HILTON, *Seismic Studies in Physical Modeling*. Boston: International Human Resources Development Corp., 1983. The book gives results of many laboratory seismic model experiments. The model structures are a set of simple shapes that give complicated patterns of reflected and diffracted waves. The data are used for two- and three-dimensional imaging studies.

NOVARINI, J. C., AND H. MEDWIN, "Computer modeling of resonant sound scattering from a periodic assemblage of wedges: comparison with theories of diffraction gratings," *J. Acoust. Soc. Am.*, 77 (1985), 1754–59. The theories had plane waves incident on a sawtooth grating. In using the Biot-Tolstoy wedge diffraction solution for each wedge, Novarini and Medwin found that reflections from facets should be ignored, because the dimensions of each facet are finite, while the Fresnel zone dimensions are infinite. Assuming that the reflection is proportional to the ratio of the areas, a finite facet area over an infinite area gives zero reflection. The wedge approximation gave numerical agreement with exact numerical evaluations within a few percent.

DETRICK, R. S. (introducer), "Introduction to mapping the sea floor" and the following 12 papers, *J. Geophys. Res.*, 91 (1986), 3331–3520. These papers are a special section on mapping the sea floor. They include papers on Deep Tow, SeaBeam, MARC I and II, GLORIA, and Seasat. Ten of the papers show the power of using multichannel high-resolution systems to give high-resolution images or pictures of the sea floor. Most of the signals are due to backscattered sound from rough features. The color illustrations are excellent. The last two papers on Seasat demonstrate the use of a satellite altimeter and gravitational theory.

Migration, Seismic Holography, and Imaging of Seismic Data

Migration and imaging techniques are closely related to diffraction and scattering phenomena. The purposes of imaging techniques are to focus on or enhance signals from diffractors and to put the image in the right place. The exploration literature uses reverse-time migration as a name for seismic holography.

SCHNEIDER, W. A., "Developments in seismic data processing and analysis (1968–70)," *Geophysics*, 36 (1971), 1043–73. This review paper gives an excellent discussion of normal moveout velocity analysis, common midpoint stack, and migration. In 1971 Schneider called it a new imaging process. He shows comparisons of CMP stacks and "raw" migration stacks.

HOOVER, G. M., "Acoustical holography using digital processing," *Geophysics*, 37 (1972), 19. Theory, numerical examples, and seismic model results for holographic imaging.

CLAERBOUT, J., *Fundamentals of Geophysical Data Processing: With Applications to Petroleum Prospecting.* New York: McGraw-Hill, 1976. This is Claerbout's first book. It develops his version of wave theory and goes to wave equation migration. An advanced text.

PHINNEY, R. A., AND D. M. JURDY, "Seismic imaging of deep crust," *Geophysics*, 44 (1979), 1637–60. The authors use steered and focused beams as a generalization of the common midpoint stack. These beams are migrated and gathered to form images of subsurface structure. The results are impressive.

BERKHOUT, A. J., *Seismic Migration: Imaging of Acoustic Energy by Wave Field Extrapolation.* Developments in Solid Earth Geophysics 12, New York: Elsevier, 1980. An advanced research monograph on migration. He uses propagation and scattering matrices in his development.

BAYSAL, E., D. D. KOSLOFF, AND J. W. C. SHERWOOD, "Reverse time migration," *Geophysics*, 48 (1983), 1514–24. Uses numerical simulations to describe imaging methods using reverse time migration or seismic holography.

LOEWNTHAL, D., AND I. R. MUFTI, "Reversed time migration in spatial frequency domain." *Geophysics*, 48 (1983), 627–35. Uses time reversed migration to image structure.

ROBINSON, E. A., *Migration of Geophysical Data.* Boston: International Human Resources Development Corp., 1983. Robinson gives the physical basis and limitations of the various migration methods. The book has excellent discussions and relatively little mathematics. I recommend the book as an introduction to migration theory.

CLAERBOUT, J., *Imaging the Earth's Interior.* Oxford: Blackwell, 1985. A witty and useful book on the propagation of waves in inhomogeneous media. It is a good introduction to finite grid techniques and includes the physics, mathematics, craft or art, and programs. The book is the result of a teacher's interacting with students.

GARDNER, G. H. F., ED., *Migration of Seismic Data.* Geophysics Reprint Series No. 4. Tulsa, Okla.: Society of Exploration Geophysics, 1985. The book contains editors' introductions and reprints of the more innovative and important papers that were published in *Geophysics* from 1937 to 1984. Rieber's paper from 1937 was decades ahead of its time. The papers range from compass-and-ruler techniques to elegant mathematics. Many papers give synthetic, scale-model, and field examples of migration processing.

McMECHAN, G. A., J. H. LUETGERT, AND W. D. MOONEY, "Imaging of earthquate sources in Long Valley Caldera, California 1983." *Bulletin Seis. Soc. of Amer.*, 75 (1985), 1005–1020. Uses recorded signals from earthquakes on an array of seismometers. Time reversed signals were numerically transmitted from the array to image the earthquake sources.

STOLT, R. H., AND A. K. BENSON, *Seismic Migration: Theory and Practice.* Amsterdam: Geophysical Press, 1986. Stolt made major contributions to migration theory and practice. This book gives a mathematical development for three dimensions. Stolt and Benson discuss the problems of migration before and after stack.

CHANG, W. F., AND G. A. McMECHAN, "Elastic reverse-time migration." *Geophysics*, 52 (1987), 1365–75. Uses reverse-time migration to image simulations and real data.

Tomography

The following papers are in *Early Geophysical Papers*, Tulsa, Okla.: Society of Exploration Geophysics, 1947.

These papers demonstrate the use of tomographic concepts for the measurement of large inhomogeneities. The early seismologists used a "fan-shooting" technique to find salt domes. The shot was fired at the center of a circle of geophones. Shot-to-geophone distances were roughly 6 to 10 km. The seismic velocity in salt is about 4 to 5 km/s, and the seismic velocities in the surrounding sediments are about 2 km/s. Waves traveling through the salt domes had large travel-time anomalies. The seismologists used transmissions between many source and receiver positions to locate and outline the salt domes. The fan-shooting phase of geophysics lasted from about 1925 through the early 1930s.

McCollom, B., and W. W. LaRue, "Utilization of existing wells in seismograph work," ibid., 119–27 (originally 1931).

DeGolyer, E., "Notes on the early history of applied geophysics in the petroleum industry," *Early Geophysical Papers*, 245–54 (originally 1932).

Rosaine, E. E., and O. C. Lester, Jr., "Seismological discovery and partial detail of Vermilion Bay Salt Dome, Louisiana," ibid., 381–89 (originally 1935).

Recent Geophysical Applications of Tomography and Inverse Methods

Herman, G. T., ed, *Image Reconstruction from Projections: Implementations and Applications.* New York: Springer-Verlag, 284 pp., 1979.

Munk, W., and C. Wunsch, "Ocean acoustic tomography: a scheme for large-scale monitoring," *Deep Sea Res.*, 26A (1979), 123–61.

Herman, G. T., ed, *Image Reconstruction from Projections: The Fundamentals of Computerized Tomography.* New York: Academic Press, 316 pp., 1980.

The Ocean Tomography Group: D. Behringer, T. Birdsall, M. Brown, B. Cornuelle, R. Heinmiller, R. Knox, K. Metzger, W. Munk, J. Spiesberger, R. Spindel, D. Webb, P. Worchester, and C. Wunsch, "A demonstration of acoustic tomography," *Nature*, 299 (1982), 121–25. This was a large multiinstitutional experiment, so correspondingly it has 13 authors. They used acoustic transmissions to study mesoscale oceanic structure. They had to measure travel-time anomalies that were far smaller than those measured by the fan shooters. They divided the ocean into cells and used inverse methods to go from data to properties of the cells.

Bishop, T. N., K. P. Burke, R. T. Cutler, R. T. Langan, P. L. Love, J. R. Resnick, R. T. Shuey, D. A. Spindler, and H. W. Wyld, "Tomographic determination of velocity and depth in laterally varying media," *Geophysics*, 50 (1985), 903–23.

Cornuelle, B., and the rest of the Ocean Tomography Group, "Tomographic maps of ocean mesoscale. P. 1: Pure acoustics," *J. Phys. Oceanography*, 15 (1985), 133–52. Gives the result of several more years of analysis of the data.

Ivanssen, S., "A study of methods for tomographic velocity estimation in the presence of low-velocity zones," *Geophysics*, 50 (1985), 969–88.

Carrion, P., *Inverse Problems and Tomography in Acoustics and Seismology.* Atlanta: Penn Publishing Company, 303 pp., 1987.

Postscript

The history of seismic methods illustrates what may be a natural life cycle. It may also be an example of evolution and survival of the fittest. Geophysicists have had to make useful innovations, improve resolution, and find interesting structures or go out of business (die).

Early seismologists located earthquakes. The data from a few seismic stations spread about the earth gave a crude picture of the interior structure of the earth. With the discovery of petroleum, people tried everything to locate oil-bearing rocks a few thousand meters beneath the surface. After trials using electrical, gravity, magnetic, and seismic refractions, reflection seismology became the method of choice. Geophysicists soon learned that more data channels gave better resolution and images of subsurface structures. The explorationist with the best images and imaginative management found the most oil. The survival urge has driven exploration geophysicists to improve their technology, resolution, and images.

A few observations about present and future seismic methods follow. First, the seismic reflection methods for imaging structures in sedimentary basins are mature as evidenced by the number of monographs on migration in the selected readings. Conventional reflection and refraction data are processed by methods that enhance signals reflected from large plane fragments of interfaces. Scattered components from rough interfaces and inhomogeneities are suppressed. Minor changes in the algorithms can improve the display of scattered components. These images show the interfaces or boundaries of layers and inhomogeneities where the physical properties of the rocks change discontinuously at a boundary. If the changes of properties are gradual and the reflected and scattered components are small, then, as an alternative, transmissions through a medium can be used to find inhomogeneities.

The art of measuring the transmissions of waves through an inhomogeneous medium to determine the structure is called *tomography*. The word means the graph of a slice or section. The discovery and development of computerized X-ray tomography earned a Nobel prize for Allan Cormack and Godfrey Hounsfield in 1979.

In seismic tomography the transmission amplitudes and travel times of waves through a region are measured from many source positions to many receiver positions. For analysis the region is divided into cells that have their own parameters or rock properties. The computations use inverse matrix methods and the data to determine the parameters for each cell. Seismic tomography requires careful design of experiments, a lot of data, good forward wave transmission algorithms, and a fast big computer.

William Menard, geologist and observer of scientists, has said that one can judge the maturity of a field and the difficulty of doing something new by looking at the standard references in recent papers. The word *tomography* is not in dictionaries published before 1970. The reference list starts with examples from the 1930s and then jumps to the late 1970s and 1980s. Seismic tomography fits Menard's criterion and is likely to be the science child that is the fruit of a generation of exploration seismology.

Appendix A

A1.1 A FEW MATHEMATICAL METHODS AND THE THREE-DIMENSIONAL WAVE EQUATION

Three techniques from advanced mathematics can greatly simplify our derivations. Depending on the reader's background this is either a minimal introduction or a brief review of complex numbers, multiple variables, and the separation of variables. These have been abstracted from standard texts on applied mathematics. Our discourse will use formulas from Appendix D.

A1.1.1 Complex Numbers

The solution of the quadratic equation shows the necessity of introducing the so-called imaginary number i

$$i \equiv \sqrt{-1} \qquad i^2 = -1 \tag{A.1}$$
$$\frac{1}{i} = -i \qquad i^3 = -i$$

(The engineering literature uses $j = \sqrt{-1}$.) The quadratic equation has two roots, z_+ and z_-

$$az^2 + bz + c = 0 \tag{A.2}$$

$$z_+ = \frac{-b + (b^2 - 4ac)^{1/2}}{2a} \tag{A.3}$$

$$z_- = \frac{-b - (b^2 - 4ac)^{1/2}}{2a} \tag{A.4}$$

213

If b^2 is greater than $4ac$, then both roots are real. If b^2 is less than $4ac$, then we factor $\sqrt{-1}$ and write the roots as

$$z_+ = \frac{-b + i(4ac - b^2)^{1/2}}{2a} \tag{A.5}$$

$$z_- = \frac{-b - i(4ac - b^2)^{1/2}}{2a} \tag{A.6}$$

The roots are complex, and the real and imaginary components are

$$\left. \begin{array}{ll} x_+ = \dfrac{-b}{(2a)} & x_- = \dfrac{-b}{(2a)} \\[2mm] iy_+ = \dfrac{i(4ac - b^2)^{1/2}}{2a} & iy_- = \dfrac{-i(4ac - b^2)^{1/2}}{2a} \end{array} \right\} \tag{A.7}$$

The roots are graphed as shown in Figure A.1. The plane is known as the *complex plane*. A point z on the complex plane has the real value x and the imaginary value iy,

$$z \equiv x + iy \tag{A.8}$$

The *complex* conjugate of a number or variable is obtained by replacing *every* i with $-i$. A common notation for the complex conjugate of z is z^*,

$$z^* \equiv x - iy \tag{A.9}$$

The complex roots of the quadratic equation have a complex conjugate relationship because $z_- = z_+^*$.

The magnitude of complex numbers and variables is measured by taking the *absolute square*

$$\left. \begin{array}{l} |z|^2 = zz^* = (x + iy)(x - iy) \\[2mm] |z|^2 = x^2 + y^2 \end{array} \right\} \tag{A.10}$$

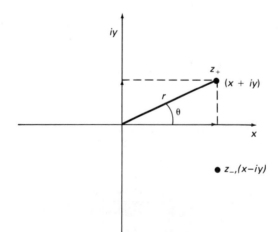

Figure A.1 Complex plane. z_+ and z_- are the roots of the equation $az^2 + bz + c = 0$.

The absolute square is real. The absolute square can be extended to complicated functions. For example, let a function $f(z)$ be

$$f(z) = a + ib + z^2 = a + ib + x^2 - y^2 + 2ixy$$

then

$$\left.\begin{aligned} f(z) &= (a + x^2 - y^2) + i(b + 2xy) \\ f(z)^* &= (a + x^2 - y^2) - i(b + 2xy) \end{aligned}\right\} \qquad (A.11)$$

The absolute square is the sum of the squares of the real and imaginary components,

$$|f(z)|^2 = (a + x^2 - y^2)^2 + (b + 2xy)^2 \qquad (A.12)$$

Polar coordinate expressions are very useful for dealing with products of complex numbers. From Figure A.1, x and y are

$$x = r \cos \theta \qquad y = r \sin \theta \qquad (A.13)$$

$$r^2 = x^2 + y^2 \qquad \theta = \arctan \frac{y}{x} \qquad (A.14)$$

Note that r^2 is the absolute square of $z = x + iy$. Euler (1707–83) gave an elegant way to write the polar expressions, namely,

$$\left.\begin{aligned} z &= r(\cos \theta + i \sin \theta) = re^{i\theta} \\ z^* &= r(\cos \theta - i \sin \theta) = re^{-i\theta} \end{aligned}\right\} \qquad (A.15)$$

The angle θ is called the *argument*, and r is called the *modulus*.

The products of $z_1 z_2$ and $z_1 z_2^*$,

$$z_1 = r_1 e^{i\theta_1} \qquad z_2 = r_2 e^{i\theta_2} \qquad (A.16)$$

are computed by multiplying r_1 and r_2 and adding the exponential

$$\left.\begin{aligned} z_1 z_2 &= r_1 r_2 e^{i(\theta_1 + \theta_2)} \\ z_1 z_2^* &= r_1 r_2 e^{i(\theta_1 - \theta_2)} \end{aligned}\right\} \qquad (A.17)$$

Powers and roots of z are

$$z^n = r^n e^{in\theta} \qquad z^{1/m} = r^{1/m} e^{i\theta/m} \qquad (A.18)$$

Taking the $1/m$ root gives m roots. For example, there are three roots of $1^{1/3}$. Write 1 as follows (where $\cos \theta = \cos 2\pi = \cos 4\pi = \cdots = 1$ and $\sin \theta = \sin 2\pi = \sin 4\pi = \cdots = 0$):

$$1 = e^{i0} = e^{i2\pi} = e^{i4\pi} \qquad (A.19)$$

We take roots of the multiple values of $2n\pi$,

$$1^{1/3} = 1, \qquad e^{i2\pi/3}, \qquad e^{i4\pi/3} \qquad (A.20)$$

We can use Euler's formula to compute the real and imaginary components and then cube each root to verify that its cube is 1.

Repeated differentiations and integrations of exponential functions are often simpler than the corresponding operations on sine and cosine functions. For example, let

$$z = e^{ia\theta}$$

$$\frac{dz}{d\theta} = iae^{ia\theta} = iaz \qquad \frac{d^2z}{d\theta^2} = -a^2 z \qquad (A.21)$$

$$\int e^{ia\theta} \, d\theta = \frac{1}{ia} e^{ia\theta} + C = \frac{z}{ia} + C \qquad (A.22)$$

To take advantage of this, Euler's formula can be used to replace cosine and sines with

$$\cos \theta = \frac{1}{2}(e^{ia\theta} + e^{-ia\theta}) \qquad \sin \theta = \frac{1}{2i}(e^{ia\theta} - e^{ia\theta}) \qquad (A.23)$$

Calculations are done using the separate expressions for $e^{ia\theta}$ and $e^{-ia\theta}$ and then recombined to form solutions for $\cos a\theta$ or $\sin a\theta$.

A1.1.2 Multiple Variables, Partial Differentiation, and Vectors

The need to use the variables x, y, and z to describe surfaces is obvious in geology and geophysics because we deal with three-dimensional structures and surfaces all the time. A second, and perhaps less obvious, use of multiple variables is in the estimation of parameters that are commonly used to describe a set of data.

The basic concept is that the function $u(x, y, z)$ or simply u is a function of the independent variables x, y, and z. The values chosen for x, as an example, do not affect the values that can be chosen for y and z. Since the x-, y-, and z-coordinates are orthogonal, a displacement of a point from a position (x_1, y_1, z_1) to a position (x_2, y_1, z_1) does not require any changes in y_1 and z_1. Similarly, displacements from y_1 to y_2 or z_1 to z_2 do not affect the other coordinates. It is easy to see the independence or orthogonality of x-, y-, and z-coordinates in three dimensions. Here the variables x, y, and z are orthogonal, because the normal projections of x on the coordinates y and z are zero. The projections of y on x and z are zero, and the projections of z on x and y are zero.

It takes more imagination to see that the concept of three independent and orthogonal coordinates can be extended to n-dimensions. Physicists and mathematicians did this more than a century ago. Defining or designating the variables $q_1, q_2, q_3, \ldots, q_n$ as being independent or orthogonal, we can write a function $u(q_1, q_2, \ldots, q_n)$. Of course, we can assign the variables any sequence symbols, such as $x_1, x_2, \ldots, x_n$ or a, b, c, d, $\ldots$, and so forth. The methods that mathematicians use to define or derive a set of n-dimensional orthogonal variables can be subtle.

Returning to three-dimensional space and x-, y-, and z-coordinates, we ask, What are the slopes of a surface $u(x, y, z)$ along the x-, y-, and z-directions? Recalling the elementary differentiations of $y = f(x)$,

$$\frac{dy}{dx} = \lim_{\Delta x \to 0} \frac{f(x + \Delta x) - f(x)}{\Delta x} \qquad (A.24)$$

dy/dx is the slope or gradient of y along the x-direction. Since the dependence of $u(x, y, z)$ on x is independent of y and z, we can take a corresponding limit,

$$\frac{\partial[u(x, y, z)]}{\partial x} = \lim_{\Delta x \to 0} \frac{u(x + \Delta x, y, z) - u(x, y, z)}{\Delta x} \tag{A.25}$$

where ∂ and ∂x are the symbols for a partial derivative. The variables y and z are held constant for the operation $\partial/\partial x$. The partial derivatives for the other variables are

$$\frac{\partial[u(x, y, z)]}{\partial y} = \lim_{\Delta y \to 0} \frac{u(x, y + \Delta y, z) - u(x, y, z)}{\Delta y} \tag{A.26}$$

$$\frac{\partial[u(x, y, z)]}{\partial z} = \lim_{\Delta z \to 0} \frac{u(x, y, z + \Delta z) - u(x, y, z)}{\Delta z} \tag{A.27}$$

These give the slopes or rates of change along the x-, y-, and z-directions.

Gradient

Instead of writing "the slope along the x-, y-, or z-direction is," we use indicators of direction, the *unit vectors*. In x-, y-, and z-coordinate systems (Figure A.2) the unit vectors are commonly **i**, **j**, and **k**. The magnitudes or lengths of **i**, **j**, and **k** are 1. The boldface type indicates that the symbol is a vector. The gradient is the vector sum of the sum of the slopes along the **i**-, **j**-, and **k**-directions. Thus, the gradient of u, or simply **grad** u, is

$$\mathbf{grad}\ u(x, y, z) \equiv \mathbf{i}\,\frac{\partial[u(x, y, z)]}{\partial x} + \mathbf{j}\,\frac{\partial[u(x, y, z)]}{\partial y} + \mathbf{k}\,\frac{\partial[u(x, y, z)]}{\partial z} \tag{A.28}$$

Grad $u(x, y, z)$ is a vector. This operation has been used so much that it is given a special symbol ∇, the *del operator*:

$$\nabla \equiv \mathbf{i}\,\frac{\partial}{\partial x} + \mathbf{j}\,\frac{\partial}{\partial y} + \mathbf{k}\,\frac{\partial}{\partial z} \tag{A.29}$$

The term *operator* means that one does the differentiation operations indicated. For example,

$$\mathbf{grad}\ u(x, y, z) = \nabla\ u(x, y, z) \tag{A.30}$$

means that one uses ∇ to compute Equation A.28. Expressions for ∇ in other coordinate systems can be found in Appendix D.

Scalar products

The product of the projection of vector **A** on vector **B** by **B** gives the scalar product. The scalar product of **A** times itself is A^2. The scalar products or dot products of **i**, **j**, and **k** are

$$\mathbf{i} \cdot \mathbf{i} = 1 \qquad \mathbf{j} \cdot \mathbf{j} = 1 \qquad \mathbf{k} \cdot \mathbf{k} = 1 \tag{A.31}$$

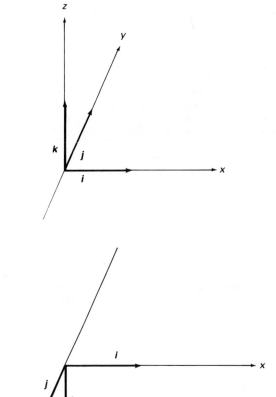

Figure A.2 Rectangular coordinates and unit vectors **i**, **j**, and **k**. Sketches show coordinates for z positive upward and z positive downward.

and since the projection of **i** on **j** is zero, and so forth,

$$\mathbf{i} \cdot \mathbf{j} = 0 \qquad \mathbf{i} \cdot \mathbf{k} = 0 \qquad \mathbf{j} \cdot \mathbf{k} = 0 \tag{A.32}$$

The unit vectors are used to facilitate the multiplication of vectors in general. For example, let **A** and **B** be the vectors

$$\left. \begin{array}{l} \mathbf{A} = \mathbf{i}\, a_x + \mathbf{j}\, a_y + \mathbf{k}\, a_z \\ \mathbf{B} = \mathbf{i}\, b_x + \mathbf{j}\, b_y + \mathbf{k}\, b_z \end{array} \right\} \tag{A.33}$$

The dot or scalar product $\mathbf{A} \cdot \mathbf{B}$ is

$$\mathbf{A} \cdot \mathbf{B} = a_x b_x + a_y b_y + a_z b_z \tag{A.34}$$

and both the right and left sides of the equation are scalars.

The del operation on a vector gives a scalar. Let $\mathbf{f}(x, y, z)$ be a vector having the components $u(x, y, z)$, $v(x, y, z)$, and $w(x, y, z)$; then

$$\mathbf{f}(x, y, z) = \mathbf{i}\, u(x, y, z) + \mathbf{j}v(x, y, z) + \mathbf{k}\, w(x, y, z) \tag{A.35}$$

The operation $\nabla \cdot \mathbf{f}(x, y, z)$ gives the so-called divergence or div,

$$\operatorname{div} \mathbf{f}(x, y, z) = \nabla \cdot \mathbf{f}(x, y, z) \tag{A.36}$$

$$= \frac{\partial[u(x, y, z)]}{\partial x} + \frac{\partial[v(x, y, z)]}{\partial y} + \frac{\partial[w(x, y, z)]}{\partial z} \tag{A.37}$$

Because $\mathbf{i} \cdot \mathbf{i} = 1$, and $\mathbf{i} \cdot \mathbf{j} = 0$, and so forth.

Laplacian

The special operation of $\nabla \cdot \nabla$ gives the Laplacian operator ∇^2

$$\nabla^2 \equiv \nabla \cdot \nabla = \frac{\partial^2}{\partial x^2} + \frac{\partial^2}{\partial y^2} + \frac{\partial^2}{\partial z^2} \tag{A.38}$$

The Laplacian operator is a result of the derivation of the three-dimensional wave equation. The Laplacian in other coordinate systems is in Appendix D.

Least squares estimation of parameters

Partial differentiations and the old calculus problem of determining the minimum of a function are combined in the least squares method. The key assumptions are that the parameters are independent, and the slopes of the test function are zero at the minimum.

For a specific example, we calculate the parameters that fit a polynomial to a set of data. Suppose that $y_0, y_1, y_2, \ldots, y_{n-1}$ are data points observed at $x_0, x_1, x_2, \ldots, x_{n-1}$. We wish to fit the data into the polynomial in x

$$f(x) = a_0 + a_1 x + a_2 x^2 \tag{A.39}$$

by adjusting a_0, a_1, and a_2 to make the sum of the squared errors as small as possible. The sum of the squared errors is

$$s \equiv \sum_{j=0}^{n-1} [y_j - (a_0 + a_1 x_j + a_2 x_j^2)]^2 \tag{A.40}$$

where s is a function of a_0, a_1, and a_2. Presumably n is much greater than 3, so the problem is overdetermined. With the assumption of independence of a_0, a_1, and a_2, we compute the slopes along a_0, a_1, and a_2 and then set the slopes to zero

$$\frac{\partial s}{\partial a_0} = 0 \qquad \frac{\partial s}{\partial a_1} = 0 \qquad \frac{\partial s}{\partial a_2} = 0 \tag{A.41}$$

giving

$$\sum_{j=0}^{n-1} [y_j - (a_0 + a_1 x_j + a_2 x_j^2)] = 0 \tag{A.42}$$

$$\sum_{j=0}^{n-1} x_j [y_j - (a_0 + a_1 x_j + a_2 x_j^2)] = 0 \tag{A.43}$$

$$\sum_{j=0}^{n-1} x_j^2 [y_j - (a_0 + a_1 x_j + a_2 x_j^2)] = 0 \tag{A.44}$$

By placing the a_0, a_1, and a_2 terms on the left side and the other terms on the right side, we obtain a set of three equations having three unknowns,

$$\left. \begin{array}{l} n\, a_0 + a_1 \,\Sigma x_j + a_2 \,\Sigma x_j^2 = \Sigma y_j \\ a_0 \,\Sigma x_j + a_1 \,\Sigma x_j^2 + a_2 \,\Sigma x_j^3 = \Sigma x_j y_j \\ a_0 \,\Sigma x_j^2 + a_1 \,\Sigma x_j^3 + a_2 \,\Sigma x_j^4 = \Sigma x_j^2 y_j \end{array} \right\} \tag{A.45}$$

Solution by elimination of variables or determinate gives a_0, a_1, and a_2.

A1.1.3 Separability of Variables

The method of separation of variables is a way of solving partial differential equations such as the wave equation. We use this as an example of the method. The separation method can be used when the solution can be written as the product of functions of the variables or as sums of separated functions of the variables. Our interest is in the products of functions. The function X_+ is an example:

$$X_+ = e^{i2\pi f(t - x/c)} = e^{i2\pi ft}\, e^{-i2\pi fx/c} \tag{A.46}$$

where X_+ is the product of a function of t and a function of x. Changing the notation, we write

$$X_+ = X\,T \tag{A.47}$$

$$X \equiv e^{-i2\pi fx/c} \tag{A.48}$$

$$T \equiv e^{i2\pi ft} \tag{A.49}$$

Substitution of XT in the wave equation

$$\frac{\partial^2 X_+}{\partial x^2} = \frac{1}{c^2}\frac{\partial^2 X_+}{\partial t^2} \tag{A.50}$$

gives

$$\frac{\partial^2 (XT)}{\partial x^2} = \frac{1}{c^2}\frac{\partial^2 (XT)}{\partial t^2} \tag{A.51}$$

Since the partial derivatives with respect to x operate only on X, and the partial derivatives with respect to t operate only on T, Eq. (A.50) becomes

$$T \frac{\partial^2 X}{\partial x^2} = \frac{X}{c^2} \frac{\partial^2 T}{\partial t^2} \tag{A.52}$$

or

$$\frac{1}{X} \frac{\partial^2 X}{\partial x^2} = \frac{1}{Tc^2} \frac{\partial^2 T}{\partial t^2} \tag{A.53}$$

All the x-dependent functions are on one side of the equal sign, and all the t-dependent functions are on the other. The left side equals the right side for *all values* of x and t. The equality condition means that each side is equal to a constant. Knowing the goal, we call the constant k^2 and write

$$\frac{1}{X} \frac{\partial^2 X}{\partial x^2} = k^2 \tag{A.54}$$

$$\frac{1}{Tc^2} \frac{\partial^2 T}{\partial t^2} = k^2 \tag{A.55}$$

Instead of solving the pair of equations, we verify the solution by substituting X and T in the wave equation and evaluating the partial derivatives. The results give an expression for the constant,

$$k^2 = \frac{4\pi^2 f^2}{c^2}$$
$$k = \pm \frac{2\pi f}{c} \tag{A.56}$$

The constant k^2 is a function of the parameters frequency f and sound velocity c. It does not depend on x or t. Hence, our assertion that Eq. (A.53) is true for all values of x and t is valid.

Our choice of k for $2\pi f/c$ or the wave number follows common usage. The wave number k is the spatial analogue of the angular frequency $\omega = 2\pi f$.

A1.1.4 Three-Dimensional Wave Equation

The wave equations for the components of vector displacement, velocity, and acceleration follow directly from the one-dimensional derivation. Let the vector displacement be

$$\mathbf{R} = X\mathbf{i} + Y\mathbf{j} + Z\mathbf{k} \tag{A.57}$$

where **i**, **j**, and **k** are unit vectors along the x-, y-, and z-coordinates. The vector sum of the one-dimensional wave equations for X, Y, and Z gives

$$\frac{\partial^2 X}{\partial x^2}\mathbf{i} + \frac{\partial^2 Y}{\partial y^2}\mathbf{j} + \frac{\partial^2 Z}{\partial z^2}\mathbf{k} = \frac{\rho}{B}\left(\frac{\partial^2 X}{\partial t^2}\mathbf{i} + \frac{\partial^2 Y}{\partial t^2}\mathbf{j} + \frac{\partial^2 Z}{\partial t^2}\mathbf{k}\right) \tag{A.58}$$

or using Eq. (A.57),

$$\nabla^2 \mathbf{R} = \frac{\rho}{B} \frac{\partial^2 \mathbf{R}}{\partial t^2} \tag{A.59}$$

where the Laplacian operator (Eq. (A.38)) is

$$\nabla^2 \equiv \frac{\partial^2}{\partial x^2} + \frac{\partial^2}{\partial y^2} + \frac{\partial 2}{\partial z^2} \tag{A.60}$$

There is a similar equation for pressure; however, deriving it requires more effort. The vector sum of the components of Newton's law for a continuous medium is

$$\frac{\partial p}{\partial x}\mathbf{i} + \frac{\partial p}{\partial y}\mathbf{j} + \frac{\partial p}{\partial z}\mathbf{k} = -\rho \left(\frac{\partial^2 X}{\partial t^2}\mathbf{i} + \frac{\partial^2 Y}{\partial t^2}\mathbf{j} + \frac{\partial^2 Z}{\partial t^2}\mathbf{k} \right) \tag{A.61}$$

or again using Eq. (A.57) and the **grad** operator ∇ Eq. (A.29), we have

$$\nabla p = -\rho \frac{\partial^2 \mathbf{R}}{\partial t^2} \tag{A.62}$$

$$\nabla \equiv \mathbf{i}\frac{\partial}{\partial x} + \mathbf{j}\frac{\partial}{\partial y} + \mathbf{k}\frac{\partial}{\partial z} \tag{A.63}$$

where ∇p is a vector.

We take the $\partial^2/\partial t^2$ operation on both sides of Eq. (A.59) and substitute $-\nabla p/\rho$ for $\partial^2 \mathbf{R}/\partial t^2$,

$$\nabla^2 (\nabla p) = \frac{\rho}{B} \frac{\partial^2}{\partial t^2} (\nabla p) \tag{A.64}$$

We move the common operation ∇ ahead of the other operations,

$$\nabla \left(\nabla^2 p - \frac{\rho}{B} \frac{\partial^2 p}{\partial t^2} \right) = 0 \tag{A.65}$$

For compressional waves the ∇ operation requires that the quantity within the parentheses be a constant, and we choose zero. Thus,

$$\nabla^2 p = \frac{\rho}{B} \frac{\partial^2 p}{\partial t^2} \tag{A.66}$$

Scalars such as p are easier to handle mathematically, so people usually solve a particular problem for the pressure and then use ∇ to compute the vector accelerations. Time integrations give the vector velocity and displacement.

Examples of using plane-wave solutions to the wave equation follow. The first example uses solutions and boundary conditions to derive Snell's law. The plane waves have particle velocities along the directions of propagation of v_i, v_r, and v_t for the incident, reflected, and transmitted waves. The wave numbers and their components are

$$\left. \begin{array}{l} k_1 = \dfrac{\omega}{c_1} \\[4mm] k_2 = \dfrac{\omega}{c_2} \end{array} \right\} \tag{A.67}$$

$$k_{1z} = k_1\cos\theta_1$$
$$k'_{1z} = k_1\cos\theta'_1$$
$$k_{2z} = k_2\cos\theta_2$$
(A.68)

$$k_{1x} = k_1\sin\theta_1$$
$$k'_{1x} = k_1\sin\theta'_1$$
$$k_{2x} = k_2\sin\theta_2$$
(A.69)

where θ_1, θ'_1, and θ_2 are the incident, reflected, and refracted angles. The equations of the particle velocities are

$$v_i = |v_i|\exp[i(\omega t - k_{z1}z - k_{x1}x)] \tag{A.70}$$

$$v_r = |v_i|R_{12}\exp[i(\omega t + k'_{z1}z - k'_{x1}x)] \tag{A.71}$$

$$v_t = |v_i|\,T_{12}\exp[i(\omega t - k_{z2}z - k_{x2}x)] \tag{A.72}$$

At the interface the vertical components of velocity are equal on each side; thus,

$$v_i\cos\theta_1 + v_r\cos\theta'_1 = v_t\cos\theta_2 \tag{A.73}$$

The substitution of Eqs. (A.70) to (A.72) and reduction by letting $z = 0$ and factoring out $\exp(i\omega t)$ gives

$$\cos\theta_1\,e^{-ik_{x1}x} + R_{12}\cos\theta'_1\,e^{-ik'_{x1}x} = T_{12}\cos\theta_2\,e^{-ik_{x2}x} \tag{A.74}$$

This expression is true for all values of x. Since the coefficients or angles do not depend on x, we conclude that the exponentials must be equal, or

$$k_{x1} = k'_{x1} = k_{x2} = k_x \tag{A.75}$$

This is Snell's law. The substitution of Eq. (A.69) in Eq. (A.75) gives Snell's law in its usual form,

$$\frac{\sin\theta_1}{c_1} = \frac{\sin\theta'_1}{c_1} = \frac{\sin\theta_2}{c_2} \tag{A.76}$$

This example also shows that the incident and reflected angles, θ_1 and θ'_1, are equal.

The second example is the penetration of a wave into medium 2, when θ_1 is beyond the critical angle. It uses Snell's law and the solutions. For angles beyond critical we write

$$\sin\theta_2 = u = \frac{c_2}{c_1}\sin\theta_1 \tag{A.77}$$

The trigonometric identity $\cos^2\theta + \sin^2\theta = 1$ is general, and we can use it to calculate $\cos\theta_2$ as follows:

$$\cos^2\theta_2 = (1 - u^2) \tag{A.78}$$

When u is greater than 1, as for reflection beyond the critical angle, we can write

$$\cos^2\theta_2 = -(u^2 - 1) \tag{A.79}$$

and then the square root gives

$$\cos\theta_2 = \pm i(u^2 - 1)^{1/2} \tag{A.80}$$

The vertical component of the wave number, k_{z2}, is

$$k_{z2} = k_2\cos\theta_2 = \pm ik_2(u^2 - 1)^{1/2} \tag{A.81}$$

The substitution of Eq. (A.81) in Eq.(A.72) gives

$$v_t = |v_i|\, T_{12}\, \exp[i(\omega t - k_x x)]\exp[-k_2(u^2 - 1)^{1/2}z] \tag{A.82}$$

where we choose the minus sign in Eq. (A.80) to attenuate the downtraveling wave.

The reflection coefficient R_{12} for reflections beyond critical has a magnitude of 1 and a phase shift. To show this, we replace $\cos\theta_2$ in Eq. (5.11) with $-i(u^2 - 1)^{1/2}$, then

$$R_{12} = -\frac{\rho_2 c_2 \cos\theta_1 + i\rho_1 c_1 (u^2 - 1)^{1/2}}{\rho_2 c_2 \cos\theta_1 - i\rho_1 c_1 (u^2 - 1)^{1/2}} \tag{A.83}$$

The numerator is the complex conjugate of the denominator, and R_{12} has the absolute value of unity. On defining the phase ϕ_{12},

$$\phi_{12} \equiv \arctan\left[\frac{\rho_1 c_1 (u^2 - 1)^{1/2}}{\rho_2 c_2 \cos\theta_1}\right] \tag{A.84}$$

R_{12} becomes

$$R_{12} = -\exp(i2\phi_{12}) \tag{A.85}$$

for the reflection beyond critical.

A1.2 DIPPING INTERFACE, INTERVAL VELOCITY MEASUREMENTS, AND THE RMS VELOCITY

A1.2.1 Dipping Interface

Before giving the derivation of the Dix-Dürbaum expression for interval velocity measurements, we briefly consider the effect of dip on the single-layer reflection. The image geometry is shown in Figure A.3. By the Pythagorean theorem, the travel time from the image to the receiver is

$$t_1^2(x) = \frac{(2h_1 - x \sin\phi_1)^2}{c_1^2} + \frac{(x\cos\phi_1)^2}{c_1^2} \tag{A.86}$$

or

$$t_1^2(x) = \frac{4h_1^2}{c_1^2} - x\frac{4h_1\sin\phi_1}{c_1^2} + \frac{x^2}{c_1^2} \tag{A.87}$$

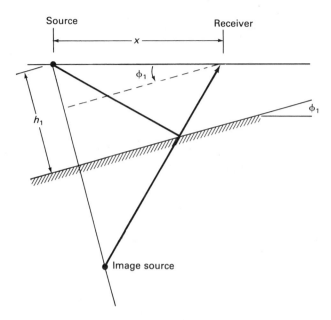

Figure A.3 Image construction for a dipping interface.

Dip introduces a first-order term in the time-squared equation. It is a common practice to measure reflection times in both $+x$ and $-y$ directions, the so-called split spread of geophones. The sum of $t_1^2(x)$ and $t_1^2(-x)$ gives

$$\frac{1}{2}\left[t_1^2(x) + t_1^2(-x) \right] = \frac{4h_1^2}{c_1^2} + \frac{x^2}{c_1^2} \tag{A.88}$$

and the difference gives

$$\frac{1}{2}\left[t_1^2(x) - t_1^2(-x) \right] = - x\frac{4h_1\sin \phi_1}{c_1^2} \tag{A.89}$$

A1.2.1 Interval Velocity Measurements

It seems reasonable to use seismic reflection data and start at the top of a horizontally layered half-space and work down to determine the thickness and velocities of the layers. A derivation of the Dix-Dürbaum equation for doing this follows. As an intermediate step we obtain the root mean square (RMS) velocity. The RMS velocity, c_{RMS} (or V_{RMS}), appears throughout the seismic exploration literature.

There are several ways to derive the interval velocity formula; we choose the power series expansion in x. The Taylor expansion about $x = 0$ is

$$f(x) = f(0) + xf'(0) + \frac{x^2}{2!}f''(0) + \cdots \tag{A.90}$$

where the primes indicate d/dx. The exact expression for reflection time squared, Eq. (A.87), is the first three terms of the Taylor expansion because it has x^0, x^1, and x^2 terms.

For horizontal layers, $f(x) = f(-x)$, and the coefficients of odd powers of x are zero. Thus, the second-order approximation of $f(x) = t_n^2(x)$ is

$$t_n^2(x) \simeq t_n^2(0) + A_n x^2 \tag{A.91}$$

$$A_n \equiv \frac{1}{2} \frac{d^2 t_n^2(x)}{dx^2} \bigg|_{x=0} \tag{A.92}$$

where $t_n(x)$ is the reflection time from the nth interface (or bottom of the nth layer) at x. The source is at $x = 0$. The calculation of A_n requires quite a bit of manipulation, and higher-order terms require much more. The derivatives of $t_n^2(x)$ are

$$\frac{d}{dx}\left(t_n^2(x) \right) \bigg|_0 = 2\, t_n(0) \frac{dt_n(x)}{dx} \bigg|_0 = 0 \tag{A.93}$$

$$\frac{1}{2} \frac{d^2}{dx^2}\left(t_n^2(x) \right) \bigg|_0 = \left(\frac{dt_n(x)}{dx} \right)^2 \bigg|_0 + t_n(0) \frac{d^2(t_n(x))}{dx^2} \bigg|_0 \tag{A.94}$$

and

$$A_n = t_n(0) \frac{d^2(t_n(x))}{dx^2} \bigg|_0 \tag{A.95}$$

We use the exact parametric expressions for $t_n(p)$ and $x_n(p)$ to compute Eq. (A.95).

$$t_n(p) = 2 \sum_{i=1}^{n} \frac{h_i}{c_i\,(1 - p^2\, c_i^2)^{1/2}} \tag{A.96}$$

$$x_n(p) = 2p \sum_{i=1}^{n} \frac{h_i c_i}{(1 - p^2\, c_i^2)^{1/2}} \tag{A.97}$$

The implicit differentiation formula is

$$\frac{d}{dx}\left[\quad \right] = \frac{\partial}{\partial p}\left[\quad \right] \frac{\partial p}{\partial x} + \frac{\partial}{\partial x}\left[\quad \right] \tag{A.98}$$

As x goes to zero, p goes to zero. After the implicit differentiation of $t_n(p)$ is done twice, Eq. (A.95) becomes

$$A_n = t_n(0) \left(\frac{\partial^2 t_n(p)}{\partial p^2} \right) \left(\frac{\partial x_n(p)}{\partial p} \right)^{-2} \bigg|_0 + \left(\frac{\partial t_n(p)}{\partial p} \right) \bigg|_0 \left\{ \text{other terms} \right\} \bigg|_0 \tag{A.99}$$

where the second expression is zero.

$$\frac{\partial^2 t_n(p)}{\partial p^2} \bigg|_{p=0} = 2 \sum_{i=1}^{n} h_i c_i \tag{A.100}$$

$$\frac{\partial x_n(p)}{\partial p} \bigg|_{p=0} = 2 \sum_{i=1}^{n} h_i\, c_i \tag{A.101}$$

$$A_n = \frac{t_n(0)}{2 \sum\limits_{i=1}^{n} h_i\, c_i} \tag{A.102}$$

We change A_n to the customary form by using

$$2h_i = c_i \tau_i \qquad (A.103)$$

where τ_i is understood to be the vertical incidence travel time in the ith layer.

$$A_n = \frac{t_n(0)}{\displaystyle\sum_{i=1}^{n} c_i^2 \, \tau_i} \qquad (A.104)$$

A1.2.3 RMS Velocity

By analogy to the reflection from a single layer, we write

$$t_n^2(x) \simeq t_n^2(0) + \frac{x^2}{c_{n,\text{RMS}}^2} \qquad (A.105)$$

$$c_{n,\text{RMS}}^2 \equiv \frac{\displaystyle\sum_{i=1}^{n} c_i^2 \, \tau_i}{t_n(0)} = \frac{1}{A_n} \qquad (A.106)$$

$$t_n(0) \equiv \sum_{i=1}^{n} \tau_i \qquad (A.107)$$

The RMS velocity comes from the second-order approximation to the reflection time squared for a multilayered half-space. It is useful for x small compared with the depth to interface n.

To determine the interval velocity for experimental seismic reflection data, we fit the experimental data to second-order expressions like Eq. (A.91) to determine A_n and A_{n+1} for the nth and $(n + 1)$st reflections. We can use $[t_n^2(x) + t_n^2(-x)]/2$ and $[t_{n+1}^2(x) + t_{n+1}^2(-x)/2$ to reduce the effects of small dips. For the $(n - 1)$st layer, $c_{n+1,\text{RMS}}$ is

$$c_{n+1,\text{RMS}}^2 = \frac{\displaystyle\sum_{i=1}^{n} c_i^2 \, \tau_i + c_{n+1}^2 \, \tau_{n+1}}{t_n(0) + \tau_{n+1}} \qquad (A.108)$$

We divide the numerator and denominator by $t_n(0)$ and identify $c_{n,\text{RMS}}^2$

$$c_{n+1,\text{RMS}}^2 = \frac{c_{n,\text{RMS}}^2 + c_{n+1}^2 [t_{n+1}(0) - t_n(0)]/t_n(0)}{1 + \dfrac{t_{n+1}(0) - t_n(0)}{t_n(0)}} \qquad (A.109)$$

Solution for c_{n+1}^2 gives the Dix-Dürbaum formula,

$$c_{n+1}^2 = \frac{c_{n+1,\text{RMS}}^2 \, t_{n+1}(0) - c_{n,\text{RMS}}^2 \, t_n(0)}{t_{n+1}(0) - t_n(0)} \qquad (A.110)$$

where c_{n+1} is the interval velocity of the $(n + 1)$st layer. Geophysicists use graphs of t^2 versus x^2 or least squares fits to determine the velocities $c_{n+1,\text{RMS}}^2 = 1/A_{n+1}$ and

$c_{n,\text{RMS}}^2 = 1/A_n$ and the intercept times $t_{n+1}^2(0)$ and $t_n^2(0)$. Primary reflections and the initial times for the signals must be carefully chosen.

A1.3 HEAD-WAVE OR REFRACTION EQUATIONS FOR MULTIPLE LAYERS, AND THE PHASE-PATH TECHNIQUE

The refraction equation for a horizontally layered subsurface (Figure A.4a) follows directly from the ray-tracing equations in Chapter 3. The point source is at A, and we choose to follow the ray at the critical angle. The steps use the form of a computational algorithm and carry from one step to the next.

The head wave from the nth layer has the ray parameter $p = 1/c_n$. The displacement and travel time to the critical reflection point are

$$t_{n-1} = \sum_{i=1}^{n-1} \frac{h_i}{c_i(1 - p^2 c_i^2)^{1/2}} \tag{A.111}$$

$$x_{n-1} = \sum_{i=1}^{n-1} \frac{h_i c_i p}{(1 - p^2 c_1^2)^{1/2}} \tag{A.112}$$

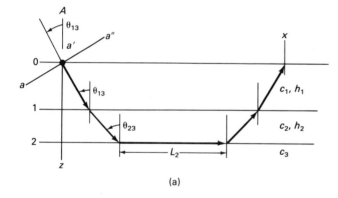

(a)

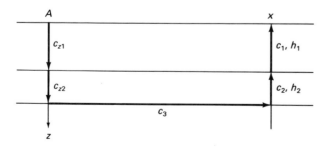

(b)

Figure A.4 Ray and phase paths for a head wave.

The horizontal distance along the interface is L_{n-1}. The displacement and travel time of the up-going headwave are t_{n-1} and x_{n-1}. The travel time for a head wave from the top of the nth layer is

$$t_n(x) = 2\,t_{n-1} + \frac{L_{n-1}}{c_n} \tag{A.113}$$

$$x = 2\,x_{n-1} + L_{n-1} \tag{A.114}$$

where

$$p = \frac{\sin\theta_{in}}{c_i} = \frac{1}{c_n} \tag{A.115}$$

$$c_n > c_1$$

The substitution of $L_{n-1} = x - 2x_{n-1}$ in Eq. (A.113) and algebraic manipulation gives

$$t_n(x) = 2\sum_{i=1}^{n-1} \frac{h_i\,(1 - p^2 c_i^2)^{1/2}}{c_i} + \frac{x}{c_n} \tag{A.116}$$

where

$$(1 - p^2 c_i^2)^{1/2} = \cos\theta_{in} \tag{A.117}$$

This is the same result as in Section 4.1 for two layers. The refraction or head-wave equations appear in this form in many papers and books. $t_n(0)$ is the intercept of the head-wave arrival line at $x = 0$, and the slope of the line is $1/c_n$. The determination of the minimum distance for the head-wave arrival requires calculation of $2\,x_{n-1}$, Eq. (A.112).

It appears that the terms in Eq. (A.116) represent something different from the ray-tracing terms in Eqs. (A.113) to (A.115). We suggest that the terms in Eq. (A.116) represent the *phase tracing* of a plane-wave front. Imagine that the medium above the 0 interface is c_1, and a plane wave is traveling downward at the angle θ_{1n}. In Figure A.4a the plane-wave front a—a'' intercepts the z-axis at a'. a' moves along the z-axis at the phase velocity $c_{Z1} = c_1/\cos\theta_{1n}$. The travel time of the wave front through the first layer is h_1/c_{Z1} or $h_1\cos\theta_{1n}/c_1$. The latter is the first term in the summation, Eq. (A.116). The ith term is $h_i\cos\theta_{in}/c_i$. The phase travels vertically downward to the top of the nth layer. Here refraction causes the plane wave to travel horizontally, and its phase velocity along the $(n-1)$st interface is c_n. The *phase travels* the distance x along the interface and then travels vertically upward to the receiver.

The algebraic manipulation in going from Eq. (A.113) to (A.115) to (A.116) significantly changes the problem. The first solution gives the travel time and displacement for a *ray from a point source* at the critical angle. The second solution gives the travel time for the *phase of a plane wave* along a prescribed path that includes refraction parallel to the $(n-1)$st interface.

The phase-path computation of the head-wave equation for dipping layers is much simpler than the ray-path computation, Figure A.5. Of course, the equation does not give the minimum distance for the existence of the head-wave arrival. For simplicity we use a three-layer problem. It is reasonably obvious how to extend the computation to n layers.

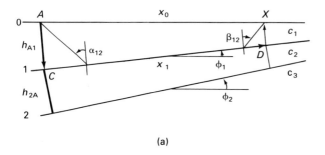

(a)

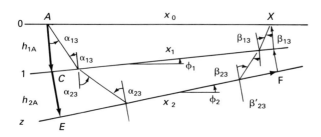

(b)

Figure A.5 Dipping-layer geometry for refractions at (a) interface 1; (b) interface 2.

The source is at A, and we wish to calculate the phase-path travel time to a receiver at x_0. The phase paths are $ACDX$ for layer 2 and $ACEFDX$ for layer 3. From the top and the source position A, the distances and normal thickness are

$$\left. \begin{array}{l} x_1 = x_0 \cos \phi_1 \\ x_2 = x_1 \cos (\phi_2 - \phi_1) \end{array} \right\} \tag{A.118}$$

$$\left. \begin{array}{l} h_{x1} = h_{A1} - x_0 \sin \phi_1 \\ h_{x2} = h_{A2} - x_1 \sin (\phi_2 - \phi_1) \end{array} \right\} \tag{A.119}$$

The program saves these results.

Computations of the angles require repeated use of Snell's law, the arcsine operation, and dip correction. We start at the bottom of the first layer and work down. At layer 2 the critical angle α_{12} is

$$\alpha_{12} = \arcsin \left(\frac{c_1}{c_2} \right)$$

$$\beta_{12} = \alpha_{12} \tag{A.120}$$

The phase-path times are, from Figure A.5b,

$$t_{AC} = \frac{h_{A1}\cos\alpha_{12}}{c_1} \tag{A.121}$$

$$t_{CD} = \frac{x_1}{c_2} \tag{A.122}$$

$$t_{DX} = \frac{h_{x1}\cos\beta_{12}}{c_1} \tag{A.123}$$

and the head-wave equation is

$$t_2(x) = t_{AC} + t_{DX} + t_{CD} \tag{A.124}$$

The angles for a head wave along layer 3 or interface 2 are, from Figure A.5a

$$\alpha_{23} = \arcsin\left(\frac{c_2}{c_3}\right) \tag{A.125}$$

$$\alpha'_{23} = \alpha_{23} + \phi_2 - \phi_1 \tag{A.126}$$

$$\alpha_{13} = \arcsin\left(\frac{c_1\sin\alpha'_{23}}{c_2}\right) \tag{A.127}$$

$$\beta_{23} = \alpha_{23} \tag{A.128}$$

$$\beta'_{23} = \beta_{23} - (\phi_2 - \phi_1) \tag{A.129}$$

$$\beta_{13} = \arcsin\left(\frac{c_1\sin\beta_{23}}{c_2}\right) \tag{A.130}$$

The phase travel times for the path ACEFDX are

$$t_{AC} = \frac{h_{A1}\cos\alpha_{13}}{c_1} \tag{A.131}$$

$$t_{CE} = \frac{h_{A2}\cos\alpha_{23}}{c_2} \tag{A.132}$$

$$t_{EF} = \frac{x_2}{c_3} \tag{A.133}$$

$$t_{FD} = \frac{h_{x2}}{c_2}\cos\beta_{23} \tag{A.134}$$

$$t_{DX} = \frac{h_{x1}}{c_1}\cos\beta_{13} \tag{A.135}$$

$$t_3(x) = t_{AC} + t_{CE} + t_{EF} + t_{FD} + t_{DX} \tag{A.136}$$

We can write expressions for the reversed profile and a source at B by letting the h_{Ai} become h_{Bi}, ϕ_{in} become $-\phi_{in}$, and x become x' as measured left from B.

A1.4 COMPUTATIONS OF ALGEBRAIC EXPRESSIONS AND A RAY-TRACING ALGORITHM

The decision to use a computer requires the choice of a computational language and a computer. To some extent these choices are fads. Scientific programming has requirements that depend on the task. Algebraic formulas or algorithms must be changed into computer code. Common computer languages are FORTRAN, BASIC, PASCAL, and C. Contrary to some of the myths, bad code can be written in any language, elegant code can be written in any language, and scientific computations can be performed in any language.

Instructions to the computer, or code, are given in abbreviated and sometimes cryptic English words and phrases. In numerical calculations the code is a sequence of instructions that can closely resemble algebraic algorithms. For geophysical computations and data processing the language must include easy access to input and output operations on files of data. Good graphics displays are essential. Geophysicists commonly use a graphics workstation and display each processing step. Most microcomputers can perform these tasks and have many hardware options. A small machine such as the Apple IIe is adequate for most of the computations in this book, but it is too small for geophysical data processing and display. Programming is easier on a computer having a larger memory and array address limits. The FORTRAN and compiled BASIC programs were written on a Macintosh plus.

We give BASIC versions of the algorithms because many (personal) computers come with a BASIC system. Usually the furnished systems use interpreters to convert the code to machine instructions for the computer. These systems require that each line of code have a statement number. An interpreter reads one line of code, converts it, and then goes to the next line of code. In interpreter mode the debugging of code is easy because the computer executes one line of code at a time. A BASIC compiler can be used to make a program run much faster. A compiler creates a set of machine instructions for the whole program. It does not waste time by going back and looking for the next instruction. The BASIC compiler does not require statement numbers.

The FORTRAN language has evolved as computer science and computers have grown. FORTRAN uses a compiler to convert the code into machine instructions. It has a huge library of tested mathematical subroutines. The great advantages of FORTRAN are the uses of separate subroutines. We can write and compile small modules or subroutines and then link them together in a main program. To facilitate the transfer of algorithms to larger problems, we also give FORTRAN 77 versions of some algorithms. Our programs for a BASIC compiler can be changed to FORTRAN by changing for-next loops to do loops and including the variable types.

A1.4.1 Instructions

Superficially, BASIC and FORTRAN instructions or program steps look like common algebraic expressions. Many computers use capital letters to represent numbers. Without being rigorous, we tend to use lowercase letters for the number and the uppercase letter to represent the computer's name of the number. For example, x is the number stored at location X.

Equal sign
The statement

$$Y = M * X \qquad (A.137)$$

instructs the computer to go to the location X and read x, go to the location M and read m, do the product mx, and store the result at the location named Y. $Y = M * X$ is not an algebraic equation in the usual sense. *The equal sign means store at location Y the result of computational operations on the right side of the equal sign.* The act of storing a number erases the old number. Algebraically the statement

$$X = M * X \qquad (A.138)$$

looks strange, but it does not confuse the computer. The computer reads the number x at location X, reads the number m at location M, multiplies m times x, and stores the result at location X. The statement is logical to the computer. Of course, the old value of x is lost.

Summations and products
We want an algorithm to duplicate the algebraic summation

$$s = \sum_{n=1}^{n1} a_n \qquad (A.139)$$

We use the computer's interpretation of the equal sign to write the algorithm. For this illustration, the a_n are $a_1 = 1, a_2 = 2, \ldots, a_n = n$ or

$$s = \sum_{n=1}^{n1} n \qquad (A.140)$$

The name of the result is S, and the first instruction is

$$\begin{aligned} N1 &= n1 \\ S &= 0 \end{aligned} \qquad (A.141)$$

The FOR-NEXT instruction loop tells the computer to repeat a series of instructions or loop a specified number of times. The summation algorithm is, letting n_1 be N1 and n be N,

$$\begin{aligned} &\text{FOR N} = 1 \text{ TO N1} \\ &\quad S = S + N \\ &\text{NEXT N} \end{aligned} \qquad (A.142)$$

On the first loop the computer reads $S = 0$, adds 1, and stores 1 in S. On the second loop, the computer sets $N = 2$, reads $S = 1$, adds 2, and stores $S = 3$. It proceeds until $N = N1$. After the statement NEXT N we add a print statement

$$\text{PRINT S} \tag{A.143}$$

to print the result. The reader can test a program for $n_1 = 5$ and get $s = 15$.

The product algorithm uses a similar logic. The product statement

$$p = \prod_{n=1}^{n_1} a_n \tag{A.144}$$

means form the product of $a_1 \cdot a_2 \cdot a_3 \ldots a_{n_1}$. The factorial function $n!$ is the product $1 \cdot 2 \cdot 3 \ldots n$. The algebraic statement to evaluate $n_1!$ is

$$p = \prod_{n=1}^{n_1} n \tag{A.145}$$

The corresponding algorithm is

$$\text{P} = 1 \tag{A.146}$$

$$\begin{array}{l} \text{FOR N = 1 TO N1} \\ \quad \text{P} = \text{P} * \text{N} \\ \text{NEXT N} \end{array} \tag{A.147}$$

$$\begin{array}{l} \text{PRINT P} \\ \text{END} \end{array} \tag{A.148}$$

Statement (A.146) initializes P to 1. Statements (A.147) are the FOR-NEXT loop. The print statement is (A.148). The reader can test this algorithm for $n_1 = 5$ and get $p = 120$.

Arrays

To store a sequence of numbers in memory, we use a computer language analogy of subscripts. For example a_0 is A(0), a_1 is A(1), . . . , a_j is A(J). Variables having two subscripts, such as a_{ij}, become A(I, J). A strong reason for using this notation is ease of using the A(J) types of names inside FOR-NEXT loops.

The computer requires that space be set aside for arrays before it starts computing. We do this with a dimension statement. The statement

$$\begin{array}{l} \text{REM DIMENSION ARRAY A(N) FOR N = 0 TO 100} \\ \\ \text{DIM A(100)} \end{array} \tag{A.149}$$

reserves the locations of 101 floating-point words having the names A(0), A(1), . . . , A(100). BASIC allows array subscripts to be positive integers starting at 0. FORTRAN 77 is more flexible. The array A($-100:100$) lets $i = -100$ to 100. An example of an array A(I,J) dimension is A(2,100) or A(0:2, 0:100) in FORTRAN, where I and J range from I = 0 to 2 and J = 0 to 100.

Remarks or comments

We explain what a program is doing by inserting explanations. It is crucial to insert explanations so that you or other users can follow the too frequently twisted logic of programs.

REM   EXPLAIN WHAT YOU ARE DOING.

c   Explain what you are doing.

A1.4.2 Code Formats and the Cases of Letters

Originally computers had strict rules about the format of instructions and the case of letters. Each line of a BASIC program had to start with a statement number. This is still true for the interpreter BASIC that comes in many personal computers such as Apple IIs and IBM-PC clones. FORTRAN also required upper case letters and statements had to be indented to the seventh space. Some line numbers were required. Improvements of the original forms of BASIC and FORTRAN let us write in upper or lower case letters. In both BASIC and FORTRAN the user can set the compiler to be insensitive to the case of letters. Common BASIC compilers such as the Microsoft BASIC for the MacIntosh do not require line numbers. Some of our BASIC and FORTRAN programs are written in upper case letters and some are in lower case letters. We set the compiler to be insensitive to case. We use line numbers for GOTO and subroutine calls within a program.

The ray trace algorithm is used to demonstrate programs written for an interpreted BASIC, compiled BASIC, and compiled FORTRAN 77. The three sets of codes are not the same on line by line comparisons, however the algorithms are the same. We believe that the compiled BASIC and FORTRAN versions are simpler to read and are much less cluttered than the interpreted BASIC version.

Open files, close files, and write to file instructions are very specific to a computer and its operating system. The device codes are used to send information to a screen, a line printer, or a disk file. FORTRAN 77 on a MacIntosh uses the device numbers of 9 for the screen, 10 for the disk, and 6 for the printer. WRITE (9,*) writes unformatted numbers or characters to the screen. WRITE (6,*) puts them on the printer. In Microsoft BASIC, the command PRINT sends output to the screen and LPRINT sends output to the printer. The utilities include examples of graphics, make file, and read file in BASIC and FORTRAN for MacIntosh computers. The reader will notice that a given graphic task usually requires less lines of code in BASIC than in FORTRAN.

A1.4.3 Example: Multilayer ray trace

The purpose of the calculation is to compute the x- and z-coordinates of the ray path and the travel time. The algebraic equations are (3.22) to (3.24), and for the ith layer

$$p = \frac{\sin \theta_1}{c_1} \tag{A.150}$$

$$a_i = p\, c_i \tag{A.151}$$

$$b_i = (1 - a_i^2)^{1/2} \tag{A.152}$$

$$t_i = \frac{h_i}{(c_i\, b_i)} \tag{A.153}$$

$$x_i = a_i \frac{h_i}{b_i} \tag{A.154}$$

The final results are the summations of x_i, t_i, and z_i. The computation of b_i, Eq. (1.152) assumes that $|a_i|$ is less than 1; programs must include an escape statement when $|a_i| \geq 1$.

For N1 $= n_1$ layers the velocities and thicknesses of the layers are in the arrays $C(I) = c_i$ and $H(I) = h_i$. For a maximum of $N = 10$, the dimension statement is

$$\text{DIM C(10),H(10),X(10),Z(10)} \tag{A.155}$$

For simplicity, we input a value of I1 $= \theta_1$ in radians and then calculate X(i), T(i), and Z(i). Equation (A.150) becomes

$$\text{P} = \text{SIN (I1) / C(1)} \tag{A.156}$$

Prior to the summations we set the sums to zero.

$$\text{T} = 0 : \text{Z} = 0 : \text{X} = 0 \tag{A.157}$$

The ray-trace loop begins

```
FOR I = 1 TO N1
A = P * C(I)                                                          (A.158)
IF ABS(A) >= 1 GOTO :REM put a statement number here
```

This is the escape for incident angles beyond the critical angle. It avoids program crashes. Continuing,

$$\text{B} = \text{SQR (1 - A * A)}$$

and the computations and summations are

```
Z = Z + H(I)
Z(I) = Z
X = X + A * H(I) / B
X(I) = X                                                             (A.159)
T = T + H(I) / (B * C(I))
T(I) = T
NEXT I
```

The outputs are Z(I), X(I), and T(I). Instructions to list the output follow:

```
PRINT "I"; TAB( 5);"Z(I)"; TAB( 11);"X(I)"; TAB( 24);"T(I)"
PRINT
```

(A.160)

```
FOR I = 1 TO N1
    PRINT I; TAB( 5);Z(I); TAB( 11);X(I); TAB( 24);T(I)
NEXT I
```

These statements are the bare minimum of a ray-trace program. In the example program that follows, they are statements 540–580, 620–920, 960–1060, and 1080. The print and output statements are 1100–1200. You will notice that the actual computation is about half of the program.

Statements 100–500 explain what is happening, dimension the array, and input data. The use of "condition indicator" Q in statements 620, 740–760, and 1000–1060 needs special explanation. After the input of an incident angle, statement 620 sets $Q = -1$. As long as Q is less than 1, and the angle is less than critical, the ray trace proceeds through 760 to 940 and then jumps to NEXT I at 1080. If the angle is greater than critical, the jump from 760 to 960 causes the output to print BEYOND CRITICAL:LAYER I. To avoid misinterpretation, X(I) and T(I) are set to 9999999, meaningless numbers. Q is set to 1 so that the program skips at statement 760 to 960 for all layers beneath the first layer having a critical reflection. These aids to the user are sometimes called "bells and whistles." Some programs have END statements and some do not. The need for an END statement depends on the particular interpreter or compiler. The final GOTO 480 causes the program to return to 480 for a new incident angle.

```
100 REM   PRGM RAYTRACE. CSC
120 REM PRGM HAS EXTRAS TO HANDLE INCIDENT ANGLES BEYOND CRITICAL
130 REM WITHOUT HANGUPS. OUTPUT NUMBERS FOR INTERFACES BEYOND
140 REM  CRITICAL ARE MEANINGLESS. X(I)=9999999, T(I)=9999999
160 PRINT
180 PRINT "RAYTRACE: AFTER LAYER INPUTS,"
200 PRINT "PRGM RETURNS FOR NEW INCIDENT"
220 PRINT "ANGLES"
240 PRINT
260   DIM  C(10),H(10),X(10),Z(10)
280 REM   COMPUTE PI=4*ATN(1) AND CONVERSION DEG TO RAD FACTOR, DR
300 DR = 4 *  ATN (1) / 180
320 PRINT "INPUT NUMBER OF LAYERS=";
340 INPUT N1
360 FOR I = 1 TO N1
380   PRINT "LAYER ";I;" INPUT C(I),H(I)=";
400   INPUT C(I),H(I)
420 NEXT I
440 PRINT
460 REM COMPUTE RAY PARAMETER
```

```
480  PRINT "INPUT INCIDENT ANGLE=";
500  INPUT ANG
520  PRINT
540 T = 0
560 Z = 0
580 X = 0
600  REM  SET Q=-1 FOR INCIDENT ANGLE LESS THAN CRITICAL
620 Q =  - 1
640 I1 = ANG * DR
660 P =  SIN (I1) / C(1)
680  REM  RAYTRACE PRGM
700  FOR I = 1 TO N1
720 A = P * C(I)
740  REM  TEST FOR BEYOND CRITICAL REFL
760  IF Q > 0 THEN 1000
780  IF  ABS (A) >  = 1 THEN 960
800 B =  SQR (1 - A * A)
820 Z = Z + H(I)
840 Z(I) = Z
860 X = X + A * H(I) / B
880 X(I) = X
900 T = T + H(I) / (B * C(I))
920 T(I) = T
940  GOTO 1080
960  PRINT "BEYOND CRITICAL:LAYER ";I
980  PRINT
1000 X(I) = 9999999
1020 T(I) = 9999999
1040  REM  SET Q FOR BEYOND CRITICAL
1060 Q = 1
1080  NEXT I
1100   PRINT "I"; TAB( 5);"Z(I)"; TAB( 11);"X(I)"; TAB( 24);"T(I)"
1120  PRINT
1140  FOR I = 1 TO N1
1160   PRINT I; TAB( 5);Z(I); TAB( 11);X(I); TAB( 24);T(I)
1180  NEXT I
1200  PRINT
1220  REM  RETURN FOR NEW ANGLE
1240  PRINT "FOR NEW ANGLE INPUT 'Y'"
1260  INPUT C$
1280  IF C$ = "Y" THEN 480
1300  END
```

'A1.4.3 Raytrace BASIC version 2

```
      PRINT "RAYTRACE: AFTER LAYER INPUTS,"
      PRINT "PRGM RETURNS FOR NEW INCIDENT ANGLE"

       DIM  C(10),H(10),X(10),Z(10)
REM   COMPUTE PI=4*ATN(1) AND CONVERSION DEG TO RAD FACTOR, DR
      DR = 4 *  ATN (1) / 180
      PRINT "INPUT NUMBER OF LAYERS=";
      INPUT N1

      FOR I = 1 TO N1
          PRINT "LAYER ";I;" INPUT C(I),H(I)=";:INPUT C(I),H(I)
      NEXT I

      PRINT
400   PRINT "INPUT INCIDENT ANGLE=";: INPUT ANG

REM  GOTO RAYTRACE SUBPROGRAM
      GOSUB 800

      PRINT "I";  TAB( 5);"Z(I)";  TAB( 11);"X(I)";  TAB( 24);"T(I)"
      PRINT
      FOR I = 1 TO N1
          PRINT I;  TAB( 5);Z(I);  TAB( 11);X(I);  TAB( 24);T(I)
      NEXT I

      PRINT "HARD COPY 'y' or 'n'";: INPUT HC$
      IF HC$="N" OR HC$="n" GOTO 500
  LPRINT
  LPRINT "INCIDENT ANGLE = ";ANG
  LPRINT
      LPRINT"I";TAB(5);"C(I)";TAB(12);"H(I)";TAB(20);"Z(I)";TAB(27);
      LPRINT  "X(I)";TAB( 40);"T(I)"
      FOR I = 1 TO N1
    LPRINT  I;TAB(5);C(I);TAB(12);H(I);TAB(20);Z(I);
        LPRINT   TAB(27);X(I); TAB(40);T(I)
      NEXT I

  REM       RETURN FOR NEW ANGLE
  500       PRINT "FOR NEW ANGLE INPUT 'y'"
      INPUT C$
      IF C$ = "Y" OR C$ = "y" THEN 400
600
      END
```

800 REM RAYTRACE SUBPROGRAM

```
REM SET INITIAL T, Z, X = 0
T = 0: Z = 0: X = 0
I1 = ANG * DR
P = SIN (I1) / C(1)
FOR I = 1 TO N1
   A = P * C(I)
   REM TEST FOR BEYOND CRITICAL REFL
   IF ABS (A) < 1 AND Q < 1 THEN
     Q = -1
     B = SQR (1 - A * A)
     Z = Z + H(I)
     Z(I) = Z
     X = X + A * H(I) / B
     X(I) = X
     T = T + H(I) / (B * C(I))
     T(I) = T
   ELSEIF ABS (A) >= 1 THEN
     Q = 1
     PRINT "BEYOND CRITICAL:LAYER ";I
     X(I) = 9999999
     T(I) = 9999999
   END IF
NEXT I


RETURN
```

```
c     A1.4.3 Ray traces for a set of angles. FORTRAN 77


      real   c(10),h(10),xa(0:10,0:50),za(0:10,0:50),ta(0:10,0:50)
      real   a,b,t,x,z,p,pi,rfd,ai,di,ri
      integer i,imax,n,nmax
      pi=4*atan(1.0)
      rfd=pi/180.

c     prgm
c
c     input data
c     default
      nmax=20
```

```
c     ai is angle step in deg
      ai=.8


c     input data

      write(9,*) 'input  number  of  layers'
      read (9,*) imax
      do 10 i=1 ,imax
         write(9,*) 'layer ',i,'  input h,c'
         read (9,*) h(i),c(i)
10    continue


c     calculate
      di=ai*rfd
      do 20 n=0, nmax
      t=0.
      x=0.
      z=0.
        p=sin(n*di)/c(1)
         do 30 i = 1,imax
           a=p*c(i)
          if (abs(a).lt.1) then
             b=sqrt(1.-a*a)
           z=z+h(i)
             x=x+a*h(i)/b
             t=t+h(i)/(b*c(i))
           ta(i,n)=t
           xa(i,n)=x
          else
             xa(i,n)=999999.
             ta(i,n)=999999.
          endif
            za(i,n)=z
30         continue
20    continue


c     raytrace complete. 999999 is beyond critical angle

      write(9,*)'  interface depth          xdist          time'
      do 40 i=1,imax
         write(9,*) 'interface=',i
      pause
       do 50 n=0,nmax
             print '(i4,2x,3(2x,f11.3))',n,za(i,n),xa(i,n),ta(i,n)
```

```
50      continue
40    continue
      pause
      end
```

' A1.4.3 Refl times T(x) for a set of stations. BASIC

```
DIM  C(10),H(10),Z(10),X(500),T(500)


PI = 4 *  ATN (1)
DR = PI / 180


PRINT "COMPUTE REFLECTION TIMES FOR BOTTOM INTERFACE"
PRINT "MAX NUMBER OF LAYERS = 10"
PRINT "INPUT NUMBER OF LAYERS=";: INPUT N1


FOR I = 1 TO N1
    PRINT "LAYER ";I;" INPUT  C(I),H(I)=";
    INPUT C(I),H(I)
NEXT I


PRINT " SOURCE AT X = 0"
PRINT "INPUT NUMBER OF GEOPHONE STATIONS = ";:INPUT JM
PRINT "INPUT STATION SPACINGS = ";: INPUT AX
REM  ESTIMATE ANGLE INCREMENT AND COMPUTE CH AND
REM  TN FOR DURBAUM-DIX APPROXIMATION
REM  EQS. A.91, A.102, AND A.106. DEEP LOW VELOCITY
REM  LAYERS CAN CAUSE RAY PATHS TO FALL SHORT
REM  OF MAX X. POOR COMPARISONS TO DIX-DURBAUM
REM  AT LARGE X INDICATES TROUBLE.


S = 0: CH = 0: TN = 0 :ANGR = 0
FOR I = 1 TO N1
   S = S + H(I)
   TN = TN + 2 * H(I) / C(I)
   CH = CH + 2 * C(I) * H(I)
   A =  ATN (JM * AX / (2 * S))
  IF ANGR<A THEN ANGR = A
   Z(I) = S
NEXT I
```

```
REM  RMS VELOCITY = CRMS. EQ.A.106
     CRMS(N1) =  SQR (CH / TN)


1    REM LET NUMBER OF ANGLE INCREMENTS BE 20 * JM
     IM = 20 * JM

       FOR J = 0 TO IM+1
         X(J) = 0
         T(J) = 0
       NEXT J


   REM  DO A REFLECTION RAYTRACE FOR EACH ANGLE INCREMENT
   REM  Q IS A CONTROL. Q=-1 FOR REFLECTIONS WITH IN RANGE.
   REM  PRGM SETS Q=1 FOR VERY LARGE X AND BEYOND CRITICAL
   REM  ANGLE. Q CAN BE RESET IN RAYTRACE SUBPRGM.

     Q = -1
       FOR J = 0 TO IM+1
         IF Q < 0 THEN
           I1 = J * ANGR / (.5 * IM)
            GOSUB 800
            REM RETURN FROM REFLECTION RAYTRACE SUBPRGM
           ELSE
           REM DISTANCES ARE TOO LARGE. SKIP TO NEXT J.
           END IF
         NEXT J


   REM  DETERMINE RAY NUMBERS THAT BRACKET X(J)

       FOR J = 1 TO JM
         X = J * AX
           FOR I = 0 TO IM
             IF X(I)<=X AND X<X(I + 1) THEN N(J)= I
           NEXT I
       NEXT J

     PRINT
   PRINT "REFLECTION TIME - DISTANCE FOR INTERFACE ";N1
     PRINT
     PRINT "DIST X";TAB(10);"EXACT";TAB(20);"DIX-DURBAUM APPROX"
     PRINT

       FOR J = 0 TO JM
          F = (J * AX - X(N(J))) / (X(N(J) + 1) - X(N(J)))
          DT = T(N(J) + 1) - T(N(J))
          T = T(N(J)) + DT * F
```

```
        T =  INT (10000 * T + .5) / 10000
        TD =   INT(10000*SQR(TN^2+(AX*J/CRMS(N1))^2)+.5)/10000
       PRINT AX * J; TAB( 10);T; TAB( 20);TD
     NEXT J


2    INPUT Q$: REM PAUSE


     END



800  REM RAYTRACE SUBPROGRAM - REFLECTED PATHS


REM MODIFY FOR ARRAYS X(J) AND T(J) AT SURFACE
REM MODIFY FOR REFLECTED PATHS. T(J)=2*T AND X(J)=2*X
REM GET I1 IN RADIANS FROM MAIN
REM SET INITIAL T, Z, X = 0


     T = 0: Z = 0: X = 0
     P =  SIN (I1) / C(1)


     FOR I = 1 TO N1
        A = P * C(I)
       REM TEST FOR BEYOND CRITICAL REFL
       IF  ABS (A) < 1 AND Q < 1 THEN
         Q = -1
         B =  SQR (1 - A * A)
         Z = Z + H(I)
         Z(I) = Z
         X = X + A * H(I) / B
         T = T + H(I) / (B * C(I))
       ELSE
         Q = 1 :REM BEYOND CRITICAL
           X(J) = 9999999
           T(J) = 9999999
       END IF
     NEXT
REM  FOR REFLECTED PATHS AT SURFACE
     X(J) = 2 * X
     T(J) = 2 * T
     IF X(J)> (JM+4)*AX THEN Q = 1
     REM NO NEED TO COMPUTE FOR X>> JM*AX
     RETURN
```

A1.4.4 Utility Programs

The ray-tracing programs are self-contained. The user enters a few numbers, the program runs, and a few numbers appear on the screen. Most geophysical computations involve a lot more numbers and more complicated programs. In FORTRAN, subprograms or subroutines are linked together by a main program. Data enter through file-reading subroutines. Output numbers are usually written to files. In BASIC we usually write programs that do only a few operations. For example, a program might (1) read a data file, (2) compute, (3) graph results, and (4) write results to file. Read files, graph, and write files are utility programs. We give simple (generic) versions because it is easy to adapt them and paste them in most programs.

Read file and write file

File instructions for writing and reading must match exactly. We usually write the file maker and then duplicate it for a mock-up of the file reader, then we use the editor to convert the filewriter to a file reader. The instructions for file operations vary from computer to computer. A common sequence of file-writing operations is: go to a disk, open a file, delete to clean the file, write the file, close the file, and return to program. Read instructions are: go to a disk, open the file, read the file into computer memory, close the file, and return to program. Examples of file operation programs are given for the Macintosh computers FORTRAN and BASIC.

Graphics

Elementary graphics start by defining the terminal screen using x- and y-coordinates. Each pixel has a coordinate. Commonly, the (0, 0) pixel is at the upper left corner. On an Apple IIe, the lower left corner is (0,159); the upper right corner is (269,0); the lower right corner is (269,159). The Macintosh has (0,0) at the upper left corner and (492,300) at the lower right corner. Graphics instructions turn a pixel to the desired color. In monochrome the colors are black and white. The program for the Apple IIe is in BASIC. The Macintosh FORTRAN program uses toolbox calls for the graphics calls. Common graphics instructions are: go to (x,y) and draw a line to $(x1, y1)$. Both programs use these instructions to plot a sine wave. One can spend a lot of time making graphs pretty.

```
rem  example of a file maker prgm BASIC complier

dim  y(100)
n1=10

for i=0 to n1
     y(i)=exp(-i/20)
     print i,y(i)
  next i
```

```
rem      make file

    print"give file name":input n$
    open n$ for output as #1

    write #1, n1

    for i=0 to n1
        write #1,y(i)
    next i

    close #1

    end


rem      example of a file reader prgm

    dim  y(100)

    print"give file name":input n$
    open n$ for input as #1

    input #1, n1

    for i=0 to n1
        input #1,y(i)
    next i

    close #1

rem  print the file

    for i=0 to n1
        print i,y(i)
    next i

    input q$: rem pause

    end



c     file maker FORTRAN 77
c     unit 9 is screen. unit 10 is disk
c     * means write in any format (integer, real, or character)
```

```
      real   x(0:20),y(0:20)
      integer i,ni
      character*10 name

      write(9,*)'enter name of file='
      read(9,*) name
c     compute a set of x and y

      ni=11
      do 10 i=0,ni
         x(i)=i
         y(i)=i*i
         write(9,*) x(i),'       ',y(i)
10    continue

c     make file

      open(unit=10,file=name,status="new")
      write (10,*) ni

      do 30 i=0,ni
         write (10,*) x(i)
         write(10,*) y(i)
30    continue

      pause

      end



c     filereader FORTRAN77

c     reads file format from file maker

      real   x(0:20),y(0:20)
      integer i,ni
      character*10 name

      write(9,*)'enter name of file to be read='
      read(9,*) name

c     read file format must match the write file exactly.
c     I usually make file write, save it, and
c     then use edit to change the writer to a reader.

      open(unit=10,file=name,status="old")
```

```
    read (10,*) ni

    do 40 i=0,ni
       read (10,*) x(i)
       read (10,*) y(i)
40  continue

    do 50 i = 0 , ni
       write (9,*) x(i),'    ',y(i)
50  continue

    pause

    end
```

'A1.4.4 Graph a sinewave BASIC

```
REM WRITTEN FOR APPLE MACINTOSH AND MICROSOFT BASIC COMPILER
REM PRGM USES A SINE WAVE TO SHOW THE BASICS OF GRAPHING
REM THE GRAPH MAKES A PIXEL MAP
REM SCREEN IS 492 BY 300 PIXELS
REM 0,0 IS THE UPPER LEFT CORNER OF THE SCREEN
REM COMPILER AND MAC CLIP LINES OUT SIDE SCREEN.

REM ONE CAN USE APPLE KEY-SHIFT-3 TO MAKE MacPaint FILES AS PIXELS.
REM OR THE APPLE KEY-SHIFT-4 TO MAKE A SCREEN DUMP IN PIXELS.


REM      COMPUTE Y(N)=SIN(2*PI*N/NP)
    DIM Y(400),X(400)
    PI = 4 * ATN (1)
    NM = 100
    NP = 25

    FOR N = 0 TO NM
       X(N) = 2 * PI * N / NP
       Y(N) = SIN (X(N))
    NEXT N

REM      MAKE GRAPH

REM      SCREEN DIMENSIONS
    XL = 492
    YL = 300
```

```
REM       SET SCALES AND COMPUTE SOME INTEGER VALUES (%)
     X0 = 20
     X0% = INT(X0)
     XL% = INT(XL)
     XS = (XL-X0) / X(NM): REM  X(NM) IS MAXIMUM VALUE OF X
     Y0 = YL / 2: REM  TO PUT Y=0 IN MIDDLE
     Y0% = INT(Y0)
     YS = YL / 4: REM   AMPLITUDE FACTOR FOR THE PLOT.
     YS% = INT(YS)

REM       CALCULATE X% AND Y% AND THEN PLOT TO X1% AND Y1%.
REM    GRAPH CALLS

     CLS: REM  CLEAR SCREEN

REM DRAW LINE (X%,Y%) TO (X1%,Y1%). % INDICATES INTEGER

     FOR N = O TO NM - 1
        X% =  INT (XS * X(N)+X0): REM  MAKE INTEGERS
        X1% =  INT (XS * X(N+1)+X0)
        Y% =  INT (Y0 - YS * Y(N)): REM   MAKE PLOT + UP
        Y1% =  INT (Y0 - YS * Y(N + 1))
        LINE  (X%,Y%)-(X1%,Y1%)
     NEXT N

     LINE (X0%,Y0%) - (XL%,Y0%)            :rem  X-axis

REM  PUT TICS ON THE X-AXIS

     FOR N = 0 TO N STEP NP
        X% =  INT (XS * X(N) + X0)        :rem  locate tics
        Y% = INT (Y0 + 10)                :rem  make tics
        Y1% = INT (Y0 - 10)
        LINE (X%,Y%) - (X%,Y1%)           :rem  draw tics
        CALL MOVETO (X%,Y%+24) : PRINT N
        REM move to a spot and print N
     NEXT N

     CALL MOVETO (20,280) : PRINT "enter a lable": INPUT L$

     PRINT " INPUT A KEY STROKE TO END"
     INPUT Q$
     CLS

     END
```

```
c     A1.4.4 graph. FORTRAN 77



c     The plot numbers are (xp1,yp1) and (xp2,yp2).
c     MOVETO goes to (xp1,yp1).LINETO draws a line from point 1 to 2.
c     If the line crosses the rectangle, CLIPRECT removes points
c     outside the bounds of the rectangle set by SETRECT. The
c     intersects are computed and the line is drawn from xp1,yp1
c     to xp2,yp2 by LINETO. Mac plots from 0,0
c     at upper left corner to 492,300 in lower right corner.Program
c     uses Mac toolbox calls.All toolbx sub's must be out of the folder.
c     all numbers that enter in toolbx graphics must be integers
c     lable is a graph title. DRAWSTRING stops writing for any 'space'


      character*256  lable
      character*3q
      real   xtic,ytic,nin,pi,xs,ys
      real   x(-400:400),y(-400:400)
      integer  xp1,yp1,xp2,yp2,x0,y0,xmax,ymax,idx,idy
      integer  n,np,nm,m,i,mmax, nmax, step,ndg


c     Set the rectangle and clip

      include QUICKDRAW.INC
      integer*2  rect(4)
      integer*4  toolbx,mypic
      call toolbx   (SETRECT,rect,0,0,492,300)
      call toolbx  (CLIPRECT,  rect)


c     coordinates of origin and graph tics

      x0=246
      y0=150
      xtic=20.
      ytic=20.
      xmax=492
      ymax=300
      mmax=int(x0/xtic)
      nmax=int(y0/ytic)
      nm=100
      np=25
      pi=4.*atan(1.0)
```

```
100  continue

c    enter a lable and coordinates x1,y1 and x2,y2

     call toolbx  (MOVETO,0,10)
     write(9,*)' enter a lable. instead of spaces enter".'"
     read (9,*) lable
     lable=char(32)//lable
c    compute a sine wave as a demo plot.
     do 10 n=-nm,nm
          x(n)=pi*n/np
          y(n)=sin(x(n))
10   continue

c    scale in x and y
     xs=xmax/(2*x(nm))
     ys=y0/2

c    Compute plotting values of xp1,yp1, etc for graph. Origin is
c    at x0,y0 and y is + in upward direction. x is + to right.

c    Erase rect first and then frame it

         call toolbx  (ERASERECT,rect)
         call toolbx  (FRAMERECT,rect)
         mypic =  toolbx (OPENPICTURE,rect)

     do 20 n=-nm ,nm-1
          xp1=int(xs*x(n)+x0)
          yp1=int(-ys*y(n)+y0)
          xp2=int(xs*x(n+1)+x0)
          yp2=int(-ys*y(n+1)+y0)

c    Plot line x1,y1 to x2,y2. plot relative to x0,y0

          call toolbx (MOVETO,xp1,yp1)
          call toolbx (LINETO,xp2,yp2)
20   continue
c    Plot coordinates

     call toolbx (MOVETO,0,y0)
     call toolbx (LINETO,xmax,y0)
     call toolbx (MOVETO,x0,0)
     call toolbx (LINETO,x0,ymax)

c    Put tics on graph
```

```
        do 30 m=-mmax,mmax,1
           idx=int(m*xtic+x0)
          call toolbx (MOVETO,idx,y0-2)
          call toolbx (LINETO,idx,y0)
30    continue

        do 50 n=-nmax,nmax,1
           idy=int(-n*ytic+y0)
          call toolbx (MOVETO,x0-2,idy)
          call toolbx (LINETO,x0,idy)
50    continue

c     Move for legend

        call toolbx  (MOVETO,40,270)
        call toolbx  (DRAWSTRING,lable)
c     end of graph operation

        call toolbx (CLOSEPICTURE)
        call toolbx  (DRAWPICTURE,mypic,rect)
        pause
c     pause gives time to look. Next calls erase picture.
        call toolbx  (ERASERECT,rect)
        call toolbx  (KILLPICTURE,mypic)
        call toolbx  (MOVETO,0,290)

        end
```

'A1.4.4 Printing graphs

'The Macintosh computers have three ways of printing hard
'copies of graphs. A brief review of how the Macintosh
'makes images on the screen follows because the first two
'methods use the screen image. The third method makes
'PICT types of instructions and files.

'The graphs and text that appear on the screen are bit or
'pixel maps. For example, a line from (x1,y1) to (x2,y2)
'appears as the string of pixels closest to the line.
'The pixel map of a diagonal line is irregular unless the
'line is exactly along a line of pixels. A character is a
'collection of dots or pixels. The quality of the image is
'limited by the dimensions of the graphic screen in pixels.
'The Macintosh Plus and SE computers have a screen of 492 by

'300 pixels. The utility programs use pixel map graphics
'because they are more robust than PICT graphic methods.

'1) SCREEN DUMP TO PRINTER: A screen dump sends the pixel
'map that is on the screen to a printer. Even the mouse pointer
'appears when it is on the screen. The Macintosh II needs
'to be in the black and white mode. The Macintosh SE/30 also
'requires the "Caps Lock" key to be down. The Apple or
'clover leaf key is used in the following command:
'Apple key-Shift-4.

'2) SCREEN DUMP TO FILE and MacPaint IMAGES: The dump to
'file operation creates a MacPaint image on the disk. The
'images are called snap shots and appear as Screen0,
'Screen1 etc. We use MacPaint or equivalent programs to read
'and alter the image files. The command follows:
'Apple key-Shift-3.

'3) MAKE A MacDraw or PICT FILE: MacDraw stores the
'data and the PICT instructions needed to draw a map.
' For example, a MacDraw file contains the PICT instructions to
'draw a line from (x1,y1) to (x2,y2) and to create a pixel map
'for the printer. A laser printer draws the line at over
'100 pixels/cm and gives high resolution images. The number of
'PICT instructions can become very large for complicated images
'such as multitrace seismograms. If the instruction set is too
'big, the computer develops system errors and has to be rebooted.
'This can happen when a program is drawing screen images and
'"make PICT file" is in the program. For big graphs,
'make several small files and then paste them together in
'MacDraw. After you have your program and the pixel map
'graphics running, paste the "make PICT file" program in it.
'A make PICT file program follows:

```
REM make PICT file

    CLS         :' CLS      clears the screen
    PICTURE ON  :' PICTURE ON puts graphic instructions in memory.
    SHOWPEN     :' SHOWPEN   puts the graph on the screen

REM     Make a graph here
            'Example for tests

    LINE (20,20) - (300,200)
    CALL MOVETO (20, 200) : PRINT" PICT graphic test"

REM End of graphics
```

```
PICTURE OFF : 'PICTURE OFF ends graphics operations.

CALL MOVETO (20,280)
PRINT " input 'y' to make a PICT file "
INPUT Q$
IF Q$ <> "y"  goto 100

pic$ = PICTURE$  : 'PICTURE$ is name of stored picture.

CALL MOVETO (50, 25)        :' name the file
PRINT   "The picture is in pic$ ("; LEN (pic$); ")"
pictFile$ = FILES$ (0, "Enter name for PICT file:")
PRINT "PICT file name is:"; pictFile$

'Save the file in "PICT" formate.
OPEN pictFile$ FOR OUTPUT AS #1

'For-next loop makes a header for PICT file formate.

FOR i = 1 TO 512
    PRINT  #1, CHR$ (0);
NEXT i

PRINT  #1, pic$

CLOSE  :'the picture 'pic$' is stored as a text file.

'Change the file from text to PICT
NAME pictFile$ AS pictFile$, "PICT"

'Use MacDraw to read the file. Then, it can be saved
'as a MacDraw drawing.

100  CLS    : 'clear screen and end

END
```

A1.5 REFRACTION TRAVEL-TIME PROGRAM
FOR DIPPING LAYERS

The program uses the phase-velocity method (Section A1.3) to calculate the intercepts of the head waves. In the phase-velocity method, we do not get the minimum distance for the head-wave arrival; we get the travel-time intercepts on a graph of travel time versus distance. The parameters refer to source A, and the layer thicknesses are under source A. The program computes the thicknesses under layer B.

Choose the number of layers, N, first. Enter L to pick a layer and then enter seismic velocity C, thickness Z, and dip P. The distance between sources is AB. PAR gives the parameters. RUN runs the program. The program asks for the refraction interface. It computes reversed intercept times for the interface. We repeat RUN for each interface. A sample run follows the program. Figure A.6 shows graphs of the intercepts for interface 2 and interfaces 2 and 3.

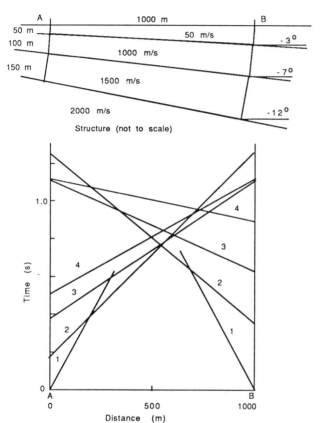

Figure A.6 Structure and refraction travel-time intercepts. The numbers 1, 2, 3, and 4 indicate the refractions from the corresponding layers.

The algorithms follow those in D. Palmer, "The generalized reciprocal method of seismic refraction interpretation," Tulsa, Okla.: Society of Exploration Geophysicists, (1980), 3–4.

```
REM     REFRACT FROM N LAYERS WITH DIPS
REM   EXAMPLE OF INTERACTIVE MENU DRIVEN PROGRAM


       DIM  X(10)
       DIM  A(10),B(10),C(10),PHI(10),Z(1,10)
       DIM  TA(1,10),TB(1,10),PD(11)
       DR = 4 *  ATN (1) / 180
```

```
      NK=1000: REM MULT OR DIVIDE BY 1000
      P(0) = 0

      PRINT " REFRACTION FOR N LAYERS"
      PRINT "MAXIMUM NUNBER OF LAYERS = 10"
      PRINT "RUN ALL LAYERS BEFORE MAKING A FILE"

200 PRINT "  MENU --- RETURNS ARE TO  -----GO-----"
      PRINT TAB( 5);"N'";    TAB( 12);"FOR NUMBER OF LAYERS"
      PRINT TAB( 5);"L'";    TAB( 12);"FOR LAYER NUMBER"
      PRINT TAB( 5);"AB'";   TAB( 12);"FOR A-B DIST'"
      PRINT TAB( 5);"PAR'";  TAB( 12);"FOR PARAMETERS"
      PRINT TAB( 5);"C'";    TAB( 12);"COMPUTE"
      PRINT TAB( 5);"R'";    TAB( 12);"DISPLAY RESULTS"
      PRINT TAB( 5);"M'";    TAB( 12);"MENU"
      PRINT TAB( 5);"MF'";   TAB( 12);"MAKE FILE"
      PRINT TAB( 5);"QUIT'"; TAB( 12);"Q"
300   PRINT "-----GO-----";: INPUT Q$

      IF Q$ = "N"   THEN 1000
      IF Q$ = "L"   THEN 1100
      IF Q$ = "PAR" THEN 1200
      IF Q$ = "AB"  THEN 1300
      IF Q$ = "C" THEN 2000
      IF Q$ = "R" THEN 2100
      IF Q$ = "Q" THEN 4000
      IF Q$ = "M"   THEN 200
      IF Q$ = "MF"  THEN 5000
      GOTO 300
1000 PRINT "INPUT NUMBER OF LAYERS=";: INPUT N
      IF N > 10 THEN 1000
      GOTO 300

1100 PRINT "INPUT LAYER L=";: INPUT L
      IF L = N    THEN 1110
      IF L > N    THEN 1100
      PRINT "OLD  C(L),Z(L),PHI(L)=";
      PRINT C(L),Z(0,L),PD(L)
      PRINT "NEW  C(L),Z(L),PHI(L)=";: INPUT C(L),Z(0,L),PD(L)
      P(L) = DR * PD(L)
      GOTO 300

1110      PRINT "C(N)=";: INPUT C(N)
      P(N)=0
      P(0)=0
      GOTO 300
```

```
1200        PRINT : REM PRINT PARAMETERS
        PRINT " LAYER C(L)   Z(L)   PHI(L)"
        FOR J = 1 TO N - 1
            PRINT  TAB(3);J;TAB(7);C(J);TAB(14);Z(0,J);TAB(23);PD(J)
        NEXT J
         PRINT   TAB( 3);J; TAB( 7);C(N)
        PRINT
         PRINT "AB=";X(0)
        PRINT
        GOTO 300


1300        PRINT "AB=";:INPUT  X(0)
        GOTO 300


REM  RUN REFRACTION CALCULATION. INPUT REFRACTED INTERFACE RI

2000        PRINT "REFRACTION INTERFACE="; : INPUT RI
        IF RI > N THEN 2000: REM FIX A BAD ENTRY

REM    CALULATE LAYER 1 DIRECT ARRIVAL
        TA(0,1) = 0
        TA(1,1) = X(0) / C(1)
        TB(1,1) = 0:TB(0,1) = TA(1,1)


    IF RI>1 THEN GOSUB 3000: REM DO REFRACTION CALCULATION

2100     REM PRINT RESULTS AFTER RETURN FROM SUBPRGM


    PRINT "AB =";X(0)
     PRINT "J      TA         TB"
    FOR J = 1 TO N
        FOR I = 0 TO 1
            A(J) =  INT (NK * TA(I,J)+.5) / NK
            B(J) =  INT (NK * TB(1 - I,J)+.5) / NK
             PRINT J;TAB(5);A(J);TAB(14);B(J)
        NEXT I
    NEXT J


    GOTO 300: REM RETURN TO GO

4000     END



3000  REM CALC FOLLOWS D. PALMER "THE GENERALIZED RECIPROCAL
REM    METHOD OF SEISMIC REFRACTION INTERPRETATION"
```

```
REM   S.E.G.1980.PP 3-4. CALCULATE LENGHTS OF X(J) FOR EACH
REM   INTERFACE AND THEN CALCULATE THE Z(1,J) AT B.

    P(0) = 0 :REM SLOPE OF INTERFACE 0 = 0
    FOR J = 1 TO N - 1
      P = P(J) - P(J - 1)
      CP =  COS (P)
      SP =  SIN (P)
      X(J) = X(J - 1) * CP
      Z(1,J) = Z(0,J) - X(J - 1) * SP
    NEXT J


REM  START AT THE REFRACTING LAYER, RI, AND WORK UPWARDS


REM  REFRACTION FROM RI
      U = C(RI- 1) / C(RI)
    IF U < 1  THEN
      A(RI - 1) =  ATN (U /  SQR (1 - U * U))

      B(RI - 1) = A(RI - 1)


REM  COMPUTE REFRACTED ANGLES IN THE NEXT LAYER UP.
      AP = A(RI - 1) + P(RI - 1)-P(RI-2)
      BP = A(RI - 1) - P(RI - 1)+P(RI-2)


REM   TRAVEL OF THE WAVEFRONT FROM INTERFACE RI TO THE SURFACE
REM  CALCULATE THE ANGLES FOR "A" AND "B" LOCATIONS

      FOR J = 2 TO RI - 1
         U = C(RI - J) *  SIN (AP) / C(RI - J + 1)
         A(RI - J) =  ATN (U /  SQR (1 - U * U))
         AP = A(RI - J) + P(RI - J) - P(RI - J-1 )
         U = C(RI - J) *  SIN (BP) / C(RI - J + 1)
         B(RI - J) =  ATN (U /  SQR (1 - U * U))
         BP = B(RI - J) - P(RI - J) + P(RI - J-1)
      NEXT J


REM    TRAVEL TIME CALCULATIONS USE PHASE VELOCITY METHOD

    T1 = 0
    T2 = 0
    FOR J = 1 TO RI - 1
        T1 = T1+ Z(0,J) *  COS (A(J)) / C(J)
        T2 = T2+ Z(1,J) *  COS (B(J)) / C(J)
    NEXT J
```

```
      T3 = X(RI - 1) / C(RI)
      TA(0,RI) = 2 * T1
      TB(1,RI) = 2 * T2
      TA(1,RI) = T1 + T2 + T3
      TB(0,RI) = TA(1,RI)


  ELSE

    PRINT "NO REFR. C(";RI;") < C(";RI-1;")"
      TA(0,RI) = 0
      TB(1,RI) = 0
      TA(1,RI) = 0
      TB(0,RI) = 0


  END IF


  RETURN


5000      REM MAKE FILE.
REM  REFRACTION CALCULATIONS ARE NEEDED FOR ALL LAYERS

    print"give  file  name":input n$
    open n$ for output as #1


    write  #1,n

      write #1, " LAYER C(L)   Z(L)   PHI(L)"
    FOR J = 1 TO N - 1
         write #1,  J,C(J),Z(0,J),PD(J)
    NEXT J
     write #1,  J,C(N)
     write  #1,"AB  =",X(0)
      write #1, " Layer TA        TB"

      for j = 1 to N
        FOR I = 0 TO 1
            A(j) =  INT (NK * TA(I,j)) / NK
            B(j) =  INT (NK * TB(1 - I,j)) / NK
              write #1,j,  A(j),B(j)
        NEXT I
      next j
      close #1
      goto 300
```

A1.6 PHYSICAL PROPERTIES OF ROCKS

Extensive tables of the properties of rocks can be found in the handbook by Robert S. Carmichael, ed., *Physical Properties of Rocks*: Vol. I Mineral Composition (1982), Vol. II Seismic Velocities (1982), and Vol. III Density of Rocks and Minerals (1984). Boca Raton, Fla.: CRC Press Inc. The physical properties include elastic, seismic velocities, seismic attenuation, electrical, magnetic, thermal, and engineering properties of rocks. The handbook also shows the dependence of physical properties on fluid saturation, pressure, and temperature.

The data in Figures A.7 and A.8 show gross relationships of rock types, seismic velocities, and densities. These data are from S. P. Clark, *Handbook of Physical Constants*. Geological Society of America, Memoir 97 (1966), and F. D. Stacey, *Physics of the Earth*. New York: John Wiley, 1977.

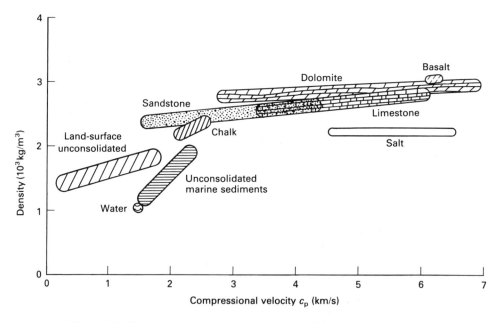

Figure A.7 Density and compressional wave velocity for various rocks. (Data from S. P. Clark, *Handbook of Physical Constants Memoir 97*. New York: The Geological Society of America, 1966.)

A1.7 AMPLITUDE COMPUTATIONS USING RAY-TRACING METHODS

The theory and equations are in Section 5.3 and the geometry is in Figure 5.5a. The program computes particle velocity amplitudes for the down-going tube of rays. It gives the incident particle velocity along the ray at interface I. For simplicity the medium above interface 0 is the same as the medium in layer 1.

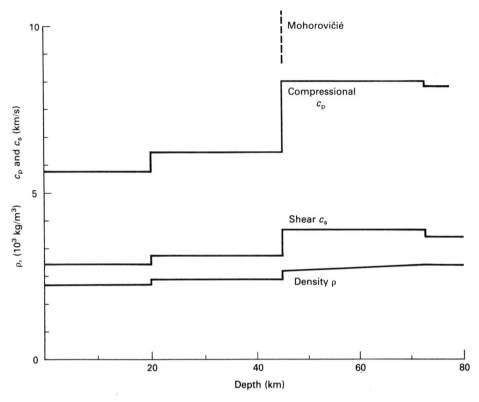

Figure A.8   Density, compressional and shear wave velocities versus depth for continental structure. (Calculated from Table G2 of F. D. Stacey, *Physics of the Earth*, (2d ed.) New York: John Wiley and Sons, 1977.)

```
REM  A1.7 RAYTRACE AMPLITUDE

REM  PRGM HAS EXTRAS TO HANDLE INCIDENT ANGLES BEYOND CRITICAL
REM  WITHOUT HANGUPS. OUTPUT NUMBERS FOR INTERFACES BEYOND
REM   CRITICAL ARE MEANINGLESS : X(I)=99999, T(I)=99999

     DIM  C(21),D(21),H(21),VR(20)
     DIM  X(20),Z(20),L(20),T(20),RF(20),TR(20)
REM COMPUTE PI=4*ATN(1) AND CONVERSION DEG TO RAD FACTOR, DR
     PI = 4 *  ATN (1): DR = PI / 180
     PW = 10000: REM   INPUT POWER DEFAULT

     PRINT      "RAYTRACE AMPLITUDES"
     PRINT   TAB( 5);"INPUT NUMBER OF INTERFACES"
     PRINT   TAB( 5);"INPUT DATA FOR LAYERS"
     PRINT   TAB( 5);"INPUT INCIDENT ANGLE"
     PRINT   TAB( 5);"INPUT INTERFACE FOR"
```

```
        PRINT    TAB( 7);"AMPLITUDE CALC."
      PRINT    "PRGM RETURNS FOR NEW INCIDENT ANGLE"
        PRINT

REM       USE RAY TRACE ALGORITHM SECTION A1.4.2
REM       ON INTERFACE I, X(I) IS MEAN X DISPLACEMENT FROM X = 0
REM       L(I) IS DIFFERENCE OF THE X-DISPLACEMENTS OF A PAIR OF RAYS.
REM       USE REFLECTION AND TRANSMISSION COEFFFICIENTS, SECTION 5.1
REM       USE RAYTRACE SIGN CONVENTION IN SECTION 5.1.2
REM       DESIGNATE R12 AS RF(1) ETC
REM       DESIGNATE T12 AS TR(1) ETC
REM       AMPLITUDE CALC FROM SECTION 5.2

     PRINT "OMNIDIRECTIONAL SOURCE ON"
     PRINT  TAB( 7);"INTERFACE 0"
     PRINT  TAB( 7);"STARTING AT 0"
     PRINT "INPUT NUMBER OF INTERFACES=";: INPUT Ni
     PRINT " ENTER LAYERS"

     FOR i = 1 TO Ni
        PRINT " INPUT H(";i;"),   C("i"), D("i") = ";
        INPUT H(i), C(i), D(i)
     NEXT i

     PRINT " INPUT BOTTOM, C("Ni+1"), D("Ni+1") = ";
     INPUT  C(Ni+1),D(Ni+1)
     H(Ni + 1) = 999999
100  PRINT "INPUT 'Q' TO QUIT, ANY KEY CONTINUES"
     INPUT Q$: IF Q$ = "Q" THEN 3000
     PRINT

     PRINT  TAB( 5);"INPUT INCIDENT ANGLE=";: INPUT IANG
     PRINT

REM  COMPUTE INCIDENT PARTICAL VELOCITY, V1, AT INTREFACE 1
     PK =  SQR (PW / (4 * PI * C(1) * D(1)))
     R1 = H(1) /  COS (IANG * DR)
     VI = PK / R1

REM  GOTO SUBROUTINE FOR DUAL RAYTRACE

GOSUB 2000

PRINT"i";TAB(4);"Z(i)";TAB(14);"X(i)";TAB(25);"L(i)";TAB(35);"T(i)"
PRINT
FOR i= 1 TO Ni
     ZC=INT(100*Z(i))/100:  XC=INT(100*X(i))/100
     LC=INT(10000*L(i)+.5)/10000:  TC=INT(1000*T(i))/1000
```

```
        PRINT i;TAB( 4);ZC;TAB( 14); XC;TAB( 25); LC ;TAB( 35);TC
NEXT i

PRINT
PRINT "i"; TAB( 4);"REFL(i,i+1)"; TAB( 18);"TRNS(i-1,i)";TAB(29);
        PRINT "TO i";TAB(38);"VEL ALONG R"
PRINT
TP = 1
FOR i= 1 TO Ni
    TP = TP * TR(i-1)
    VR(i) = VI *  SQR (L(1) * X(1) / (L(i) * X(i))) * TP
        PRINT  i;TAB(4);RF(i);TAB(18);TR(i-1);TAB(28);TP;TAB(38);VR(i)
NEXT i

PRINT "POWER IN=";PW;" WATTS"
PRINT
GOTO 100


3000      END

2000      REM  DUAL RAYTRACE CALCULATION

    DA = .2 : REM ANGLE INCREMENT FOR DUAL RAYS
    IF IANG > DA THEN
      I0 = (IANG - DA) * DR: REM   INNER RAY
      IC = IANG * DR     : REM   CENTER RAY
      I1 = (IANG + DA) * DR: REM   OUTTER RAY
    ELSE
      I0 = 0 : IC = DA * DR
      I1 = 2 * DA * DR
    END IF
    P0 =  SIN (I0) / C(1): REM  DUAL RAY PAREMETERS
    PC =  SIN (IC) / C(1)
    P1 =  SIN (I1) / C(1)


REM    TEST FOR BEYOND CRITICAL REFL. USE A1 <.9 BECAUSE
REM    RAYTRACE AMPLITUDE CALCULATIONS ARE POOR
REM    FOR INCIDENT ANGLES NEAR CRITICAL.

    Q = - 1 :   REM  SET Q=-1 FOR INCIDENT ANGLE LESS THAN CRITICAL
    TR(0) = 1 :   REM TRANSMISSION IN LAYER 1 IS  1
    T = 0: X0 = 0: X1 = 0: Z = 0

    FOR i = 1 TO Ni
      A0 = P0 * C(i)
```

```
         AC = PC * C(i)
         A1 = P1 * C(i)
      IF Q < 0 AND A1 < .9 THEN
         B0 =  SQR (1 - A0 * A0)
         BC1 =  SQR (1 - AC * AC)
         B1 =  SQR (1 - A1 * A1)
        Z = Z + H(i)
        Z(I) = Z
         X0 = X0 + A0 * H(i) / B0
         X1 = X1 + A1 * H(i) / B1
        L(i) = X1 - X0
         X(i) = (X0 + X1) / 2
         T = T + H(i) / (BC1 * C(i))
         T(i) = T
      REM REFLECTION AND TRANSMISSION COEF FOR MEAN RAYPATH
        AC2 = PC * C(i + 1)
        IF AC2 < 1 THEN
           BC2 = SQR(1 - AC2*AC2)
           N1 = D(i) * C(i) * BC2
           N2 = D(i + 1) * C(i + 1) * BC1
           RF(i) = (N2 - N1) / (N1 + N2): REM  ALONG RAY PATH
           TR(i) = 2 * D(i) * C(i) * BC1 / (N1 +N2)
        ELSE
           PRINT "BEYOND CRITICAL AT ";i;" - ";i+1;"  INTERFACE"
           TR(I) = 0
           RF(I) = 1
        END IF
      ELSE
        PRINT "TOO NEAR CRITICAL: LAYER ";i
         X(I) = 99999 : T(I) = 99999 :L(I) = 99999
        Q = 1 : REM  SET Q FOR BEYOND CRITICAL
      END IF
     NEXT i




     RETURN
```

A1.7 Sample calc for raypath amplitudes

Transmission amplitude calculation for
2 layers over a half space, C(3) = 3000, D(3) = 3000
Incident angle = 10 deg

Layer	H(i), m	C(i), m/s	D(i), kg/m^3
1	1000	1000	1000
2	1000	2000	2000

Interface	Depth,m	X(i),m	L(i),m	Time,s
1	1000	176.32	3.60	1.015
2	2000	546.68	11.94	1.548

Interface	R(i,i+1)	T(i-1,i)	Vel. along R
1	0.615	1.0	2.78E-05
2	0.424	0.404	3.51E-06

ANSWERS TO PROBLEMS—UNIT 1

Chapter 1

1. (a) $f(t, x) = \sin[2\pi f(t - x/c)]$
  (b) $f(t, x) = \sin[2\pi f(t + x/c)]$
4. $f = 1000$ Hz: $X = 10^{-6}$ m, $v_x = 2\pi \times 10^{-3}$ m/s,
  $a_x = -4\pi^2$ m/s^2; $p = 3\pi \times 10^3$ n/m^2.

Chapter 2

1. $\alpha = 2.6 \times 10^{-5}$ f, m^{-1}. 10 Hz, $a = 0.77$; 100 Hz, $a = 0.074$.
2. $f = 10$ Hz: $R = 100$ M, $a = 0.99 \times 10^{-2}$; $R = 10^4$ m, $a = 0.42 \times 10^{-4}$.
4. Pa · m or N/m

Chapter 3

1. $\sin \theta_1/c_1 = \sin \theta i_2/c_2$. $\theta_2 = \arcsin (c_2 \sin \theta_1/c_1)$
  (a) $\theta_1 = 30°$, $\theta_2 = 41.8°$
  (b) $\theta_1 = 30°$, $\theta_2 = 14.5°$

2. (a) Incident 10°: $\theta_{1p} = 10°$, $\theta_{1s} = 5°$
    Incident 60°: $\theta_{1p} = 60°$, $\theta_{1s} = 25.7°$
  (b) Incident 20°: $\theta_{2p} = 30.9°$, $\theta_{2s} = 22°$
  (c) Critical angle p_1p_2, $\theta_{1p2p} = 41.8°$
    Critical angle p_1s_2, $\theta_{1p2s} = 65.4°$
  (d) Yes
3. PRGM PRINTOUT

Let incident angle be I0 in degrees and I1 in radians. The refracted angle is I2. Snell's law gives

c   in FORTRAN77

```
u = C2 * sin( I1) / C1
I2 = asin(u)
```

Many small computers take only the arc tangent. Use $\arcsin(u) = \arctan[u/(1 - u^2)^{1/2}]$.

```
REM            PRGM  SNELL'S LAW

        PRINT "SNELL'S LAW:  RAY INCIDENT IN"
        PRINT "  LAYER 1, REFRACTED INTO"
        PRINT "  LAYER 2
        PRINT
        DR = 4 *  ATN (1) / 180
        PRINT "INPUT C1,C2="; : INPUT C1,C2
        U = C1 / C2

REM TEST : IF U = C1/C2 <1 THEN TOTAL REFLECTION CAN EXIST.
REM IF U > 1 NO CRITICAL ANGLE.

        IF U > = 1 THEN 100
        IC =  ATN (U /  SQR (1 - U * U))
        PRINT "CRITICAL ANGLE=";IC / DR
        PRINT

100 PRINT "INPUT INCIDENT ANGLE,DEG="; : INPUT IANG
        PRINT

REM CONVERT DEG TO RAD
        I1 = IANG * DR
        P =  SIN (I1) / C1
        A2 = P * C2

REM     TEST FOR BEYOND CRITICAL REFRACTION.
        IF A2 > = 1 THEN 200

        B2 =  SQR (1 - A2 * A2)
        I2 =  ATN (A2 / B2)

        PRINT "REFRACTED ANGLE=";I2 / DR
        PRINT
        PRINT "INPUT 'Y' FOR ANOTHER ANGLE ";: INPUT C$
        IF C$ < > "Y" THEN 300
        PRINT
        GOTO 100

200     PRINT "BEYOND CRITICAL ANGLE"
        PRINT
        PRINT "INPUT 'Y' FOR ANOTHER ANGLE ";: INPUT C$
        IF C$ < > "Y" THEN 300
        PRINT
```

                    GOTO 100

300   END

        SNELL'S LAW: RAY INCIDENT IN
            LAYER 1, REFRACTED INTO
            LAYER 2
        INPUT C1,C2=? 1000,2000
        CRITICAL ANGLE= 30
        INPUT INCIDENT ANGLE,DEG=? 20
        REFRACTED ANGLE= 43.16018

        INPUT 'Y' FOR ANOTHER ANGLE ? Y

        INPUT INCIDENT ANGLE,DEG=? 40
        BEYOND CRITICAL ANGLE

        SNELL'S LAW: RAY INCIDENT IN
            LAYER 1, REFRACTED INTO
            LAYER 2

        INPUT C1,C2=? 2000,1500
        INPUT INCIDENT ANGLE,DEG=? 40
        REFRACTED ANGLE= 28.822
        INPUT 'Y' FOR ANOTHER ANGLE ? N

Chapter 4

1. $\theta_{1p2p} = 48.6°$, minimum distance $= 2\,h_1\tan\theta_{1p2p} = 79.4$ km. At vertical incidence p_1p_1 two-way reflection time $= 11.67$ s. The $p_1p_2p_1$ intercept is 7.72 s. At 100 km $p_1p_2p_1$ arrives at 20.22 s.

2. The problem simplified from M. Blaik, J. Northrop, and C. S. Clay "Some seismic profiles on shore and off-shore Long Island, New York," *J. Geophys. Res.*, 64 (1959), 231–39. Layer 1 has a trapped wave or ground roll having a velocity of about 300 m/s and slower. The vertical incidence two-way reflection times are Interface 2: 0.264 s, Interface 3: 0.614 s.

3. Start at the top layer. Use the ratio c_1/c_2 to compute θ_{12}, then use the intercept at $x = 0$ and $t_n(0) = 2\,h_1\cos\theta_{12}/c_1$ to compute h_1. Use c_i, c_2 and c_3 to compute θ_{13} and θ_{23}. The intercept of the c_3 line is $2\,h_1\cos\theta_{13}/c_1 + 2\,h_2\cos\theta_{23}/c_2$. Subtract the first-layer term to evaluate h_2. Solutions for deep layers follow the same procedures.

Chapter 5

1. In Eqs. (5.1) and (5.2) eliminate v_i and v_r using $v = \pm\, p/(\rho c)$.

$$R_{12}^p = \frac{\rho_2 c_2 \cos \theta_1 - \rho_1 c_1 \cos \theta_2}{\rho_2 c_2 \cos \theta_1 + \rho_1 c_1 \cos \theta_2}$$

$$T_{12}^p = \frac{2 \rho_2 c_2 \cos \theta_1}{\rho_2 c_2 \cos \theta_1 + \rho_1 c_1 \cos \theta_2}$$

$R_{12}^p = -R_{12}$, where superscript is for pressure.

2. This program has a few extra statements to assist in its use.

```
REM  Unit 1, Chapter 5, Problem 2

    REM  PRGM 'REFL COEF' FOR PARTICLE VELOCITY
    PRINT "COMPUTE REFL COEF, TRANS COEF"
    PRINT "REFRACTED ANGLE"
    PRINT " MED 1 TO MED 2"
    PRINT
100 PRINT "INPUT C1,C2";
    INPUT C1,C2
    PRINT "INPUT D1,D2";
    INPUT D1,D2
    PRINT

    REM  DEGREES TO RADIANS
    DR = 4 *  ATN (1) / 180
    PRINT "INPUT IANG,DEG="; : INPUT IANG
    PRINT

    I1 = IANG * DR
    A1 =  SIN (I1)
    B1 =  COS (I1)
    A2 = A1 * C2 / C1
    IF  ABS (A2) > = 1 THEN 200

    B2 =  SQR (1 - A2 * A2)
    N1 = D1 * C1 * B2
    N2 = D2 * C2 * B1
    R = (N1 - N2) / (N1 + N2)
    T = 2 * B1 * C1 * D1 / (N1 + N2)
    I3 =  ATN (A2 / B2) / DR

    PRINT "R12=";R
    PRINT "T12=";T
    PRINT "REFRACTED ANGLE=";I3
    PRINT
    PRINT "INPUT 'Y' TO DO AGAIN";: INPUT C$
```

```
        IF C$ < > "Y" THEN 300
        GOTO 100


200   PRINT "BEYOND CRITICAL"
        PRINT "R12=1,  T12=0"
        I3 =  ATN ((C1 / C2) / (1 - (C1 / C2) ^ 2) ^ .5) / DR
        PRINT "CRITICAL ANGLE=";I3
        PRINT
        PRINT "INPUT 'Y' TO DO AGAIN";: INPUT C$
        IF C$ = "Y" THEN 100


300   END



        COMPUTE REFL COEF, TRANS COEF
          AND, REFRACTED ANGLE
         FROM MEDIUM 1 TO MEDIUM 2

        INPUT C1,C2 = 1000, 2000
        INPUT D1,D2 = 1, 2
        INPUT INCIDENT ANGLE, DEG = 0

        R12 = -.6
        T12 = .4
        REFRACTED ANGLE = 0 DEG

        INPUT 'Y' TO DO AGAIN? Y

        INPUT C1,C2 = 1000, 2000
        INPUT D1,D2 = 1, 2
        INPUT INCIDENT ANGLE, DEG = 20

        R12 = -.6749512
        T12 = .4187378
        REFRACTED ANGLE = 43.16018 DEG
```

3. $\theta_{12} = 30°$. At vertical incidence, $\theta_1 = 0$,
   (a) $R_{12} = -0.6$, $T_{12} = 0.4$
   (b) $\theta_1 = \theta_2$. At all angles of incidence
       $R_{12} = \frac{1}{3}$, $T_{12} = \frac{2}{3}$
   (c) No critical angle.
   (d) At $\theta_1 = 0$, $R_{12} = 0$ and $T_{12} = 1$. There are reflections for other angles of incidence. In general, $\rho_1 c_1 \cos \theta_2 = \rho_2 c_2 \cos \theta_1$ gives no reflection and complete transmission. The corresponding θ_1 is sometimes called the *angle of intromission*.

Appendix B

B1.1 GENERATING FUNCTIONS, Z-TRANSFORMS, AND FOURIER SERIES

There are two ways of describing signals, the time domain and frequency or spectral domain. In the time domain the signal is a function of time. We use a polynomial or a generating function to express a sequence of signal amplitudes. Robinson introduced these techniques to the geophysical literature in the 1950s. Generating functions are particularly convenient for digital computations on many seismic problems.

The frequency-domain description gives the output of the system for a continuous sine-wave input. That is, the amplitude and phase response of the system are measured as a function of the frequency of the input sine wave. The performance specifications of practically all analog electronic equipment are measured and given in the frequency domain. Second, most procedures for solving the wave equation start by assuming the source to be a sine-wave signal. In geophysics the result is the amplitude and phase response of the earth to a sine-wave source.

The time- and frequency-domain descriptions are analogous to two languages because both can describe the same phenomenon. The Fourier transformation is analogous to the translator's dictionary; it enables us to go between domains. Geophysicists generally become bilingual.

B1.1.1 Generating Functions

The generating functions of Laplace can be constructed by choosing an arbitrary set of coefficients a_j to form

$$A(z) \equiv a_0 + a_1 z + \cdots + a_n z^n \tag{B.1}$$

where z is indeterminate with the range

$$0 \leq |z| \leq |z_0|$$

and where the a_j and $|z_0|$ are finite. The polynomial in z must satisfy convergence conditions if n is allowed to become infinite. The absolute value signs are used because z can be real or complex. For finite n, we disallow values of z that are roots of $A(z)$.

The algebraic statement that $A(z) = B(z)$ gives the proof that the coefficients of like powers of z are equal, $b_j = a_j$, and b_i is independent of b_j for $i \neq j$. The equality and its expansions are

$$A(z) = B(z) \tag{B.2}$$
$$a_0 + a_1 z + \cdots + a_n z^n = b_0 + b_1 z + \cdots + b_n z^n$$

where again, the roots of $A(z)$ and $B(z)$ are not allowed values of z. Rearrangement of the terms gives

$$(a_0 - b_0) + (a_1 - b_1)z + \cdots + (a_n - b_n)z^n = 0 \tag{B.3}$$

Equation (B.3) must be satisfied for an infinite set of values of $|z| \leq |z_0|$.

We use a first-order polynomial, $n = 1$, to demonstrate a method of proof. Since the values of z in Eq. (B.1) are indeterminate, we choose a pair of values, $z = 2$, and $z = 5$. Substitutions in Eqs. (B.2) and (B.3) give two equations,

$$\left.\begin{array}{l} (a_0 - b_0) + 2(a_1 - b_1) = 0 \\ (a_0 - b_0) + 5(a_1 - b_1) = 0 \end{array}\right\} \tag{B.4}$$

If we regard $(a_0 - b_0)$ and $(a_1 - b_1)$ as unknowns, the only solutions are

$$\left.\begin{array}{l} (a_0 - b_0) = 0 \\ (a_1 - b_1) = 0 \end{array}\right\} \tag{B.5}$$

because the determinate of the coefficients is not zero, that is,

$$\begin{vmatrix} 1 & 2 \\ 1 & 5 \end{vmatrix} = 3, \qquad \text{not } 0 \tag{B.6}$$

If Eq. (B.6) is zero, we can choose different values of z. Thus, the coefficients of z^0 are equal, $a_0 = b_0$, and the coefficients of z^1 are equal, $a_1 = b_1$. The value of b_0 depends only on a_0; the value of b_1 depends only on a_1; thus, b_0 and b_1 are independent. We can use the same procedure to extend the proof to the nth-order polynomial and obtain $b_j = a_j$ and b_i independent of b_j.

Polynomial multiplication demonstrates an application. The product of the generating functions $A(z)$ and $B(z)$ gives $C(z)$,

$$C(z) = A(z)B(z) \tag{B.7}$$

Again choosing short polynomials for an example, we have

$$\left.\begin{array}{l} A(z) = a_0 + a_1 z \\ A(z) = b_0 + b_1 z + b_2 z^2 \end{array}\right\} \tag{B.8}$$

The product gives

$$C(z) = a_0 b_0 + (a_0 b_1 + a_1 b_0)z + (a_0 b_2 + a_1 b_1)z + a_1 b_2 z^3$$

$$C(z) = c_0 + c_1 z + \cdots$$

$$c_0 = a_0 b_0$$

$$c_1 = a_0 b_1 + a_1 b_0$$ $\qquad$ (B.9)

$$c_2 = a_0 b_2 + a_1 b_1$$

$$c_3 = a_1 b_2$$

where the coefficients c_j are the result of equating the coefficients of like powers of z. All coefficients of $C(z)$ for n greater than 3 are zero because the highest power of z on the right side is 3.

B1.2 ALTERNATIVE FORMS OF CONVOLUTION EXPRESSIONS

The purpose of this section is to show the relation of our derivation of the convolution to other ways of expressing the convolution operation. The derivation is heuristic.

We begin by generalizing the z-transforms to include negative exponents, z^{-j}. Since z^j gives time steps in the positive time direction from $t = 0$, z^{-j} gives time steps in the negative time direction from $t = 0$. Assuming that all the series converge, we write, using infinite limits,

$$E(z) = \sum_{i=-\infty}^{\infty} e_i z^i \qquad (B.10)$$

$$A(z) = \sum_{j=-\infty}^{\infty} a_j z^j \qquad (B.11)$$

$$C(z) = \sum_{n=-\infty}^{\infty} c_n z^n \qquad (B.12)$$

When we move the summation signs next to each other, the convolution, Eqs. (6.30) or (6.31), gives

$$\sum_{n-\infty}^{\infty} c_n z^n = \sum_{i=-\infty}^{\infty} \sum_{j=-\infty}^{\infty} e_i a_j z^{i+j} \qquad (B.13)$$

We let

$$n = i + j$$

$$\text{or} \qquad\qquad\qquad (B.14)$$

$$i = n - j$$

and change the summation on i to one on n. Since the limits are $-\infty$ to $+\infty$, the infinite limits remain infinite, and

$$\sum_{n=-\infty}^{\infty} c_n z^n = \sum_{n=-\infty}^{\infty} \sum_{j=-\infty}^{\infty} e_{n-j} a_j z^n \qquad (B.15)$$

If we move the right side to the left side and take the common summation on n outside, Eq. (B.15) becomes

$$\sum_{n=-\infty}^{\infty} \left(c_n - \sum_{j=-\infty}^{\infty} e_{n-j} a_j \right) z^n = 0 \qquad (B.16)$$

This is true for any finite value of z; thus, the ( ) must be zero and

$$c_n = \sum_{j=-\infty}^{\infty} e_{n-j} a_j \qquad (B.17)$$

By replacing j with $n - i$ in Eq. (B.13), we could have obtained

$$c_n = \sum_{i=-\infty}^{\infty} e_i a_{n-i} \qquad (B.18)$$

These are common expressions of the convolution as a summation. The corresponding infinite integral expressions are

$$c(t) = \int_{-\infty}^{\infty} e(t - t') a(t') dt' \qquad (B.19)$$

or

$$c(t) = \int_{-\infty}^{\infty} e(t') a(t - t') dt' \qquad (B.20)$$

Our derivation of the convolution and expressions (6.31) and (6.39) follow directly from the addition of a signal at a set of time delays. Explanations of the meaning of Eqs. (B.17) to (B.20) are usually quite convoluted because the signals have to be reversed in time and then summed or integrated. A convolution algorithm follows.

```
rem Example for test follows:
dim  e(100),a(100),c(100)
n1= 3 : j1 = 10
a(0)=5: a(1)=3: a(2)=2 : a(3) =1
e(4)=1:  e(8)=-2

rem  convolution using Eq B.17

for n = 0 to n1 +j1
     s = 0
         for j = 0 to j1
         ea = 0
```

$$(B.21)$$

```
                    if n-j>=0 then ea=e(n-j)*a(j)
                        s = s + ea
                      next  j
                    c(n) = s
                  next n

                for n = 0 to j1 + n1
                    print n,c(n)
                  next n

                input q$
                end
```
(B.21)

where M1 is the number of terms of A(M), and N1 is the number of terms of E(N). The algorithm Eq. (B.21) does the same operations as Eqs. (6.45 and 6.46).

B1.2.1 Generating Functions and the Engineering z-Transform

The generating function $A(z)$ as given in Eq. (B.1) allows for an infinite number of choices of numbers that can be substituted for z. In the generating function context the coefficients are the sampled amplitudes at a sequence of times. A specific definition or method of choosing the numerical values of z leads to another dimension of application.

Choosing z to be a complex number

$$z \equiv e^{-iw}$$
(B.22)

and substituting in $A(z)$ gives

$$A(z) = a_0 + a_1 e^{-iw} + a_2 e^{-2iw} + \cdots + a_n e^{-niw}$$
(B.23)

The coefficients $a_0, a_1, \ldots$ still retain their identification as the signal amplitudes at 0, $t_0, 2t_0, \ldots$. We can include negative time steps and write a new $A(z)$,

$$A(z) = \cdots + a_{-1} e^{iw} + a_0 + a_1 e^{-iw} + \cdots$$

$$A(z) = \sum_{-\infty}^{\infty} a_n e^{-inw}$$
(B.24)

This is the form needed for comparison with the Fourier series. For comparison, we give standard versions of the Fourier series.

B1.2.2 Fourier Series and z-Transforms: Comparison

Fourier used the sums of sines and cosines having a harmonic set of frequencies to represent functions in his study of heat. Since then, Fourier methods have been applied to many problems, including both transient signals and periodic signals. The reader is referred to the numerous texts on Fourier theory for details. We use periodic signals in this section. A periodic function $f(t)$,

$$f(t) = f(t + nT), \qquad n = 0, \pm 1, \pm 2, \ldots \tag{B.25}$$

has the expansion in the complex form

$$f(t) = \sum_{m=-\infty}^{\infty} F_m \, e^{i2\pi mt/T} \tag{B.26}$$

$$F_m \equiv \frac{1}{T} \int_0^T f(t) \, e^{-2\pi imt/T} \, dt \tag{B.27}$$

The integral is over one period and can be shifted by any fraction of T for mathematical convenience. F_m has the same units as $f(t)$.

For digital data sets the integrations become finite summations. The time steps are t_0, so

$$t = n \, t_0 \tag{B.28}$$

$$dt \rightarrow t_0 \tag{B.29}$$

The period T becomes Nt_0, and

$$\frac{t}{T} = \frac{n}{N}$$
$$t_0 = \frac{T}{N} \tag{B.30}$$

The substitution of n/N for t/T in Eq. (B.27) and change of the integral to a summation gives

$$\left. \begin{array}{c} F_m \rightarrow \dfrac{1}{N} \displaystyle\sum_{n=0}^{N-1} f_n \, e^{-i2\pi nm/N} \\[2mm] f_n \equiv f(n \, t_0) \end{array} \right\} \tag{B.31}$$

The values of f_0 and f_N are equal because f_n has the period N.

It is customary to move the $1/N$ factor to the summation on F_m, Eq. (B.26), and write the transformation pair for finite data sets as follows:

$$f_n = \frac{1}{N} \sum_{m=0}^{N-1} F_m \, e^{i2\pi nm/N} \tag{B.32}$$

$$F_m = \sum_{n=0}^{N-1} f_n \, e^{-i2\pi nm/N} \tag{B.33}$$

Equations (B.32) and (B.33) are the standard forms of the finite or discrete Fourier transformation (FFT or DFT) and inverse discrete Fourier transformation (IFFT or IDFT). Most people use the fast Fourier transformation algorithm, given at the end of this section, to compute the DFT and IDFT.

Both f_n and F_m are periodic, and the summations can be shifted to

$$f_n = \frac{1}{N} \sum_{m=-N/2}^{N/2-1} F_m \, e^{i2\pi mn/N} \tag{B.34}$$

$$F_m = \sum_{n=-N/2}^{N/2-1} f_n\, e^{-i2\pi mn/N} \qquad (B.35)$$

Comparison

The identification of the z-transform as being a Fourier transform follows directly by comparing Eqs. (B.24) and (B.33) or (B.35) for a finite data set. We write $A(z)$ as

$$\left. \begin{aligned} A(z) &= \sum_{n=-N/2}^{N/2-1} a_n\, z^n \\[2mm] A(z) &= \sum_{n=-N/2}^{N/2-1} a_n\, e^{-inw} \end{aligned} \right\} \qquad (B.36)$$

let w be

$$w \equiv \frac{2\pi m}{N} \qquad (B.37)$$

and evaluate $A(z)$ for a particular m:

$$\left. \begin{aligned} A(z) &= \sum_{n=-N/2}^{N/2-1} a_n\, e^{-i2\pi mn/N} \\[2mm] A_m &\equiv A(z), \quad -\frac{N}{2} \le m < \frac{N}{2} \end{aligned} \right\} \qquad (B.38)$$

Thus, the definitions of $z = e^{-iw}$ and $w = 2\pi m/N$ change the generating function or z-transform into the Fourier transformation.

The choice of $z = e^{-iw}$ gives another context of meaning to $A(z)$. Since F_m is the amplitude of the harmonic component having the frequency mw, A_m has the same meaning. The numerical evaluation of summation for $z = e^{-iw}$ gives the frequency component. We call this use of the generating function the *engineering z-transform* because it came from research in electrical engineering.

The geophysical and engineering literatures differ on the definitions. The geophysical literature usually uses the form in Eq. (B.24). The electrical engineering literature usually uses a different definition of z, namely,

$$\left. \begin{aligned} z &\equiv e^{iw} \\[1mm] A_E(z) &= a_0 + a_1 z^{-1} + a_2 z^{-2} + \cdots + a_n z^{-n} \end{aligned} \right\} \qquad (B.39)$$

If we include negative time steps, $A_E(z)$ becomes

$$A_E(z) = \cdots + a_{-1} z^n + a_0 + a_1 z^{-n} + \cdots \qquad (B.40)$$

The particular choices of $z = e^{\pm iw}$ give dual meanings to $A(z)$. The coefficients a_j still retain their identifications as being signal amplitudes at the sequence of times. Evaluation of the summation gives the amplitudes of harmonic components as a function of frequency. Both interpretations are "legal" and coexist.

B1.2.4 Properties of Fourier Series

Fourier transformations are actually a pair that enable us to go from time space to frequency space and vice versa. Repeating the expressions, the pair of transformations are

$$f_n = \frac{1}{N} \sum_{m=-N/2}^{N/2-1} F_m \, e^{i2\pi mn/N} \tag{B.41}$$

$$F_m = \frac{1}{N} \sum_{n=-N/2}^{N/2-1} f_n \, e^{-i2\pi mn/N} \tag{B.42}$$

To evaluate f_n, we choose an n and evaluate the right side of Eq. (B.41). Full evaluation requires computing the sum for each value of n from $-N/2$ to $N/2 - 1$ or 0 to $N - 1$. Evaluation of F_m is the same using a sequence of values from

$$F_m = F_{m+kN} \tag{B.43}$$

where k is any positive or negative integer, and the summation on m from $-N/2$ to $N/2$ covers the complete range of m for any value of k. The periodicity of F_m requires a special interpretation of the meaning of m/T as being frequency. The range of $m = 0$ to $N/2$ gives positive frequencies, and $m = 0$ to $-N/2$ gives negative frequencies. For both, the maximum absolute value is $N/(2T)$. By the sampling theorem, this is also the maximum allowed frequency for any component of the signal. The components for $m > N/2$ do not represent higher frequencies, and $N/(2T)$ is the folding frequency.

Most IFFT routines (Eq. (B.33)) use the range of $m = 0$ to $N - 1$. The periodic equation gives values of F_m as follows:

$$\text{Compute } F_m \text{ for } -N/2 \leq m \leq N/2. \tag{B.44}$$

For m in the range $N/2 < m < N$, use

$$F_{N-m} = F_{-m} \qquad \text{for } 0 \leq m \leq N/2 \tag{B.45}$$

When the f_n are real, the reader can derive the following relationships:

$$F^*_{-m} = F_m \tag{B.46}$$

$$\text{real } F_{-m} = \text{real } F_m \tag{B.47}$$

$$-\text{imag } F_{-m} = \text{imag } F_m \tag{B.48}$$

The real parts fold symmetrically about $N/2$. The imaginary parts fold antisymmetrically about $N/2$.

Fast Fourier transformation algorithm

This algorithm is also known as the Cooley-Tukey algorithm. The BASIC version is adapted from a FORTRAN version. Fr(M) and Fi(m) are the real and imaginary components of the f_m and F_m in Eqs. (B.32) and (B.33). The arrays Fr(0,m) and Fi(0,m) hold the input. The dft or the fft gives outputs in Fr(0,M), Fi(0,m), Fr(1,M), and Fi(1,m). The ifft has output only in Fr(1,m) and Fi(1,m). The parameter sg instructs the program to do the dft when sg $= -1$ and the idft when sg $= 1$.

This version requires that the number of terms $n_t = 2^{n_1}$. When using the dft and idft to study transients, it is necessary to add a string of zeros after the signal to isolate the transient from periodic repetition at n_t. People usually add zeros to extend the signal to 2^{n_1} terms.

WARNING:

Fourier transformation (FFT or DFT) programs are in most scientific libraries. Users should check the particular algorithms. Many of these programs were written in old versions of FORTRAN when FORTRAN did not allow the array subscript 0. Then, programmers put the time step 0 or $n = 0$ in array subscript 1 or $f(1)$. The dft operation puts frequency 0 in array subscript 1 or F(1). Input data must be shifted one time step, and output frequencies must be shifted to $m - 1$. The DFT algorithms given here use array subscripts 0. Output frequency $= 0$ has $m = 0$.

```
rem finite or discrete Fourier transformations: BASIC versions

rem The transformations use complex signals.
rem The input is in Fr(0,m) + i * Fi(0,m), where i = sqr(-1).
rem Since our Basic uses real numbers, we carry the
rem imaginary components seperately.
rem The algorithm does direct and inverse Fourier transformations.
rem The dft algorithm is in a subroutine at the end.

rem The number of terms must be nt = 2^ne, where ne is an integer.
rem The arrays use m = 0 to nt-1.

        dim  Fr(1,512),  Fi(1,512)

rem Little main program for test

        nt = 8
        A$ = "dft"

        for m = 0 to nt
            Fr(0,m) = 0
            Fi(0,m) = 0
        next m
        Fr(0,0) = 1

rem Results are Fr(1,m) = 1 and Fi(1,m) = 0.

        gosub 1000
```

rem return and display

```
        for m = 0 to nt-1
                print m, Fr(1,m), Fi(1,m)
        next m

        input q$ : rem pause

        end
```

```
1000       rem dft or fft subroutine
rem  The real input is in Fr(0,m).
rem  The imaginary input is in Fi(0,m)
rem  The real output is in Fr(1,m).
rem  The imaginary output is in Fi(1,m)
rem  The number of terms is nt = 2^ne, where ne is an integer.
rem  sg is the sign of t in the computations of sin(t) and cos(t)
rem  sg = -1 gives the forward dft. sg = 1 gives the inverse idft.

        if A$ = "dft" then sg = -1
        if A$ = "idft" then sg = 1

rem  Since nt must = 2^ne, compute ne and then recompute nt.
        ne = int ( log(nt)/log(2) +.1 )
        nt = 2 ^ ne
        n2 = nt / 2
        pf = 8 * atn(1) / nt
        for j = 1 to ne
            n3 = nt / (2^j)
            n4 = (2^j) / 2

            for i = 0 to n4-1
                i2 = i * n3
                 t = i2 * pf * sg
                rw = cos (t)
                iw = sin (t)

                for k= 0 to n3 - 1
                    b= k + i2
                    b1 = k + 2 * i2
                    b2 = b1 + n3
                    b3 = b + n2
                        Fr(1,b)  =  Fr(0,b1)+rw*Fr(0,b2)-iw*Fi(0,b2)
                        Fi(1,b)  =  Fi(0,b1)+iw*Fr(0,b2)+rw*Fi(0,b2)
```

```
                    Fr(1,b3)  =  Fr(0,b1)-rw*Fr(0,b2)+iw*Fi(0,b2)
                    Fi(1,b3)  =  Fi(0,b1)-iw*Fr(0,b2)-rw*Fi(0,b2)
                next k
            next i

            for m = 0 to nt-1
                    Fr(0,m) = Fr(1,m)
                    Fi(0,m) = Fi(1,m)
            next m
        next j
```

rem The Fourier transformation is finished.
rem By convention, the idft has the factor 1/nt.

```
            if sg > 0 then
              for m = 0 to nt-1
                    Fr(1,m) = Fr(0,m) / nt
                    Fi(1,m) = Fi(0,m) / nt
              next m
              else
```
rem   For idft only Fr(1,m) and Fi(1,m) have idft coefficients.
rem   return to main part
```
              end if


            return
```

c   B1.2.4 main dft. FORTRAN 77

c   main dft.....discrete Fourier transform using common array
c   dft is also known as fft....fast Fourier transform
c   sg=-1 gives dft. sg=1 gives inverse dft
c   The dft sub uses subscripts 0 to nt. nt is 2**ne.
c   The same sub routine does dft (forward) and idft (inverse)
c   Common instruction carries data to sub and back.

```
     complex  f(0:4094)
     integer sg, ne, nt, i
     character *4q
     common sg,ne,nt,f
```

c   input dft numbers

```
10   write (9,*) 'input ne for 2^ne, ne='
     read (9,*) ne
     nt= 2**ne
```

```
      write  (9,*)  'nt=',nt
      write  (9,*)  'if ok enter "y" '
      read (9,*) q
      if ( q.eq.'y') then
         go to 20
      else
         go to 10

      endif
20    continue

      write (9,*) 'dft or idft input either'
      read (9,*) q

      if (q.eq.'dft') then
         sg=-1
      else
         sg=1
      endif


c     enter real and imaginary components

      write (9,*) 'enter real, imag'
      do 30 i=0 , nt-1
         write (9,*) 'i=',i
         read (9,*) f(i)
30    continue


c     do discrete Fourier transform
      call dft


c     print output

      write  (9,*) q
      write  (9,*) 'i   real    imag'

      do 40 i=0 ,nt-1
         write (9,*) i,f(i)
40    continue
      pause

      end
```

```
c       B1.2.4 dft sub FORTRAN 77
        subroutine dft

c       Complex array f(m) initially holds input;after execution
c       it returns output.sg=1 for IFFT ; sg=-1 for FFT.
        complex  f(0:4094),f1(0:4094),w
        real pf,t
        integer  sg,b,b1,b2,b3,ne,nt,i,i2,j,k,m,n2,n3,n4
        common sg,ne,nt,f
        nt=2**ne
        n2=nt/2
        pf=8*atan(1.0)/nt
        do 1 j=1,ne
            n3=nt/(2**j)
            n4=(2**j)/2
          do 2 i=0,n4-1
            i2=i*n3
            t=i2*pf*sg
            w=cmplx(cos(t),sin(t))
            do 2 k=0,n3-1
               b=k+i2
                b1=k+2*i2
               b2=b1+n3
               b3=b+n2
                 f1(b)=f(b1)+w*f(b2)
                 f1(b3)=f(b1)-w*f(b2)
2       continue
          do 1 m=0,nt-1
            f(m)=f1(m)
1       continue
        if (sg.lt.0) goto 4
          do 3 m=0,nt-1
             f(m)=f(m)/nt
3       continue
4       continue
        return
        end
```

B1.3 MULTIPLE REFLECTIONS AT A THIN LAYER

Thin layer means that the layer is thin relative to the distances to the source or receiver. The solution uses the place-wave approximation and ignores the geometric spreading of ray paths. The solution is easy to continue or generalize into the plane-wave reflection coefficient for the multi-layered half-space.

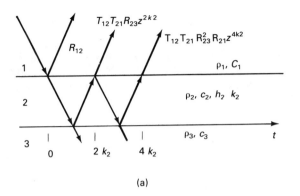

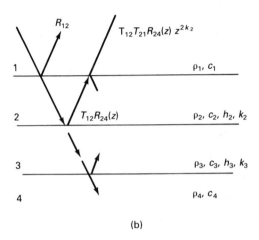

Figure B.1 Constructions for reflection at a thin layer. (a) The vertical axis is depth, and the horizontal axis is time steps. The plane waves are incident vertically. (b) Reflection from four layers.

The geometry is shown in Figure B.1. The receiver is in medium 1 and just above the first interface. For simplicity, the plane-wave front is vertically incident. To indicate time delays and the multiple reflection paths, the horizontal axis is time, and the vertical axis is depth. In Figure B.1 the up-going waves combine as a z-transform. The sum of all up-traveling wave fronts is $R_{13}(z)$,

$$R_{13}(z) = R_{12} + T_{12}T_{21}R_{23}\,z^{2k_2} + T_{12}T_{21}R_{23}^2R_{21}\,z^{4k_2} + \cdots \tag{B.49}$$

where the two-way travel time in layer 2 is $2k_2$, and R_{12}, R_{23}, T_{12}, T_{21}, and R_{21} are the reflection and transmission coefficients, Eqs. (5.11) to (5.16). The infinite series is, using $R_{12} = -R_{21}$,

$$R_{13}(z) = R_{12} + T_{12}T_{21}R_{23}\,z^{2k_2}\sum_{n=0}^{\infty}(-R_{12}R_{23}\,z^{2k_2})^n \tag{B.50}$$

We use $T_{12}T_{21} = 1 - R_{12}^2$ and the sum of the infinite geometric series to reduce Eq. (B.50) to

$$R_{13}(z) = \frac{R_{12} + R_{23}\,z^{2k_2}}{1 + R_{12}R_{23}\,z^{2k_2}} \tag{B.51}$$

where $|R_{12}R_{23}| < 1$. An algorithm for computing expressions such as Eq. (B.51) appears at the end of this section.

Extension to the multilayered half-space uses the procedure of starting at the bottom and working up layer by layer. We demonstrate the procedure by adding layer 3, which has the two-way travel time z^{2k_3} and ρ_3 and c_3, Figure B.1. At the *top* of the third layer, the reflection coefficient is

$$R_{24}(z) = \frac{R_{23} + R_{34}\, z^{2k_3}}{1 + R_{23}R_{34}\, z^{2k_3}} \tag{B.52}$$

$R_{24}(z)$ becomes the reflection coefficient at the bottom of layer 2, and $R_{14}(z)$ is

$$R_{14}(z) = \frac{R_{12} + R_{24}(z)\, z^{2k_2}}{1 + R_{12}R_{24}(z)\, z^{2k_2}} \tag{B.53}$$

$R_{14}(z)$ is the ratio of polynomials in z. Robinson and Treitel (1980, 295–320) give an equivalent result using matrix techniques.

An algorithm for computing Eq. (B.52) uses the recursive or feedback technique. We start by stating the algebraic problem first:

$$\sum_{i=0}^{\infty} c_i\, z^i = \frac{a_0 + a_1\, z^{2k}}{1 + b_1\, z^{2k}} \tag{B.54}$$

$$\sum_{i=0}^{\infty} c_i\, z^i + b_1 \sum_{i=0}^{\infty} c_i\, z^{i+2k} = a_0 + a_1\, z^{2k} \tag{B.55}$$

Solutions for like powers of z give

$$c_0 = a_0 \tag{B.56}$$

$$c_i = 0, \qquad 0 < i < 2k \tag{B.57}$$

$$c_{2k} = a_1 - b_1\, c_0 \tag{B.58}$$

$$c_i = -b_1\, c_{i-2k}, \qquad 2k < i \tag{B.59}$$

The only c_i that are not zero have the index $i = 2nk$, $n = 0, 1, \ldots$.

For the BASIC algorithm the parameters are

$$
\begin{aligned}
&\text{a0 = R12} \\
&\text{a1 = R23} \\
&\text{b1 = R12 * R23} \\
&\text{k = k2 :}
\end{aligned}
\tag{B.60}
$$

and the output c_i is R(i). Since all terms for i not equal to $2nk$ are zero, we reduce computational time by stepping the index at $2k$ steps. i_1 is the maximum number of time steps. Initialize the array R(i) = 0 if necessary.

$$
\begin{aligned}
&\text{k1 = 2 * k} \\
&\text{R(0) = a0} \\
&\text{R(k1) = a1-b1 * R(0) :}
\end{aligned}
\tag{B.61}
$$

```
for i k1 to i1 step k1
    R(i) = -b1 * R(i - k1)
next i :
```
(B.62)

We can do a number of trials by varying k_2, R_{12}, and R_{23}.

Ricker wavelet

Our examples show the convolution of the thin-layer reflection and the *Ricker wavelet*. This wavelet or signal is used in many synthetic seismograms because it is simple to compute and is a good approximation for seismic signals (Chapter 2, Problems 1 and 2). The Ricker wavelet is proportional to the second derivative of the normal function.

$$w(t) = -\frac{w_0^2}{2}\frac{d^2}{dt^2}e^{-t^2/w_0^2}$$
(B.63)

$$w(t) = \left(1 - \frac{2t^2}{w_0^2}\right)e^{-t^2/w_0^2}$$
(B.64)

$$w_0 = w_b/\sqrt{6}$$

where the parameter w_b is the wavelet breadth. The wavelet breadth is the separation of the wavelet valleys.

An algorithm for computing the Ricker wavelet follows. Let $t = i - i_0$ and w_0 be in time steps.

```
rem subroutine ricker wavelet
rem Wavelet breadth wb is the peak to peak time of the wavelet.
rem The program uses wb to calculate i0. w(i) has 2*i0 duration.

w0= wb/(sqr(6)*t0)
i0=int(4*w0)
nw=int(2*i0)
w2=w0*w0
print "i0=";i0

for i=0 to nw
    i2=(i-i0)^2
        w(i)=(1-2*i2/w2)*exp(-i2/w2)
next i

rem        return to main part of the calculation
```
(B.65)

Equation (B.63) has infinite tails. The algorithm terminates the tails at the width $2i_0$.

B1.4 POLYNOMIAL FEEDBACK ALGORITHM FOR ALGEBRAIC LONG DIVISION

Computations of the reflection coefficients for multiple layers and inverse filters often give an output that is the ratio of two polynomials in z. We begin by writing the recursive solution for a second-order polynomial in the denominator. The input signal is $A(z)$, and the output is $C(z)$,

$$\sum_{j=0}^{\infty} c_j z^j = \frac{\sum_{j=0}^{n_1} a_j z^j}{b_0 + b_1 z + b_2 z^2} \tag{B.66}$$

where the infinite limits mean carry as many terms as needed. Algebraic manipulation gives

$$b_0 \sum_{j=0}^{\infty} c_j z^j + b_1 \sum_{j=0}^{\infty} c_j z^{j+1} + b_2 \sum_{j=0}^{\infty} c_j z^{j+2} = \sum_{j=0}^{n_1} a_j z^j \tag{B.67}$$

The coefficients of like powers of z are equal on each side of Eq. (B.67) and are for $j \le n_1$

$$\left.\begin{array}{c} b_0 c_0 = a_0 \\ b_0 c_1 + b_1 c_0 = a_1 \\ b_0 c_2 + b_1 c_1 + b_2 c_0 = a_2 \\ \vdots \\ b_0 c_j + b_1 c_{j-1} + b_2 c_{j-2} = a_j, \quad j \le n_1 \end{array}\right\} \tag{B.68}$$

Notice that the sums of the subscripts of b_m and c_{j-m} are equal and that they equal the subscript of a_j on the right side. For $j > n_1$, the coefficients on the right side are zero, and

$$b_0 c_j + b_1 c_{j-1} + b_2 c_{j-2} = 0, \quad j > n_1 \tag{B.69}$$

The feedback solutions are for $j \le n_1$

$$c_0 = \frac{a_0}{b_0} \tag{B.70}$$

$$\left.\begin{array}{c} c_1 = \dfrac{a_1 - b_1 c_0}{b_0} \\[2mm] c_2 = \dfrac{(a_2 - b_1 c_1 - b_2 c_0)}{b_0} \\[2mm] c_j = \dfrac{a_j - b_1 c_{j-1} - b_2 c_{j-2}}{b_0} \end{array}\right\} \tag{B.71}$$

and for $j > n_1$

$$c_j = -\frac{b_1 c_{j-1} + b_2 c_{j-2}}{b_0} \tag{B.72}$$

The reader can generalize the computation to $B(z)$ having the order m_1,

$$c_0 = \frac{a_0}{b_0} \tag{B.73}$$

$$c_j = \frac{\left(a_j - \sum_{m=1}^{m_i} b_m c_{j-m}\right)}{b_0}, \qquad j \le n_1, j - m \ge 0 \tag{B.74}$$

$$c_j = \frac{-\left(\sum_{m=1}^{m_i} b_m c_{j-m}\right)}{b_0}, \qquad j > n_1, j - m \ge 0 \tag{B.75}$$

From Eq. (B.71) the conditional statement $j - m \ge 0$ yields $b_1 c_0$ for $j = 1$ and $m = 1$. The index $j = 2$ allows $m = 1$ and 2 and yields $b_1 c_1 + b_2 c_0$.

In writing the BASIC algorithm, we assume that the array elements A(J) are zero. The signals a_j are stored in a(j), and the filter coefficients b_m are in b(m). For the algorithm the maximum indices are m1 $= m_1$, n1 $= n_1$, and j1 $= j_1$. The choice of statement numbers is arbitrary.

```
rem   program to run polynomial division

rem for simplicity i enter data as defaults

      dim  a(100),b(100),c(100)
rem zero  arrays

      for m=0 to 100
         a(m) = 0
         b(m) = 0
         c(m) = 0
      next m

rem  data
      j1=6  :m1=2:  n1=4
      a(0)=1.0  :a(1)=-1.0:  a(2)=-5.0  :a(3)=17.0  :a(4)=-12.0
      b(0)=1.0  :b(1)=-3.0  :b(2)=4.0

rem  do polynomial division

      c(0)  = a(0)/b(0)
      for j  = 1 to j1
```

```
        s = 0.0
        for m =1 to m1
           bc = 0.0
100        if j >= m then
              bc = b(m)*c(j-m)
           end if
           s = s+bc
        next m
           c(j)=-s/b(0)
200        if j <= n1 then
              c(j)=c(j)+a(j)/b(0)
           end if
        next j


rem print results
     for j = 0 to j1
        print "c( ";j;")=";c(j)
     next j
     input q$ :rem pause
     end
```

Statement 100 causes the algorithm to compute $b(m) * c(j - m)$ only when $j \geq m$. This satisfies the conditions on the summation in Eq. (B.74). The conditional statement 200 causes the algorithm to compute Eq. (B.74). When $j \geq n_1$, the algorithm computes Eq. (B.75). We choose these statements because they avoid GOTO instructions. Incidentally, most computers give error messages in BASIC programs if a program jumps out of a FOR-NEXT loop.

We can test the algorithm and the program by computing a known $C(z)$, for example,

$$
\left.
\begin{aligned}
A(z) &= 1 - z - 5z^2 + 17z^2 - 12z^3 \\
B(z) &= 1 - 3z + 4z^2 \\
C(z) &= 1 + 2z - 3z^2
\end{aligned}
\right\}
\tag{B.76}
$$

The algorithm should give $c(j) = 0$ for all $j > 2$. Test it with $j1 = 10$.

```
c     main polynomial divide FORTRAN77

c     main poly div
c     main program to run polynomial division sub
c     for simple test, enter data as defaults

      integer  j,m,n,j1,m1,n1
      real  a(0:100),b(0:100),c(0:100)

c     data
      j1=6;  m1=2;  n1=4
```

```
        a(0)=1.0;  a(1)=-1.0;  a(2)=-5.0;  a(3)=17.0;  a(4)=-12.0
        b(0)=1.0;  b(1)=-3;  b(2)=4.0

c       call subroutine
        call  polydi(j1,m1,n1,a,b,c)

c       print results
        do 10 j=0,j1
            write(9,*) 'c( ',j,')=',c(j)
10      continue
        pause
        end


        subroutine  polydi(j1,m1,n1,a,b,c)
c       computes  c(z)=a(z)/b(z)  for a(z) of order n1, b(z)
c            of order m1, and c(z) of order j1.
c            b(z) should have nonzero constant.

        integer  j,m,n,j1,m1,n1
        real  a(0:n1),b(0:m1),c(0:j1),s,bc

        c(0)=a(0)/b(0)
        do 1  j=1,j1
            s=0.0
            do 2 m=1,m1
                bc=0.0
                if (j.ge.m) then
                    bc=b(m)*c(j-m)
                endif
                s=s+bc
2           continue
                c(j)=-s/b(0)
                if (j.le.n1) then
                    c(j)=c(j)+a(j)/b(0)
                endif
1       continue
        return
        end
```

B1.5 WIENER PREDICTION FILTER

The prediction filter is one of a general class of least-squares minimum error filters studied by Norbert Wiener (1894–1964). The prediction-error filter is powerful in its ability to simplify geophysical data and is particularly useful for the reduction of reverberation effects. G. P. Wadsworth, et al. (1953) gave the first applications of Wiener's filter theory

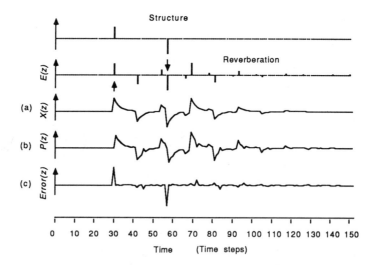

Figure B.2 Prediction filtering. The reflection signals are shown for an impulsive decaying exponential source. (a) $E(z)$ for no reverberation. (b) The input signal x_1 including reverberation. (c) The predicted signal p_i. (d) The output of the prediction filter *Error* (z). The error of the prediction represents the unexpected reflections.

to seismic data. Enders A. Robinson (1967; 1969) gives a tutorial discussion of digital seismic data processing. Robinson is the author and coauthor of many books and papers that give the mathematical details.

An example of the operation of a prediction-error filter is shown in Figure B.2. The signal consists of reflections and reverberations in a surface layer. The reflections are indicated by the arrows, and all other arrivals are reverberations of the reflection signals. Although the details of the filter are given later, it suffices here to state that the filter length of 15 time steps is longer than the reverberation interval of 12 time steps. The filter predicts the signal one time step ahead. The signal is the top trace. Before the arrival of the signal, the filter predicts zero, so that the difference (c) between the actual signal (a) and predicted signal (b) is zero. When the signal arrives, the filter predicts zero, and the signal is large, so the error is large. This indicates the presence of an unpredicted event, the reflection. The prediction error (c) is large at the reflection travel time. Continuing to step along in time, the filter uses the reflection signal to predict the arrival of the first reverberation signal (1) at 42 time steps on trace (b). The difference between the actual and predicted arrival is small. The reverberation arrival is approximately removed. The next reverberation arrival is at 54 time steps, which adds to the reflection at 57 time steps. The prediction filter predicts the reverberation arrival (2) on trace (b) and not the reflection at 57 time steps. The prediction error is large at the arrival time of the reflection at 57 time steps. After this, the prediction filter correctly predicts the reverberation arrivals, and the errors are small.

The least squares filter techniques are very powerful because we do not have to know or guess the inverse filters that are needed to deconvolve the signals. As will be shown in the derivations, the signal is used to compute the filter coefficients that deconvolve the signal to the desired form with minimum error. For simplicity we give the derivation of filter coefficients for the prediction-error filter. General filter derivations can be found in the literature references at the end of Unit 2.

Least squares prediction-error filter

To establish notation the z-transforms are input signal $X(z)$, filter $F(z)$, and output $Y(z)$. The convolution is

$$Y(z) = F(z) X(z) \tag{B.77}$$

Using Eq. (B.18) as a model, we write the convolution in the form

$$y_n = \sum_0^{j_1} f_j x_{n-j} \tag{B.78}$$

for $j_1 + 1$ filter coefficients. For prediction we want to choose a set of f_j and $x_{n-j}, \ldots ,$ x_n to predict a future estimate of x_{n+a} so that the errors are a minimum. The least squares procedure enables us to do this. Labeling the predicted value of x_{n+a} as p_{n+a} and using Eq. (B.18), we obtain the convolution equation,

$$p_{n+a} = \sum_{j=0}^{j_1} f_j x_{n-j} \tag{B.79}$$

where a is the number of future time steps.

The least squares procedure is convenient for determining f_j. We start by using the n_1 terms of the time series that are available. Assuming that the statistical behavior of the time series remains the same, we can use past behavior to make predictions. The first steps are to form $(x_{n+a} - p_{n+a})^2$ and sum over n,

$$I = \sum_{n=j1}^{n_1} (x_{n+a} - p_{n+a})^2 \tag{B.80}$$

where n_1 is much larger than j_1. Our task is to choose the set of f_j so that I is a minimum. Recall from calculus that the value of x for the minimum or maximum of a function $y = f(x)$ is determined by solving $dy/dx = 0$. Here we temporarily regard the f_j as being variables and use the partial derivatives to calculate $\partial I/\partial f_i = 0$

$$\frac{\partial I}{\partial f_i} = -2 \sum_{n=j1}^{n_1} (x_{n+a} - p_{n+a}) \frac{\partial p_n}{\partial f_i}$$

$$\frac{\partial I}{\partial f_i} = 0 \tag{B.81}$$

If we substitute Eq. (B.79) in Eq. (B.81) and use Eq. (B.79) to replace p_{n+a}, we have

$$\sum_{n=j_1}^{n_1} x_{n-i} x_{n+a} - \sum_{j=0}^{j_1} f_j \sum_{n=j_1}^{n_1} x_{n-j} x_{n-i} = 0 \tag{B.82}$$

We simplify the summations by noticing that both have the same form and are strictly properties of the time series x_n. Using the first summation as an example, we let $n' = n - i$ and

$$\sum_{n=j_1}^{n_1} x_{n-i} x_{n+a} = \sum_{n'=j_1-i}^{n_1-i} x_{n'} x_{n'+i+a} \tag{B.83}$$

We use this to define the variance r_{i+a},

$$r_{i+a} \equiv \frac{1}{n_1} \sum_{n=j_1-i}^{n_1-i} x_n x_{n+i+a} \tag{B.84}$$

Similarly, the other summation becomes

$$r_{i-j} \equiv \frac{1}{n_1} \sum_{n=j_1-i}^{n_1-i} x_n x_{n+i-j} \tag{B.85}$$

where i and j range from 0 to j_1. Equation (B.82) becomes

$$\sum_{j=0}^{j_i} f_j r_{i-j} = r_{i+a} \tag{B.86}$$

Equation (B.86) is a set of equations,

$$\left.\begin{aligned}
f_0 r_{0-0} + f_1 r_{0-1} + f_2 r_{0-2} + \cdots + f_{j_1} r_{0-j_1} &= r_{0+a} \\
f_0 r_{1-0} + f_1 r_{1-1} + \cdots + f_{j_1} r_{1-j_1} &= r_{1+a} \\
\vdots \\
f_0 r_{f_j-0} + \cdots + f_j r_{j_1-j_1} &= r_{j_1+a}
\end{aligned}\right\} \tag{B.87}$$

The next simplification involves noticing in Eq. (B.84) that the index i slides the summation along the time series, and the $(i - j)$ combination gives the displacement of x_{n+i-j} relative to x_n. Thus, for $j_1 \ll n_1$, we would expect $r_{0-0}, r_{1-1}, r_{2-2}, \ldots$ to have the same values. Similarly, r_{0-1} has the same values as $r_{1-2}, r_{2-3}, \ldots$. We can interchange i and j in Eq. (B.85) and after manipulation show that $r_{i-j} = r_{j-i}$. If we use these simplifications, Eq. (B.87) becomes

$$\left.\begin{aligned}
f_0 r_0 + f_1 r_1 + f_2 r_2 + \cdots + f_{j_1} r_{j_1} &= r_a \\
f_0 r_1 + f_1 r_0 + f_2 r_1 + \cdots + f_{j_1} r_{j_1-1} &= r_{a+1} \\
\vdots \\
f_0 r_{j_1} + f_1 r_{j_1-1} + \cdots + f_{j_1} r_0 &= r_{j_{1+a}}
\end{aligned}\right\} \tag{B.88}$$

The left side has $j_1 + 1$ unknown coefficients f_i and $j_1 + 1$ rearrangements of the j_1 variances r_i. Mathematicians call this the *Toeplitz form*, and Levinson's algorithm is an efficient way to solve for f_i (E. A. Robinson and S. Treitel, 1980, 163–69). A BASIC version of the algorithm, EUREKA, appears at the end of this section.

In comparing our special case of prediction with general optimum filter theory in the literature, the reader will find the terms on the right side, $r_a, r_{a+1}, \ldots, r_{a+j_1}$, replaced by $g_0, g_1, \ldots, g_{j_1}$. Notations vary in the literature. The g_i come from the solution of the general problem in which g_i is the cross correlation of a "desired output" d_n and the signal x_{n-i},

$$g_i = \frac{1}{n_1} \sum_{n=0}^{n_1} d_n x_{n-i}, \qquad n \geq i \qquad (\text{B.89})$$

As an example, if we desire the output to be the impulse response, then we set $g_0 = 1$ and all other $g_i = 0$. The EUREKA algorithm is written for the general case, and we choose to let $g_i = r_{i+a}$ in our examples.

Now we are ready to do what we mentioned in the first paragraphs, to make a prediction filter. Assuming prediction errors are due to the *nonpredictable* presence of new arrivals, the error indicates a signal. The output of the prediction filter is

$$e_{n+a} = x_{n+a} - p_{n+a} \qquad (\text{B.90})$$

where p_{n+a} uses the $x_{n-j_1}, \ldots, x_n$ samples of the signal to predict the value at $n + a$. Substitution of p_{n+a} using Eq. (B.79) gives

$$e_{n+a} = x_{n+a} - f_0 x_n - f_1 x_{n-1} - \cdots - f_{j_1} x_{n-j_1} \qquad (\text{B.91})$$

The "error" e_{n+a} is the prediction filtered signal. A prediction algorithm is given at the end of this section. The usefulness of the prediction filter depends on how well it works.

Figure B.3 shows the operation of a prediction filter. The structure consists of a reverberating layer over two reflecting layers, Figure B.2a. Choosing the filter length j_1 and the prediction time steps a is an art. It is hoped that the prediction is not too sensitive to the exact choice and the signal. Figure B.4 shows a set of output signals for different filter lengths and prediction time steps. The input signal and correlation function are the same as in Figure B.3. It is easy to identify the reflection interfaces and the relative magnitudes of the reflections. These comments are a minimal introduction to a very important field of practice in geophysics.

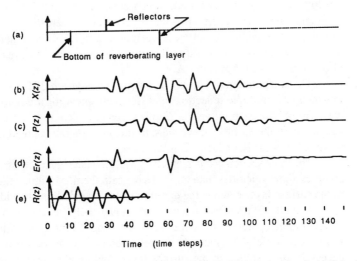

Figure B.3 Prediction filter operation for a Ricker wavelet input signal. (a) Seismic structure or $E(z)$. Depths are two-way travel times in time steps ($k = h = 6$). (b) Reflected signal without multiple reflections. Reverberation in layer 1 is included. (c) Predicted output, p_i. (d) Error, or "signal out," e_i. (e) Correlation, r_i. The filter length is $j_1 = 17$ time steps and the filter predicts $a = 5$ time steps ahead.

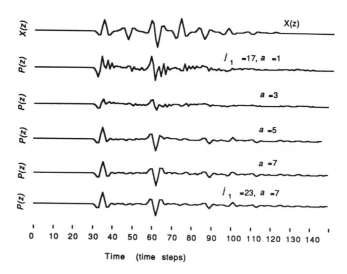

Figure B.4 Prediction filter operations for various choices of the prediction parameter *a*. The signals are the same as in Figure B.3.

B1.5.1 Seismic-Signal Utility Programs

The steps for constructing synthetic seismograms that include reverberation are numerous, and the corresponding programs tend to be long and complicated. The deconvolution programs are equally complicated. A set of subroutines or utility programs that can be put together are more useful and much easier to understand. Besides, they can be put together in many ways. The write- and read-file utilities are the glue that holds the sequences of operations together. In all these operations graphs of the signals are used at each step to follow the operations. Utility subroutines such as make file, read file, and graph appear in Section A1.4 of Appendix A.

The computation of $E(z)$ is a first step. The program E(Z) does a reflection calculation in which both multiple reflections and spherical spreading are ignored. This gives a plane-wave result. A base program for $E(z)$ is given later. These computations are for a velocity geophone on the surface. If we want to include spherical spreading, the spreading of the wavefronts is in 'RAYTRACE.AMP', A1.7.

The thick section is represented by $E(z)$. Surface layers commonly have much smaller compressional velocities than the rocks beneath them. Multiple reflections or reverberation in the surface layer extends the duration of the signal. We can add write-file instructions.

A synthetic reverberation calculation is usually treated as an operation on $E(z)$. The reverberation operation uses the algebraic long-division algorithm, Section B1.4. We would read a file of $E(z)$ and do the reverberation. One pass through the reverberation is necessary for a source in the surface layer, and another pass is necessary for a receiver in the surface layer. The reverberation utility has a single pass. We can use the write-file instruction to store the results for the next operations. Convolution with a signal is next. This utility uses the Ricker wavelet for the signal. We use a read file to enter $E(z)$ and write file to output the signal $S(j)$.

For examples and tests we include the program REVERBERATION. It has data in it and it produces a signal or time series file. The file is intended to be read by the prediction filter program WIENER. To assist the reader in testing their code, the program TEST EUREKA is included.

The prediction-filter computation is in WIENER FILTER. After reading the signal file $X(n)$, autocovariance computes the set of $R(j)$. The set of $R(j)$ enter the EUREKA program. We can choose the length of the filter and predict steps pa. The desired output $G(j)$ is set to $R(j + pa)$. EUREKA gives a set of filter coefficients $F(n)$ for the last steps.

The last step uses the filter coefficients $F(n)$ and the signal $X(n)$. Convolution of $F(j)X(n - j)$ gives the predicted signal $P(m)$. The error $ER(n)$ is $X(n) - P(n)$. The "error" is the deconvolved signal. The graphic section plots the input signal, predicted signal, the error signal or deconvolved signal, and the autocorrelation.

```
rem  Computes E(z) at vertical incidence for Chap 6, prob 5.
rem  Input layer parameters for each layer.
rem  Ignore spherical spreading for simplicity.
rem  Ignore all multiple reflections.
rem  Although not included the program is intended to write a file.

     dim  E(1000), R(10), T(10), nt(10), h(10), d(10), c(10)

rem  Data from Blaik, Northrop, and Clay,
rem  J. Geoph. Res.64,231-239 (1959). Data are mks units.

       im = 5   : rem number of layers
     h(1) = 90  :     d(1) = 2000 :    c(1) = 1620
     h(2) = 150 :     d(2) = 2200 :    c(2) = 1740
     h(3) = 270 :     d(3) = 2300 :    c(3) = 1950
     h(4) = 300 :     d(4) = 2400 :    c(4) = 2100
     h(5) = 99999:    d(4) = 3000 :     c(4) = 5000

     t0 = 0.005 :rem sampling interval

rem  Compute the reflection and two-way transmissions.
rem  Use velocity coefficients.

     for i = 2 to im
        a = d(i) * c(i)
        b = d(i-1) * c(i-1)
        R(i) = (b-a) / (b+a)
     next i

     T(1) = 1
     for i = 2 to im - 1
        T(i) = 1 - R(i) * R(i)
     next i
```

```
rem reflection times in time steps

    sum = 0
    for i = 1 to im - 1
        sum = sum + int(2 * h(i) / (c(i) * t0) +0.5)
        nt(i) = sum
    next i

rem calculate E(z)

    for i = 0 to 400
        E(i) = 0
    next i

rem Reflection for interface i
rem Note   velocity at surface = 2*incident velocity
    for i = 1 to im
        p = 1 : rem inner loop for amplitudes
        for j = 1 to i
            p = p * T(j)
        next j
        E(nt(i)) = 2 * p * R(i +1)
    next i

rem Output

    print  tab(3);"i";tab(6);"nt(i),  ts";tab(16);"E(i)"
    print
    for i = 1 to im - 1
        print i;tab(6);nt(i);tab(16);E(nt(i))
    next i
    print  tab(6);"t0  =  ";t0

    input q$ :rem pause

    end
```

'B1.5.1 Reverberation BASIC

```
REM  ASSUME THE SIGNAL IS INCIDENT ON A THIN LAYER AT THE SURFACE.
REM  THE LAYER HAS PARAMETERS HK, DK, AND CK.
REM  THE LAYER IS OVER A LAYER D(1) AND C(1).
REM  THE IMPULSE RESPONSE OF THE UPWARD TRAVELING SIGNAL IS E(Z)
REM  ALL OF THE UPWARD TRAVELING SIGNAL IS ASSUMED TO REVERBERATE
```

```
REM  IN THE LAYER. PROGRAM HAS A MENU FOR CHOICE OF WAVELET
REM  PARAMETER, GRAPHING, AND FILING.SCREEN IS 492 BY 300 PIXELS

     DIM  E(1100),CE(1100),X(1100),W(100)
     X0%=20 :Y0%=20: XL%= 480: YL%=280

REM  CLEAN ARRAYS

     FOR i = 0 TO 1100
        E(i) = 0
        CE(i) = 0
        X(i) = 0
     NEXT i

10   PRINT " Enter t0 = ";:INPUT t0

REM  CHOOSE E(Z) FROM REFLECTED SIGNAL CALCULATION.
REM  USUALLY ONE WOULD READ A FILE FOR E(Z)

REM  DEFAULT FOR DEMONSTRATION. HK = THICHKNESS,M. DK = DENSITY,
REM  CK = SOUND VELOCITY. t0 = SAMPLING INTERVAL

REM  DEFAULT E(Z) EXAMPLE
     HK = 50:DK = 1:CK = 1500
     D(1) = 2:C(1) = 1800
REM  CALCULATE TIME STEPS FOR DEPTHS OF REFLECTORS

REM  TOP LAYER
     K2 =  INT (2 * HK / (CK * T0) + .5): REM   K2=2*K
     R0 = 1: REM  FREE SURFACE VELOCITY REFLECTION
     A = D(1) *  C(1)
     B = DK * CK
     R1 = (B - A) / (B + A)
     T1 = 2 * D(1) * C(1) / (B + A)
     BR = - R0 * R1
REM  ONE GETS A REVERBERATION CONTRIBUTION EVERY K2 TIME STEPS

REM  LOWER LAYERS, FOR SIMPLICITY ASSUME AVERAGE CA=2000 m/s.
     CA = 2000
     CT=CA * t0
     ND1 = INT(2 *  500/CT)
     ND2 = INT(2 *  800/CT)
     ND3 = INT(2 *  1400/CT)
     im = ND3 + 4 * K2
     nt = ND3 + 2 * K2
```

```
        PRINT " max time steps = ";im
        IF im  > 1000 GOTO 10


        E(0) = 1:E(K2) = R1: E(ND1) = -.294:E(ND2) =.357:  E(ND3) = -.325

REM        DO THE REVERBERATION IN THE TOP LAYER
REM        USE POLYNOMIAL LONG DIVISION.

REM        CE(Z) = E(Z) * T1 / (1 - R0 * R1 * Z ^ 2 * K)
REM        CE(Z)=E(Z)*T1/(1+BR*Z^K2), SIMPLER NOTATION.
REM        USE THE SPECIAL FORM OF DIVISION FOR (1+BR*Z^K2)
REM        TO SPEED THE CALCULATION.

        FOR I = 0 TO K2 - 1
          CE(I) = E(I) * T1
        NEXT I

        FOR I = K2 TO IM + 2 * K2
          CE(I) = E(I) * T1 - BR * CE(I - K2)
        NEXT I

REM   END REVERB CALC


100  rem menu

        PRINT "enter g  for graphs"
        PRINT "        mf to make a file"
        PRINT "         c  for convolution with Ricker wavelet"
        PRINT "        so to start over"
        PRINT "         q  to quit"
        input q$
        if q$ = "g"  goto 2000
        if q$ = "mf" goto 3000
        if q$ = "c"  goto 1000
        if q$ = "so" goto 10
        if q$ = "q"  goto 4000

2000      REM DISPLAY

        YS = (YL%-Y0%) / 4
         TSC = (XL%-XO%) / (im+1)

        print"input amplitude factors for e(n) and ce(n) and x(n)";
        input eamp,ceamp,xamp

        emp = eamp * YS: cemp = YS * ceamp: xmp = xamp * YS
```

```
cls

for n = 0 to im
    T% = int (n * TSC +X0%)
    T1% = int ((n+1) * TSC +X0%)
    YP% = int (YS - emp * E(n))
    YP1% = int ( YS - emp * E(n+1))
    LINE (T%,YP%) - (T1%,YP1%)
next n

for n = 0 to im
    T% = int (n * TSC +X0%)
    T1% = int ((n+1) * TSC +X0%)
    YP% = int (2*YS - cemp * CE(n))
    YP1% = int (2*YS - cemp * CE(n+1))
    LINE (T%,YP%) - (T1%,YP1%)
next n

for n = 0 to im
    T% = int (n * TSC +X0%)
    T1% = int ((n+1) * TSC +X0%)
    YX% = int (3*YS - xmp * X(n))
    YX1% = int (3*YS - xmp * X(n+1))
    LINE (T%,YX%) - (T1%,YX1%)
next n
input Q$


GOTO 100


1000     REM CONVOLVE WITH RICKER WAVELET

rem subroutine ricker wavelet
rem Wavelet breadth wb is the peak to peak time of the wavelet.
rem The program uses wb to calculate i0. w(i) has 2*i0 duration.

  PRINT "Reflection time in top layer =";INT(1000*2*HK/CK)/1000
  PRINT " Input wavelet breadth in s = ";: INPUT wb

  w0= wb/(sqr(6)*t0)
  i0=int(4*w0)
  nw=int(2*i0)
  w2=w0*w0
  print  "i0=";i0
```

```
    for i=0 to nw
        i2=(i-i0)^2
            W(i)=(1-2*i2/w2)*exp(-i2/w2)
    next i

rem  convolve CE(z) * W(z)
    for m = 0 to nw
        for n = 0 to nt
        k = m + n
        X(k) = X(k) + W(m) * CE(n)
        next n
    next m

    GOTO 100
3000    REM     FILE MAKER

REM     Number of terms is im
REM     Data is in X(n)

    print"give file name":input n$
    open n$ for output as #1

    write #1, im

    FOR n = 0 TO im
        write #1, X(n)
    NEXT n

    close #1

    GOTO 100

4000    END



REM                 PRGM TEST EUREKA.CSC
REM         ROBINSON CALLED HIS PRGM TO DO THE INVERSION 'EUREKA'.
REM                 I LIKE THE NAME.

    DIM  R(50),G(50),GI(50),FZ(1,50),AZ(1,50),ID(50),TESTG(50)
    M = 0:N = 1: REM  ARRAYS 0 AND 1


REM   R(J)  HAS TOEPLITZ FORM.R(J) STARTS AT 0 TO LI.R(J) ARE
REM  AUTO CORRELATIONS. GI(J) ARE DESIRED OUTPUT. F(J) ARE.
REM   FILTER COEF. A(J) ARE SOLN OF A(J)*R(J,K)=(1,0,0,...)
REM         PRGM USES NORMAN LEVINSON'S RECURSIVE ALGORITHM.
```

```
REM       REF:J.F. CLAERBOUT 'FUNDIMENTALS OF DATA PROCESSING'
REM       P 53-57.BLACKWELL SCI. PUB.1985 AND E.A.ROBINSON
REM       & S.TREITEL 'GEOPHYSICAL SIGNAL ANALYSIS'
REM       P 163-169.PRENTICE-HALL INC. 1980
REM    I USE TRANSFER OF ARRAYS AZ(M,J) TO AZ(N,J) FOR EACH LOOP
REM       ON L. LAST STEP IS TO MOVE ALL ARRAYS FROM AZ(N,J) TO
REM  AZ(M,J) FOR NEXT LOOP ON L.


 PRINT "TEST OF EUREKA FOR INPUTS OF R(M) AND GI(M) "
 PRINT " R(M,N)*F(M)=G(M). TOEPLITZ FORM"
 PRINT "PRGM USES INPUT OF R(I,J) AND G(I) TO COMPUTE F(J)."
 PRINT "   IT COMPUTES A MATRIX A(IJ)=(R(I,J))^-1.
 PRINT "THE TEST COMPARES THE CALC G(M) AND INITIAL GI(M)"
 PRINT
 PRINT "INPUT DATA FOR FILTER COMPUTION"
 PRINT  TAB( 5);"INDICES 0 TO LI. LI=";: INPUT LI

     PRINT "INPUT G(J)
     FOR J = 0 TO LI
        PRINT J;:INPUT G(J)
        GI(J) = G(J)
     NEXT J
     PRINT "INPUT R(J)
     FOR J=0 TO LI
        PRINT J;:INPUT R(J)
     NEXT J

     GOSUB 1000: REM  DO SUB EUREKA

REM       OUTPUTS

     PRINT " L       A(L)      F(L)/F(0),F(0)=";FZ(M,0)

     FOR L = 0 TO LI
         PRINT L;TAB(6);AZ(M,L);TAB( 21);FZ(M,L)/FZ(M,0)
     NEXT L
     INPUT Q$ : REM PAUSE
     PRINT
     PRINT "CHECK A(I,J)*R(I,J) SHOULD BE (1,0,...)"
     PRINT
     PRINT " J      A(I,J)*R(I,J)"

     V = 0
     FOR J = 0 TO LI
     V = V + AZ(M,J) * R(J)
   NEXT J
```

```
FOR L = 0 TO LI
   S = 0
     FOR I = 0 TO LI
        K =  ABS (L - I)
         S = S + AZ(M,I) * R(K)
     NEXT I
     ID(L) = S/V
   PRINT L; TAB( 6);ID(L)
NEXT L
INPUT Q$
PRINT
PRINT "CHECK F(L). R(MN)*F(N)=G(N)"
PRINT
PRINT " L   TEST G(L)          ORIG G(L)"

FOR L = 0 TO LI
   S = 0
     FOR I = 0 TO LI
        K =  ABS (L - I)
         S = S + FZ(M,I) * R(K)
     NEXT I
     TESTG(L) = S
   PRINT L; TAB( 6);TESTG(L); TAB( 23);GI(L)
NEXT L
INPUT Q$
GOTO 3000


1000      REM  PRGM EUREKA.CSC

REM       L=0

   AZ(M,0)  =  1
   FZ(M,0)  =  G(0) / R(0)
   AL(M)  =  R(0)
   GM(M)  =  FZ(M,0) *  R(1)
   BT(M)  =  R(1)
   K(M)  =   - BT(M) / AL(M)

FOR L = 1 TO LI
   AZ(N,0)  =  AZ(M,O)

      FOR J = 1 TO L - 1
          AZ(N,J)  =  AZ(M,J)+K(M)*AZ(M,L-J)
      NEXT J
```

```
        AZ(N,L) = K(M) * AZ(M,0)
        AL(N) = AL(M) + K(M) * BT(M)
      REM  FOR BETA
        BT(N) = 0

      FOR J = 0 TO L
            BT(N) = BT(N)+AZ(N,J)*R(L+1-J)
      NEXT J

        K(N) =  - BT(N) / AL(N)
        Q = (G(L) - GM(M)) / AL(N)

      FOR J = 0 TO L - 1
            FZ(N,J) = FZ(M,J)+Q * AZ(N,L-J)
      NEXT J

        FZ(N,L) = Q * AZ(N,0)
REM          GAMMA
        GM(N) = 0

      FOR J = 0 TO L
          GM(N) = GM(N) + FZ(N,J) * R(L+1-J)
      NEXT J

REM        SLIDE COEFFICIENTS

      FOR J = 0 TO L
            AZ(M,J) = AZ(N,J)
            FZ(M,J) = FZ(N,J)
      NEXT J

REM        SLIDE
        AL(M) = AL(N)
        BT(M) = BT(N)
        K(M) = K(N)
        GM(M) = GM(N)
REM        REPEAT FOR NEW L

    NEXT L

    RETURN : REM  RETURN TO DISPLAY RESULTS

3000      REM  FILE MAKER

    print"give file name":input n$
    open n$ for output as #1
```

```
write #1, "Test of eureka: input R(m) and GI(m)"
write #1," L        A(L)        F(L)/F(0),F(0)=";FZ(M,0)

FOR L = 0 TO LI
     write#1,L,AZ(M,L),FZ(M,L) / FZ(M,0)
NEXT L

write #1,"CHECK A(I,J)*R(I,J)  SHOULD BE  (1,0,...)"
write #1," J        A(I,J)*R(I,J)"

FOR L = 0 TO LI
     write #1,L,ID(L)
NEXT L

write #1,"CHECK F(L).  R(MN)*F(N)=G(N)"
write #1," L   TEST G(L)        ORIG G(L)"

FOR L = 0 TO LI
     write #1,L,TESTG(L),GI(L)
NEXT L
close #1
END
```

Test of eureka: input R(j) and GI(j)

j	R(j)	GI(j)
0	1.	4
1	.7	-2
2	.4	7
3	-.3	-3
4	-.7	8

Output

j	A(j)	F(j)/F(0)
0	1	1
1	-3.099994	-1.372132
2	.9333293	-.4229499
3	2.566663	1.804918
4	-2.399994	-.8950827

$$F(0) = -89.70598$$

Check A(j)*R(j) should be (1,0,...)

j	A(j)*R(j)
0	1
1	5.259249E-07
2	5.259249E-07
3	0
4	0

Check F(j).   R(j)*F(j) = G(j)

j	TEST G(j)	ORIG G(j)
0	3.999996	4
1	-1.999985	-2
2	7.000015	7
3	-2.999977	-3
4	8.000015	8

```
c     B1.5.1 main eureka FORTRAN 77

c     prgm main eureka.csc
c     Data for test of subroutine are covariance r(j)
c     and input desired output g(j). Get filters f(j)

      real   r(0:100),g(0:100),fzm(0:100),azm(0:100)
      real   azn(0:100),fzn(0:100),gc(0:100),s,km,kn,v
      integer  i,j,li,k

      li=5
      r(0)=3.;r(1)=1.;r(2)=-1.;r(3)=0.;r(4)=.5;r(5)=-.2
      g(0)=-3.;g(1)=-2.;g(2)=1.;g(3)=4.;g(4)=2.;g(5)=1.

      call  eureka(li,r,g,fzm,azm)

      write (9,*) ' j        a(j)      f(j)/f(0)'

      do 10 j=0,li
          write(9,1000) j, azm(j),fzm(j)/fzm(0)
1000          format(i3,2x,f10.5,2x,f10.5)
10    continue

c     check  a(j)*r(j)=1
      do 20 j=0 ,li
          v=v+azm(j)*r(j)
```

```
20   continue

     write (9,1200) 'j','a(j)*r(j)    should be 1.0,0,...'
1200     format( a3,4x,a30)

     do 25 j=0,li
       s=0
       do 26 i=0,li
         k=abs(i-j)
         s=s+azm(i)*r(k)
26      continue
         write (9,1500) j,s/v
1500           format(i3,2x,f10.7)
25   continue

     do 30 j=0, li
       s=0
       do 40 i=0 ,li
         k=abs(j-i)
         s=s+fzm(i)*r(k)
40      continue
         gc(j)=s
30   continue

     write(9,*)' j      orig g(j)      calc g(j)'
     do 50 j=0,li
         write (9,2000) j,g(j),gc(j)
2000           format(i3,2x,f10.5,2x,f10.5)
50   continue
     pause
     end



     SUBROUTINE eureka(li,r,g,fzm,azm)

c    subroutine  eureka uses Norman Levinson's recursive algorithm.
c    r(m)f(m)=g(m)   Toeplitz form.
c    subroutine uses input of r(i,j) and g(i) to compute f(j).
c        it computes a matrix a(i,j)=(r(i,j))^-1.
c        fzm(j) contains filter coef. and
c        azm(j) the inverse matrix.
c    Memory usage is reduced by transfering arrays azm(j) to
c    azn(j)  and fzm(j) to fzn(j) for each loop on l.

     integer  li,i,j
     real   r(0:100),g(0:100),fzm(0:100),azm(0:100)
     real   azn(0:100),fzn(0:100),km,kn,gmm,gmn,btm,btn,alm,aln,q
```

```
c    initial conditions
     azm(0)=1.;fzm(0)=g(0)/r(0);alm=r(0)
     gmm=fzm(0)*r(1);btm=r(1);km=-btm/alm

     do 1 l=1,li
        azn(0)=azm(0)
        do 2 j=1,l-1
          azn(j)=azm(j)+km*azm(l-j)
2       continue
        azn(l)=km*azm(0)
        aln=alm+km*btm
c    for beta
        btn=0.
        do 3 j=0,l
          btn=btn+azn(j)*r(l+1-j)
3       continue
        kn=-btn/aln
        q=(g(l)-gmm)/aln
        do 4 j=0,l-1
          fzn(j)=fzm(j)+q*azn(l-j)
4       continue
        fzn(l)=q*azn(0)
c    for gamma
        gmn=0.
        do 5 j=0,l
          gmn=gmn+fzn(j)*r(l+1-j)
5       continue
c    slide coefficients
        do 6 j=0,l
          azm(j)=azn(j)
          fzm(j)=fzn(j)
6       continue
        alm=aln
        btm=btn
        km=kn
        gmm=gmn
c    repeat for new l
1    continue
     return
     end
```

'B1.5.1 Wiener filter BASIC

REM PROGRAM READS A FILE, COMPUTES AUTOCORRELATION, FILTER
REM COEFFICIENTS, AND THEN DOES WIENER'S OPTIMUM FILTER

REM OPERATION ON THE SIGNAL X(n). THE RESULT IS THE ERROR ER(n).
REM SCREEN IS 492 BY 300 PIXELS. USE THESE TO SET DISPLAY LIMITS.

```
DIM  X(1100),P(1100),ER(1100)
DIM  R(100),G(50),GI(50),FZ(50),FZ1(50),AZ(50),AZ1(50)
X0%=20 :Y0%=20: XL%= 480: YL%=280
```

REM ZERO ALL ARRAYS

```
for n = 0 to 1100
    X(n) = 0: P(n) = 0: ER(n) = 0
next n

for n= 0 to 100
    R(n) = 0
next n

for n = 0 to 50
    G(n) = 0: AZ(n) = 0: Az1(n) = 0
    GI(n) = 0: FZ(n) = 0: FZ1(n) = 0
next n
```

REM READ FILE

```
print"read signal file for input"
print"give file name":input n1$
open n1$ for input as #1
input #1, nt

FOR n = 0 TO nt
    input #1, X(n)
NEXT n
close #1
```

REM GOTO SUBROUTINE TO CALCULATE AUTOCORRELATION FUNCTION
```
    PRINT " INPUT MAXIMUM TIME STEP FOR AUTOCORRELATION = "
    INPUT j1
    GOSUB 2000
```

REM GOTO SUBROUTINE TO CALCULATE PREDICTION FILTER COEFF.

```
100 PRINT "MAX LENGTH CORRELATION =";j1
    PRINT  TAB( 5);"INDICES 0 TO Li. Li=";: INPUT Li
    PRINT "INPUT PREDICT AHEAD TIME STEPS=";: INPUT pa

    GOSUB 3000
```

```
REM  GOTO SUBROUTINE FOR PREDICTION FILTER OPERATION
     GOSUB 4000

REM  DISPLAY EVERYTHING
     GOSUB 5000

     END

2000     REM AUTOCORRELATION

REM  CALC AUTO CORREL, EQ.(B.85)
     PRINT " n       R(n)"

     FOR j = 0 TO j1
        sum = 0
        FOR n = 0 TO nt - j
           sum = sum + X(n) * X(n + j)
        NEXT n
        R(j) = sum
     NEXT j
     norm =  R(0)

     FOR j = 0 TO j1
        R(j) = R(j) / norm
     NEXT j

     j2 =INT( j1 / 2 + .5)
     FOR j = 0 TO j2
         PRINT j;  TAB(6);  R(j);TAB(20);j2 + j;TAB(26);  R(j2+j)
     NEXT j

REM  END AUTO CORRELATION CALC
     PRINT "PAUSE:ANY KEY TO CONTINUE"
     INPUT Q$

     RETURN

3000     REM PREDICTION FILTERS USING EUREKA
     M = 0:N = 1: REM  ARRAYS 0 AND 1 IN EUREKA SUBROUTINE

REM  PUT R(j+pa) in the array G(j). R(j+pa) is the correlation
REM  of the signal and the desired output.

     for j = 0 to Li +pa
        G(j) = R(j + pa)
     next  j
```

```
REM  L=0 TERM IS FIRST.

      AZ(0) = 1
      FZ(0) = G(0) / R(0)
      AL = R(0)
      GM = FZ(0) * R(1)
      BT = R(1)
      K =  - BT / AL

      FOR L = 1 TO Li
         AZ1(0) = AZ(O)

           FOR J = 1 TO L - 1
                AZ1(J) = AZ(J) + K * AZ(L - J)
           NEXT J

         AZ1(L) = K * AZ(0)
         AL1 = AL + K * BT
      REM  FOR BETA
         BT1 = 0

         FOR J = 0 TO L
              BT1 = BT1 + AZ1(J) * R(L + 1 - J)
         NEXT J

          K1 =  - BT1 / AL1
          Q = (G(L) - GM) / AL1

         FOR J = 0 TO L - 1
              FZ1(J) = FZ(J) + Q * AZ1(L - J)
         NEXT J

          FZ1(L) = Q * AZ1(0)
REM  GAMMA
          GM1 = 0

         FOR J = 0 TO L
              GM1 = GM1 + FZ1(J) * R(L + 1 - J)
         NEXT J

REM  SLIDE COEFFICIENTS

         FOR J = 0 TO L
              AZ(J) = AZ1(J)
              FZ(J) = FZ1(J)
         NEXT J
```

```
REM  SLIDE
        AL = AL1
        BT = BT1
        K = K1
        GM = GM1
REM         REPEAT FOR NEW L

     NEXT L

     print "filter  done"

     RETURN : REM THE PREDICTION FILTER OPERATION IS NEXT.

4000        REM DO PREDICTION AND CALCULATE ERROR

     for n = 0 to PA
        P(n) = X(n)
     next n

     for n = pa to nt
        sum = 0
        for j = 0 to Li
          FX = 0
          k = n-j
          if k >= 0 then FX = FZ(j) * X(k)
          sum = sum + FX
         next j
        ER(n+pa) = X(n+pa) - sum
        P(n + pa) = sum
     next n

REM  WIENER PREDICTION FILTER OPERATION IS FINISHED.
     RETURN

5000        REM DISPLAY

     YS = (YL%-Y0%) / 5
      TSC = (XL%-XO%) / (nt+1)

     print"input  amplitude  factors  for  X(n),  P(n),  E(n)  and  R(j)";
     input xamp, pamp,eamp, ramp

     xmp = xamp*YS: pmp = YS*pamp: emp = eamp*YS: rmp=ramp*YS

     cls
```

```
for n = 0 to nt
    T% = int (n * TSC +X0%)
    T1% = int ((n+1) * TSC +X0%)
    YX% = int (YS - xmp * X(n))
    YX1% = int (YS - xmp * X(n+1))
    LINE (T%,YX%) - (T1%,YX1%)
next n

for n = 0 to nt
    T% = int (n * TSC +X0%)
    T1% = int ((n+1) * TSC +X0%)
    YP% = int (2 * YS - pmp * P(n))
    YP1% = int (2 * YS - pmp * P(n+1))
    LINE (T%,YP%) - (T1%,YP1%)
next n

for n = 0 to nt
    T% = int (n * TSC +X0%)
    T1% = int ((n+1) * TSC +X0%)
    YE% = int (3 * YS - emp * ER(n))
    YE1% = int (3 * YS - emp * ER(n+1))
    LINE (T%,YE%) - (T1%,YE1%)
next n
for n = 0 to j1
    T% = int (n * TSC +X0%)
    T1% = int ((n+1) * TSC +X0%)
    YE% = int (4 * YS - rmp * R(n))
    YE1% = int (4 * YS - rmp * R(n+1))
    LINE (T%,YE%) - (T1%,YE1%)
next n

INPUT Q$
PRINT" NEW PLOT y or n";:INPUT Q$
IF Q$ = "y" OR Q$ = "Y" GOTO 5000
```

Appendix C

C1.1 MULTICHANNEL SEISMIC DATA, SEISMIC HOLOGRAPHIC AND STACKING ALGORITHMS AND DIFFRACTIONS

Our purpose is to give the basic algorithms that we used to make many of the figures in Chapters 13 to 15. Comments within the programs describe the operations.

A two-dimensional rough interface can be modeled in many ways for numerical studies. We use finite-width plane facets and calculate ray path reflections from the facets and diffractions from the wedge apex where a pair of facets join. The facets have infinite length. If the width of a facet is large compared to $2x_1$ in Eq. (14.29), then the ray path reflects with nearly the same amplitude as if dimensions of the facet were infinite. If the facet width is less than $2x_1$, then the amplitude of the reflection decreases roughly as (facet width)/$(2x_1)$. The discussion of the Fresnel integrals in Chapter 14 gives details. Actually, the infinite length requirement can be relaxed to require only that the length be much greater than $2y_1$ in Eq. (14.29).

For very simple examples, we give the reflection from a finite-width plane facet and the scattering by a point diffractor. The seismic pseudosections from these features can be imaged by the holographic imaging algorithm. The algorithms were written to demonstrate principles and keep the algebra simple. Since a large number of square roots are needed in the holographic algorithm, the computation is slow. References in the selected readings on imaging give much faster algorithms.

A few words of caution in the use of point diffractors follow. Point diffractors are a conceptual convenience in visualizing the construction of Huygens wavelets, however amplitudes are ignored. Actually, points have zero area and diffractions from points have zero amplitude. The diffraction from eges of planes or wedges is a difficult problem. The final equations of the Biot-Tolstoy wedge diffraction solution are given later in this section.

Here, it is sufficient to state that the sign of the diffraction signal from the edge of a plane is negative when the colocated source and receiver are over the plane and is positive when the colocated source and receiver are not over the plane.

```
REM  C.1 REFLECTION TIMES AND SYN' SEISMOGRAM

REM  REFLECTION FROM INTERFACE AT H
REM  SOURCE AT XSHOT. GEOPHONES AT X(I)
REM  MAKES DATA FILE FOR C.2 NORMAL MOVEOUT CORRECTION
REM  SIMPLE IMPULSIVE SOURCE AND SIGNAL: AMPLITUDE = 1
REM  SCREEN IS 492 BY 300 PIXELS
REM  USE INTEGER ARRAYS TO CONSERVE MEMORY

     DIM  S%(20,300),N%(300) ,E%(20,300),a(50),XR(20)
     DIM  C(10),H(10),X(300),Z(10),T(300) ,D(10),Crms(10)
     PI = 4 *  ATN (1)
     DR = PI / 180

     PRINT "MAX NUMBER OF LAYERS = 10"
     PRINT "INPUT NUMBER OF LAYERS=";: INPUT N1

     FOR I = 1 TO N1
         PRINT "LAYER ";I;" INPUT C(";I;"),H(";I;")=";
        INPUT C(I),H(I)
     NEXT I
     PRINT" SOURCE AT X = 0"
     PRINT "INPUT NUMBER OF GEOPHONE STATIONS = ";:INPUT JM
     PRINT "INPUT STATION SPACINGS = ";: INPUT AX

REM       ESTIMATE ANGLE INCREMENT
REM       COMPUTE CH AND TN FOR
REM DURBAUM-DIX APPROXIMATION EQS. A.91, A.102, AND A.106

     S = 0:CH = 0:TN = 0 :ANGR = 0
     FOR I = 1 TO N1
       S = S + H(I)
       TN = TN + 2 * H(I) / C(I)
       CH = CH + 2 * C(I) * H(I)
       a =  ATN (JM * AX / (2 * S))
       if ANGR< a then ANGR = a
       D(I) = S
       CRMS(I)= SQR (CH/TN):REM RMS VELOCITY=CRMS.EQ.A.106
     NEXT I

     X=JM*AX
     estT  = sqr(4* S^2+X^2)/CRMS(N1)
```

```
20    PRINT "estimated max time =";estT
      PRINT"input time step t0 = ";: input t0
      JH = INT(estT/t0+.5)
      PRINT " JH = ";JH
      PRINT "max JH = 300. OK 'y' or 'n'";:input q$
      if q$ = "n" goto 20
      PRINT " Input wavelet breadth in s = ";: INPUT wb


      w0=  wb/(sqr(6)*t0)
      i0=int(4*w0)
      nw=int(2*i0)
      w2=w0*w0
      print  "i0=";i0
      waf = 1000
      for i=0 to nw
          i2=(i-i0)^2
              a(i)=(1-2*i2/w2)*exp(-i2/w2)*waf
      next i
      JH=JH+nw

REM  ZERO ARRAY

      FOR i = 0 TO JM+1
        FOR j = 0 TO 300
           E%(i,j) = 0
           S%(i,j)  = 0
        NEXT j
      NEXT i
```

REM travel times. use subroutine from A1.4.2 n-layer refl
rem   and brute force ray trace.

REM START AT STATEMENT 1
REM ' REM LET NUMBER OF ANGLE INCREMENTS BE...'
REM AND COPY (PASTE) THROUGH STATEMENT 2
REM 'INPUT Q$: REM PAUSE'
REM INCLUDE 800 THE RAYTRACE SUB.

REM Brute force ray trace gives E(z) for each geophone.

```
rem  convolve  E(z)*A(z)
     for i = 0 to JM
       for  k= 0 to nw
         for  j = i0 to JH
            m=j+k-i0
              S%(i,m)=S%(i,m)+E%(i,j)*a(k)
```

```
            next j
          next k
           print" jug ";i
        next i


REM  PLOT  E%(j,k)
REM      PLOT reflections


100  PRINT "INPUT AMPLITUDE FACTOR = ";:INPUT AM
     PRINT "INPUT TIME TICS = ";:INPUT DT
     Print "input  max  time =";:input tmax
     DTJ = DT / T0
     kmax  =  INT(tmax/t0+.5)
       X0%=20 :Y0%=20: XL%= 480: YL%=270


      XSC = (XL%-XO%) / (2*JM+2)
      AF = AM*.8 * XSC
      ys = (YL%-Y0%)/kmax


      CLS          :'clear screen


       LINE (X0%,Y0%)-(XL%,Y0%)
      FOR j = 0 TO JM
          X% = INT(j * XSC + 1.8 * X0%)
           FOR k = 1 TO kmax
               YP% = INT(k*ys+Y0%)
          XP% =INT( X% + AF * E%(j,k))
          LINE (X%,YP%)-(XP%,YP%)
          LINE (XP%,YP%+1) - (X%,YP%+1)
        NEXT k
    NEXT j


FOR j = 0 TO JM
    X% = INT((j+jm+1) * XSC + 1.8 * X0%)
    FOR k = 1 TO kmax-1
        YP% = INT(k*ys+Y0%)
        YP1% = INT((k+1)*ys+Y0%)
        XP% =INT( X% + AF * S%(j,k)/waf)
        XP1% =INT( X% + AF * S%(j,k+1)/waf)
        LINE (XP%,YP%)-(XP1%,YP1%)
     NEXT k
   NEXT j
```

```
REM TIME TICS

    NZJ = INT(kmax / DTJ +0.5)
    X1% = INT(X0%/5)

    FOR J = 0 TO NZJ
        YP% = INT(J * DTJ*ys +0.5 + Y0%)
        XP% = INT(X1% + .2 * XSC)
        LINE (X1%,YP%) - (XP%,YP%)
        LINE (XP%,YP%) - (X1%,YP%)
    NEXT J
      call MOVETO(x0%,280)
    PRINT "ax=";ax;" tics=";DT ;" wavelet breadth =";wb

    'end of graph operation
    INPUT Q$          :'pause for viewing picture

    CLS    :rem clear screen

3600     PRINT" NEW PLOT y or n";:INPUT Q$
    IF Q$ = "y" OR Q$ = "Y" GOTO 100

    PRINT "Make file y or n";
    INPUT Q$
    IF Q$ ="n" OR Q$ = "N" GOTO 200

REM make file
    print"give file name":input n$
    open n$ for output as #1

    WRITE #1, JM
    WRITE #1, XSHOT
    WRITE #1, JH
    WRITE #1, n1
    WRITE #1, t0

    FOR i = 0 TO JM
        WRITE #1, XR(i)
    NEXT i

    FOR I = 0 TO JM
        FOR J = 0 TO JH
        WRITE #1, S%(I,J)
    NEXT J
  NEXT I
```

```
for n = 1 to n1
    write #1,c(n)
    write #1,h(n)
    write #1,Crms(n)
next n

    CLOSE#1

200  END

800  REM PUT RAY TRACE SUB HERE


REM  C.2 NORMAL MOVEOUT CORRECTION

REM  USE DATA FILE FROM C.1 REFLECTION TIMES
REM  REMAP S%(I,Q) TO SP%(I,J) FOR NMO CORRECTION
REM  IM = NUMBER OF GEOPHONES. JH = MAX TIME STEPS
REM  USE INTEGER ARRAYS TO CONSERVE MEMORY
REM  SCREEN IS 492 BY 300 PIXELS

    DIM  S%(40,300),SP%(40,300),X(40)
    X0%=20 :Y0%=20: XL%= 480: YL%=280

    PRINT"Give file name":input n$
    OPEN n$ for input as #1

     INPUT #1, IM
     INPUT #1, XSHOT
     INPUT #1, JH
     INPUT #1, n1
     INPUT #1, t0

    FOR I = 0 TO IM
        INPUT #1, X(I)
    NEXT I
    ax = X(1)-X(0)

    FOR I = 0 TO IM
      FOR J = 0 TO JH
          INPUT #1, S%(I,J)
      NEXT J
    NEXT I
    for n = 1 to n1
        input#1,c(n)
```

```
      input #1,h(n)
      input #1,Crms(n)
   next n

   close #1
   PRINT "FILE READ"

20   PRINT"Input a stacking velocity 'sCrms =";:input sCrms

REM  ZERO ARRAY
   FOR I = 0 TO IM +1
     FOR J = 0 TO 300
        SP%(I,J) = 0
     NEXT J
   NEXT I

REM REMAPPING OF S%(I,Q) TO SP%(I,J)

   FOR J = 0 TO JH
     FOR i= 0 TO IM
        XI =  (X(i)-XSHOT)/(sCrms*t0)
       Q = INT(SQR(J^2 + XI^2)+0.5)
       IF Q<= 300 THEN SP%(i,J) = S%(i,Q)+SP%(i,J)
     NEXT i
   NEXT J

REM    STACK IN TO SP%(IM+1)
   FOR J = 0 TO JH
     FOR I = 0  TO IM
          SP%(IM+1,J) = SP%(IM+1,J)+SP%(I,J)/(IM+1)
     NEXT I
   NEXT J

REM PLOT SP%(I,J)
REM   PLOT INTERFACES

100 PRINT "Input amplitude factor = ";:INPUT AM
   PRINT "Input time tics = ";:INPUT DT
   PRINT "input max time =";:input tmax
   DTJ = DT / T0
   kmax  = INT(tmax/t0+.5)
     X0%=20 :Y0%=20: XL%= 480: YL%=270

   XSC = (XL%-XO%) / (2*IM+3)
   waf = 1000    :'waf was used to make integer arrays in preceding prgm
   AF = AM*.8 * XSC
   ys = (YL%-Y0%)/kmax
```

```
    CLS          :' clear screen

   LINE (X0%,Y0%)-(XL%,Y0%)
  FOR j = 0 TO IM
     X% = INT(j * XSC + 1.8 * X0%)
      FOR k = 1 TO kmax-1
         XP% =INT( AF * S%(j,k)/waf)
         XP1% =INT(AF * S%(j,k+1)/waf)
         YP% = INT(k*ys+Y0%)
         YP1% = INT((k+1)*ys+Y0%)
         LINE (XP%+X%,YP%)-(XP1%+X%,YP1%)
      NEXT k
   NEXT j

  FOR j = 0 TO IM+1
     X% = INT((j+IM+1) * XSC + 1.8 * X0%)
      FOR k = 1 TO kmax-1
         XP% =INT( AF * SP%(j,k)/waf)
         XP1% =INT(AF * SP%(j,k+1)/waf)
         YP% = INT(k*ys+Y0%)
         YP1% = INT((k+1)*ys+Y0%)
         LINE (XP%+X%,YP%)-(XP1%+X%,YP1%)
      NEXT k
   NEXT j


REM TIME TICS

   NZJ = INT(kmax / DTJ +0.5)
   X1% = INT(X0%/5)

  FOR J = 0 TO NZJ
     YP% = INT(J * DTJ*ys +0.5 + Y0%)
     XP% = INT(X1% + .2 * XSC)
     LINE (X1%,YP%) - (XP%,YP%)
     LINE (XP%,YP%) - (X1%,YP%)
     NEXT J
       call MOVETO(x0%,290)
     PRINT "ax=";ax;" tics=";DT ;" stack c =";crms
     INPUT Q$

     CLS    :rem clear screen

3600     PRINT" NEW PLOT y or n";:INPUT Q$
       IF Q$ = "y" OR Q$ = "Y" GOTO 100
```

```
      PRINT" NEW stack y or n";:INPUT Q$
      IF Q$ = "y" OR Q$ = "Y" GOTO 20


   end


REM  C.3 REFLECTIONS FROM FINITE PLANE

REM MAKES DATA FILE FOR C.5 HOLOGRAPHIC IMAGING.
REM  COLOCATED SOURCE AND RECEIVER ARE OVER A FINITE WIDTH
REM  PLANE. GEOPHONES ARE AT XG(I) ON THE SURFACE Z = 0.
REM  LEFT EDGE OF THE PLANE IS AT XL AND ZL.
REM  RIGHT EDGE OF THE PLANE IS AT XR AND ZR.
REM  SIMPLE IMPULSIVE SOURCE AND SIGNAL: AMPLITUDE = 1
REM  THE IMPULSE HAS 2 TIME STEPS DURATION.
REM  USE INTEGER ARRAYS TO CONSERVE MEMORY
REM  SCREEN IS 492 BY 300 PIXELS

      DIM S%(40,282), XG(40)
      X0%=20 :Y0%=20: XL%= 480: YL%=280

      C = 2000
      T0 = 0.005
      NJ = 280

      PRINT"C = ";C;"   T0 = ";T0;"   MAX TIME STEP = ";NJ
      PRINT"INPUT NUMBER OF GEOPHONES, NG <= 40, =";: INPUT NG
      PRINT"INPUT GEOPHONE SEPERATIONS X = ";: INPUT DX
      PRINT"INPUT LOCATE LEFT EDGE, INPUT XL AND ZL = ";:INPUT XL,ZL
      PRINT"INPUT LOCATE RIGHT EDGE, INPUT XR AND ZR = ";:INPUT XR,ZR

REM  ZERO ARRAY
      FOR I = 0 TO NG
        FOR J = 0 TO NJ
          S%(I,J) = 0
        NEXT J
      NEXT I

REM  CALCULATE THE ANGLE THE PLANE MAKES WITH HORIZONTAL
REM  ANGLE PHI IS + C-CLOCKWISE

      PHI = ATN((ZL - ZR)/(XR - XL))

REM  CALCULATE LEFT AND RIGHT INTERCEPTS OF NORMALS AT THE
REM  SURFACE FOR EACH EDGE OF THE PLANE. SOURCE / RECEIVER
REM   MUST BE BETWEEN INTERCEPTS FOR A REFLECTION TO EXIST.
REM  IGNORE DIFFRACTIONS.
```

```
LNI = XL - ZL * TAN(PHI)
RNI = XR - ZR * TAN(PHI)


REM USE CONSTRUCTION ZC = ZL + (XL - XG(I)) * TAN(PHI)


    CT = C * T0


    FOR I = 0 TO NG
       XG(I) = I * DX
       IF XG(I) >= LNI AND XG(I) <= RNI THEN
         ZC = ZL + (XL - XG(I)) * TAN(PHI)
         TWT = 2 * ZC * COS(PHI) / CT
         J = INT(TWT + 0.5)
         S%(I,J) = 1
         S%(I,J+1) = 1
       ELSE
        REM NO REFLECTION
       END IF
    NEXT I

REM PLOT  S%(I,J)
REM    PLOT INTERFACES


    XSC = (XL%-XO%) / (NG+2)
    AF% = INT(.8 * XSC)


    CLS


    LINE (X0%,Y0%)-(XL%,Y0%)
    FOR I = 0 TO NG
       X% = INT(I * XSC) + X0%
        FOR J = 1 TO NJ
           YP% = J+Y0%
           XP% = X% + AF% * S%(I,J)
           LINE (X%,J+Y0%)-(XP%,YP%)
           LINE (XP%,YP%) - (X%,YP%)
        NEXT J
    NEXT I
    INPUT Q$

    PRINT "Make file y or n";
    INPUT Q$
    IF Q$ ="n" GOTO 100
```

```
REM make file
    print"give file name":input n$
    open n$ for output as #1

    WRITE #1, NG
    WRITE #1, NJ
    WRITE #1, C
    WRITE #1, T0
    FOR I = 0 TO NG
        WRITE #1, XG(I)
    NEXT I

    FOR I = 0 TO NG
        FOR J = 0 TO NJ
            WRITE #1, S%(I,J)
        NEXT J
    NEXT I

    close#1

100  END


REM  C.4 POINT DIFFRACTOR

REM MAKES DATA FILE FOR C.5 HOLOGRAPHIC IMAGING.
REM  DIFFRACTIONS FROM POINT AT AO AND H
REM  COLOCATED SOURCE AND RECEIVER
REM  SOURCE AT A0. GEOPHONES AT XG(I)
REM  SIMPLE IMPULSIVE SOURCE AND SIGNAL: AMPLITUDE = 1
REM  THE IMPULSE HAS 2 TIME STEPS DURATION.
REM  SCREEN IS 492 BY 300 PIXELS AND THIS LIMITS
REM  MAX VALUE OF VERTICAL PIXELS OR DEPTH STEPS TO 282.
REM  USE INTEGER ARRAYS TO CONSERVE MEMORY

    DIM S%(40,282), XG(40)
    X0%=20 :Y0%=20: XL%= 480: YL%=280

    C = 2000
    T0 = 0.005
    NJ = 280

    PRINT"C = ";C;"   T0 = ";T0;"   MAX TIME STEP = ";NJ
    PRINT"INPUT NUMBER OF GEOPHONES, NG <= 40, =";: INPUT NG
    PRINT"INPUT GEOPHONE SEPERATIONS X = ";: INPUT DX
    PRINT"INPUT LOCATE POINT IN X = ";: INPUT A0
    PRINT"INPUT DEPTH OF DIFFRACTOR H = ";:INPUT H
```

```
REM  ZERO ARRAY
     FOR I = 0 TO NG
       FOR J = 0 TO NJ +1
           S%(I,J) = 0
       NEXT J
     NEXT I


REM  CALCULATE SIGNALS

     CT = C * T0
     HT =2 * H / CT

     FOR I = 0 TO NG
        XG(I) = I * DX
        XT = 2 * (XG(I) - A0) / CT
        J = INT(SQR(XT^2 + HT^2) +0.5)
        IF J< 280 THEN
          S%(I,J) = 1
          S%(I,J+1) = 1
        ELSE
         REM NO REFLECTION
        END IF
      NEXT I


REM PLOT  S%(I,J)
REM    PLOT INTERFACES

     XSC = (XL%-XO%) / (NG+2)
     AF% = INT(.8 * XSC)

     CLS

     LINE  (X0%,Y0%)-(XL%,Y0%)
   FOR I = 0 TO NG
      X% = INT(I * XSC) + X0%
        FOR J = 1 TO NJ
           YP% = J+Y0%
           XP% = X% + AF% * S%(I,J)
           LINE (X%,J+Y0%)-(XP%,YP%)
           LINE (XP%,YP%) - (X%,YP%)
        NEXT J
     NEXT I
     INPUT Q$
```

```
      PRINT "Make file y or n";
      INPUT Q$
      IF Q$ ="n" GOTO 100

REM make file
      print"give file name":input n$
      open n$ for output as #1

       WRITE #1, NG
       WRITE #1, NJ
       WRITE #1, C
       WRITE #1, T0
      FOR I = 0 TO NG
         WRITE #1, XG(I)
      NEXT I

      FOR I = 0 TO NG
        FOR J = 0 TO NJ
           WRITE #1, S%(I,J)
        NEXT J
      NEXT I

      close #1

100  END


REM  C.5 HOLOGRAPHIC IMAGING

REM  USE DATA FILES FROM C.3 REFLECTIONS FROM FINITE PLANE AND
REM  C.4 POINT DIFFRACTOR. FOR SIMPLICITY USE CONSTANT VELOCITY C
REM  PRGM MAPS POINTS IN TO CIRCLES BY MOVING S%(I,Q)
REM  TO SP%(I,J). ALL SIGNALS ON CIRCLES PASSING THROUGH (K,J)
REM  ARE ADDED BY STACKING SP%(I,J). USE INTEGER ARRAYS
REM  S%(I,Q) AND SP%(I,J) TO SAVE MEMORY.

      DIM  S%(40,282),SP%(40,282),XG(40)
        X0%=20 :Y0%=20: XL%= 480: YL%=280

      print"give file name":input n$
      open n$ for input as #1

       INPUT #1, NG
       INPUT #1, NJ
       INPUT #1, C
       INPUT #1, T0
```

```
FOR I = 0 TO NG
   INPUT #1, XG(I)
NEXT I
FOR I = 0 TO NG
  FOR J = 0 TO NJ
     INPUT #1, S%(I,J)
  NEXT J
NEXT I
close #1
PRINT "FILE READ"


REM ZERO ARRAY
  FOR I = 0 TO NG
    FOR J = 0 TO NJ
       SP%(I,J) = 0
    NEXT J
  NEXT I


REM REMAPPING OF S%(I,Q) TO SP%(I,J)
REM PRGM IS SLOW SO WE WRITE K TO TELL YOU IT IS WORKING.

    FOR K = 0 TO NG
      WRITE" K = ";K
      FOR J = 0 TO NJ
        FOR I= 0 TO NG
           XT = 2 * (XG(I) - XG(K))/(C*T0)
           Q = INT(SQR(J^2 + XT^2)+0.5)
           IF Q< 280 THEN SP%(K,J) = S%(I,Q) + SP%(K,J)
        NEXT I
      NEXT J
    NEXT K

REM PLOT SP%(I,J)
REM PLOT INTERFACES

100 PRINT "INPUT AMPLITUDE FACTOR = ";:INPUT AM
    PRINT "INPUT DEPTH TICS = ";:INPUT DZ
    DZJ = 2 * DZ / (T0 * C)
    XSC = (XL%-XO%) / (NG+2)
    AF = AM * XSC/NG

   CLS

   LINE (X0%,Y0%)-(XL%,Y0%)
```

```
FOR I = 0 TO NG
    X% = INT(I * XSC + 1.8 * X0%)
      FOR J = 1 TO NJ
        YP% = J+Y0%
        XP% = X% + INT(AF * SP%(I,J))
        LINE (X%,J+Y0%) - (XP%,YP%)
        LINE (XP%,YP%+1) - (X%,YP%+1)
      NEXT J
    NEXT I

REM DEPTH TICS

    NZJ = INT(NJ / DZJ +0.5)
    X1% = INT(X0%/4)

    FOR J = 0 TO NZJ
       YP% = INT(J * DZJ +0.5 + Y0%)
       XP% = INT(X1% + .2 * XSC)
       LINE (X1%,YP%) - (XP%,YP%)
       LINE (XP%,YP%) - (X1%,YP%)
    NEXT J


    INPUT Q$
    PRINT" NEW PLOT y or n";:INPUT Q$
    IF Q$ = "y" OR Q$ = "Y" GOTO 100

    END
```

C1.2 HELMHOLTZ-KIRCHHOFF DIFFRACTION FORMULA

This section gives the steps between a standard expression of the Helmholtz-Kirchhoff integral and the Kirchhoff diffraction formula, Equation (14.5). Derivations of the Helmholtz-Kirchhoff integral can be found in standard intermediate physics texts in wave propagation and optics. Our starting place is the result of Clay and Medwin's derivation for underwater sound in section A10 of *Acoustical Oceanography* (New York: Wiley Interscience, 1977). Following their notation, $U(Q)$ is the field (here sound pressure) at point Q due to a continuous wave source. The integral is

$$U(Q) = \frac{e^{i\omega t_r}}{4\pi} \int_s \left[U \frac{\partial}{\partial n}\left(\frac{e^{-ikR_r}}{R_r}\right) - \frac{e^{-ikR_r}}{R_r}\frac{\partial U}{\partial n} \right]_s dS \tag{C.1}$$

$$k = \frac{\omega}{c}$$

The integral is evaluated on the surface S. $\partial/\partial n$ is the normal derivative at dS and is on the surface as drawn inward toward the point Q. R_r is the distance from the elemental area dS to a receiver at Q. U is the field on the surface. t_r is the travel time from dS to Q.

The integral is difficult to evaluate because we must know U and $\partial U/\partial n$ on the surface. Kirchhoff made an approximation in which U is the reflection of the incident wave field,

$$U \simeq R_{12}^p \, U_s \tag{C.2}$$

where U_s is the incident wave field from a source, and R_{12}^p is the local plane-wave reflection coefficient at dS. Because of the change of direction caused by reflection, the approximation gives

$$\frac{\partial U}{\partial n} \simeq - R_{12}^p \frac{\partial U_s}{\partial n} \tag{C.3}$$

The substitution of Eq (C.2) and (C.3) in Eq. (C.1) gives

$$U(Q) \simeq \frac{e^{i\omega t_r}}{4\pi} \int_S R_{12}^p \frac{\partial}{\partial n} \left(U_s \frac{e^{-ikR_r}}{R_r} \right) dS \tag{C.4}$$

From the geometry in Figures 14.2 and 14.4, U_s for a point source is

$$U_s = \frac{B e^{i(\omega t_s - kR_s)}}{R_s} \tag{C.5}$$

where B is a source amplitude function and R_s is the source to dS distance. The substitution of Eq. (C.5) in Equation (C.4) and replacement of $U(Q)$ with p_r gives

$$p_r \simeq \frac{B e^{i\omega t}}{4\pi} \int_S R_{12}^p \frac{\partial}{\partial n} \left[\frac{e^{-ik(R_s + R_r)}}{R_s R_r} \right] dS \tag{C.6}$$

$$t \equiv t_s + t_r$$

To take the derivative, we depress dS a distance ζ below the mean reflection interface at h. From Figure 14.2 R_s^2 and R_p^2 are

$$R_s^2 = (R_1 \sin \theta_1 + x)^2 + y^2 + (R_1 \cos \theta_1 + \zeta)^2 \tag{C.7}$$
$$R_r^2 = (R_2 \sin \theta_2 - x)^2 + y^2 + (R_2 \cos \theta_1 + \zeta)^2$$

whereas in a modification to Figure 14.2, R_1 is the distance from the source to the origin, and R_2 is the distance from the origin to the receiver. The expansions, using Eq. (14.9), give

$$R_s \simeq R_1 + x \sin \theta_1 + \zeta \cos \theta_1 + \frac{x^2}{2R_1} \left(1 - \sin^2\theta_1 \right) + \frac{y^2}{2R_1} \tag{C.8}$$
$$R_r \simeq R_2 - x \sin \theta_2 + \zeta \cos \theta_2 + \frac{x^2}{2R_2} (1 - \sin^2\theta_2) + \frac{y^2}{2R_2}$$

The sum $R_s + R_r$ is approximately

$$R_s + R_r \simeq R_1 + R_2 + x(\sin \theta_1 - \sin \theta_2) + \zeta (\cos \theta_1 + \cos \theta_2) + \frac{x^2}{2} \left(\frac{\cos^2\theta_1}{R_1} + \frac{\cos^2\theta_2}{R_2} \right) + \frac{y^2}{2} \left(\frac{1}{R_1} + \frac{1}{R_2} \right) \quad \text{(C.9)}$$

The normal derivative in Eq. (C.6) becomes, using Eckart's approximation for small slopes,

$$\frac{\partial}{\partial n} \simeq -\frac{\partial}{\partial \zeta}$$

where $\partial/\partial n$ is upward, and $\partial/\partial \zeta$ is downward,

$$\frac{\partial}{\partial \zeta} \left[\frac{e^{-ik(R_s+r)}}{R_s R_r} \right] \simeq \left[\frac{-ik(\cos \theta_1 + \cos \theta_2)}{R_s R_r} + \frac{\partial}{\partial \zeta} \frac{1}{R_s R_r} \right] e^{-ik(R_s+R_r)} \quad \text{(C.10)}$$

The second term on the right side of Eq. (C.10) gives terms of the order of $R_s^{-2}R_r^{-1}$ and $R_s^{-1}R_r^{-2}$. These terms are small compared with $kR_s^{-1}R_r^{-1}$. For R_s and R_r much greater than λ, we can ignore the second expression. The substitution of the remaining term in Eq. (C.10) gives

$$p_r \simeq \frac{ikB}{4\pi} \int_S R_{12}^p \frac{(\cos \theta_1 + \cos \theta_2)}{R_s R_r} e^{i[\omega t - k(R_s + R_p)]} \, dS \quad \text{(C.11)}$$

We can replace $(\cos \theta_1 + \cos \theta_2)$ with $(\cos \theta_s + \cos \theta_r)$ to express more accurately the normal derivative at dS when dS is not horizontal. Then Eq. (C.11) becomes the integral Equation (14.5).

C1.3 APPROXIMATION AND NUMERICAL COMPUTATION OF THE DIFFRACTION IMPULSE SOLUTION

This section is from Daneshvar and Clay (1987). The exact impulse solution of an infinitely rigid wedge is in Biot and Tolstoy (1957), Tolstoy (1973), sec. 8–4 and Tolstoy and Clay (1987), Appendix 5);

$$P_d(t) = \frac{-A_0 cv}{2\pi r r_0 \sinh \eta} \left\{ \frac{\sin [v(\pi \pm \theta \pm \theta_0)]}{\cosh v\eta - \cos [v(\pi \pm \theta \pm \theta_0)]} \right\} \quad \text{for } t > \tau_0 \quad \text{(C.12)}$$

where $\tau_0 = [(r + r_0)^2 + y^2]^{1/2}/c$

$$A_0 = \frac{\rho S}{4\pi}$$

$$v = \frac{\pi}{\theta_w}$$

$$\eta = \text{arccosh} (U)$$

$$U = [c^2 t^2 - (r^2 + r_0^2 + y^2)]/2rr_0$$

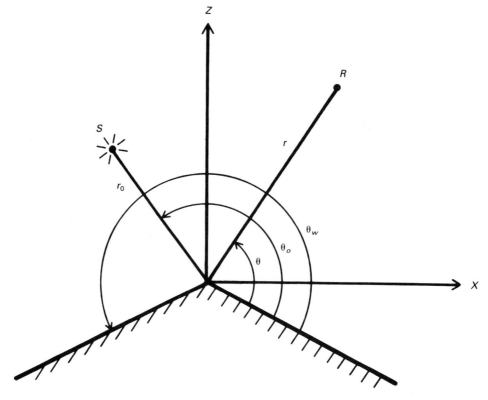

Figure C.1 Geometry for the wedge problem. The coordinate *y* is normal to the page and is not shown.

S = source strength (m^3/s)

ρ = density of the propagation medium

θ_w = wedge angle

c = sound speed

y = offset source-receiver

The angles θ_0, θ, and θ_w and the ranges r_0 and r are shown in Figure C.1. The offset y is suppressed. The braces enclose four terms obtained by using all combinations of $\pi \pm \theta \pm \theta_0$. If $sin\,[\nu(\pi \pm \theta \pm \theta_0)] = 0$ for any of the four combinations, then the corresponding term in braces will be zero.

The solution has a singularity at $t = \tau_0$, the diffraction arrival time. In numerical computations the discrete representations are used. According to the Nyquist sampling theorem, if the diffracted signal is sampled at intervals t_0, the sampling frequency is $1/t_0$, and the highest frequency present in the sampled signal is $1/2t_0$. Therefore, if the signal is low-pass filtered, below $1/2t_0$ the high-frequency components that cause difficulties in numerical computations are removed with no loss of information.

Equation (C.12) needs to be low-pass filtered with a boxcar by the following integration:

$$\frac{1}{T_b} \int_0^{t'} P_d(t)\, dt \qquad (C.13)$$

Where t' is the time after τ_0, and $Tb = 2t_0$ is the length of the boxcar filter. Equation (C.12) is not in a form suitable for integration. An asymptotic approximation to Eq. (C.12) is derived by approximating $\sinh \eta$ and $\cosh \nu\eta$. We can express the time as $t = \tau_0 + \Delta t$, where Δt is the time increment after τ_0 and is small compared with τ_0. Substituting for t, we obtain the expression

$$U = 1 + \frac{c^2 \tau_0}{r r_0} \Delta t + \frac{c^2}{2 r r_0} \Delta t^2 \qquad (C.14)$$

We use the following identity:

$$\cosh^2 \eta - \sinh^2 \eta = 1 \qquad (C.15)$$

and ignore powers of Δt higher than one. We then have

$$\sinh \eta \approx \left[\frac{2c^2 \tau_0}{r r_0} \right]^{1/2} \sqrt{\Delta t} \qquad (C.16)$$

In order to estimate $\cosh \nu\eta$ we use the identity (C.17) and power series expansions (C.18),

$$\operatorname{arccosh}(U) = \ln(U + \sqrt{U^2 - 1}) \qquad (C.17)$$

$$\ln(1 + q) = q - \frac{1}{2} q^2 + \cdots$$

$$\cosh \nu\eta = 1 + \frac{(\nu\eta)^2}{2!} + \frac{(\nu\eta)^4}{4!} + \cdots \qquad (C.18)$$

Ignoring the powers of Δt higher than one, we have

$$\eta \approx \left(\frac{2c^2 \tau_0}{r r_0} \Delta t \right)^{1/2} \qquad (C.19)$$

$$\cosh \nu\eta \approx 1 + \frac{c^2 \nu^2 \tau_0}{r r_0} \Delta t$$

The approximation to Eq. (C.12) has the following form:

$$P_d(\Delta t) \approx -\frac{A_0 \nu}{2\pi (2 r r_0 \tau_0)^{1/2} \sqrt{\Delta t}} \left\{ \frac{\sin[\nu(\pi \pm \theta \pm \theta_0)]}{1 + \dfrac{c^2 \nu^2 \tau_0}{r r_0} \Delta t - \cos[\nu(\pi \pm \theta \pm \theta_0)]} \right\} \qquad (C.20)$$

Integration of Eq. (C.13) can be applied to Eq. (C.20) by letting $x = \sqrt{\Delta t}$ and using a standard integral.

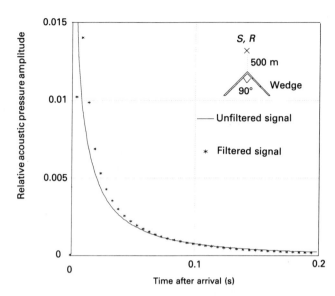

Figure C.2 Diffracted impulse response from a rigid 90° wedge. The solid curve is the exact solution. *s represents the numerical filtered diffracted impulse response. The filter is a two-step boxcar. The signals have arbitrary amplitudes. The sound speed is 1500 m/s. (Figure from M. R. Daneshvar and C. S. Clay, "Imaging of rough surfaces for impulsive and continuously radiating sources," *J. Acoust. Soc. Am.*, 82 (1987), 360–69.

$$\int \frac{dx}{a + bx^2} = \sqrt{\frac{b}{a}} \arctan\left(x \sqrt{\frac{b}{a}} \right) \tag{C.21}$$

The result is given in Eq. (C.22).

$$P_{df}(\Delta t) = -\frac{A_0}{\sqrt{2}\pi c T_b \tau_0} \left\{ \frac{\sin[\nu(\pi - \theta \pm \theta_0)]}{\sqrt{1 - \cos[\nu(\pi \pm \theta \pm \theta_0)]}} \right\}$$
$$\times \arctan \sqrt{\frac{\nu^2 c^2 \tau_0}{rr_0(1 - \cos[\nu(\pi \pm \theta \pm \theta_0)])} \Delta t} \tag{C.22}$$

Equation (C.22) is the filtered form of Eq. (C.20) and is valid for $\Delta t \ll \tau_0$. Comparison of $P_d(t)$ before and after a low-pass filter operation is shown in Figure C.2. Comparisons of the Biot-Tolstoy theory and an impulsive solution of the Kirchhoff diffraction formula are in Li and Clay (1988). The data match the Biot-Tolstoy theory. The impulsive solution of the Kirchhoff diffraction formula is inaccurate for the pressure signal diffracted and reflected at the edge of a half plane.

References to Appendix C

BIOT, M. A., AND I. TOLSTOY, "Formulation of wave propagation in infinite media by normal coordinates with an application to diffraction," *J. Acoust. Soc. Am.*, 19 (1957), 381–91.

DANESHVAR, M. R., AND C. S. CLAY, "Imaging of rough surfaces for impulsive and continuously radiating sources," *J. Acoust. Soc. Am.*, 82 (1987), 360–69.

LI, S. AND C. S. CLAY, "Sound transmission experiments from an impulsive source near rigid wedge," *J. Acoust. Soc. Am.*, 84 (1988), 2135–43.

TOLSTOY, I., *Wave Propagation*. New York: McGraw-Hill, Sec. 8–4, 339–47, 1973.

TOLSTOY, I., AND C. S. CLAY, *Ocean Acoustics*. Acoustical Society of America reprint. New York: American Institute of Physics, 1987. The appendices include the Biot and Tolstoy paper. We use their expression for $\partial\phi/\partial t$, Eq. (5.24). We replace their source $-1(t - R/\alpha)/4\pi$ by A_0. Then, the pressure $P_d(t)$ is their $\partial\phi/\partial t$.

Appendix D: Formulas

D1.1 FOURIER SERIES

Fourier series for periodic functions

$$f(t) = f(t + nT), \quad n = 0, \pm 1, \pm 2, \ldots$$

$$f(t) = \sum_{m=-\infty}^{\infty} F_m e^{i2\pi mt/T}$$

$$F_m \equiv \frac{1}{T} \int_0^T f(t) \, e^{-2\pi imt/T} \, dt$$

Finite or discrete Fourier transformation and inverse discrete transformation

$$t_0 = \frac{T}{N} \qquad\qquad \frac{t}{T} = \frac{n}{N}$$

$$f_n = \frac{1}{N} \sum_{m=-N/2}^{N/2-1} F_m \, e^{i2\pi mn/N} \qquad f_n = \frac{1}{N} \sum_{m=0}^{N-1} F_m \, e^{i2\pi nm/N}$$

$$F_m = \sum_{n=-N/2}^{N/2-1} f_n \, e^{-i2\pi mn/N} \qquad F_m = \sum_{n=0}^{N-1} f_n \, e^{-i2\pi nm/N}$$

Infinite Fourier integral

The infinite Fourier integral for transient signals appears in the geophysical literature. The transformation pairs are, from texts and handbooks,

$$\int_{-\infty}^{\infty} |f(t)|\, dt = \text{finite}$$

$$F(f) = \int_{-\infty}^{\infty} f(t)\, e^{-i2\pi ft}\, dt$$

$$f(t) = \int_{-\infty}^{\infty} F(f)\, e^{i2\pi ft}\, df$$

$F(f)$ is the spectral density. People often use the DFT to calculate the spectral coefficients F_m and then wish to compare them to the results of Fourier integral calculations. The Fourier integral gives $F(f)$ having the dimensions of $f(t)$ times time.

$$F(f) = F_m t_0$$

where we understand $F(f)$ to mean the spectral density from the integral.

D1.2 FORMULAS OF VECTOR ANALYSIS*

D1.2.1 Multiplication

Scalar product

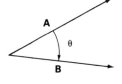

$$\mathbf{A} \cdot \mathbf{B} = AB \cos \theta$$
$$= A_x B_x + A_y B_y + A_z B_z$$

Vector product

$$\mathbf{A} \times \mathbf{B} = \mathbf{n}\, AB \sin \angle \theta$$
$$= \mathbf{i}(A_y B_z - A_z B_y) + \mathbf{j}(A_z B_x - A_x B_z) + \mathbf{k}(A_x B_y - A_y B_x)$$
$$= \begin{vmatrix} \mathbf{i} & \mathbf{j} & \mathbf{k} \\ A_x & A_y & A_z \\ B_x & B_y & B_z \end{vmatrix}$$

D1.2.2 Differentiation

Rectangular coordinates (Mutually perpendicular unit vectors i, j, k.)

Gradient $$\nabla p = \mathbf{i}\frac{\partial p}{\partial x} + \mathbf{j}\frac{\partial p}{\partial y} + \mathbf{k}\frac{\partial p}{\partial z}$$

Divergence $\nabla \cdot \mathbf{A}$ $$= \frac{\partial A_x}{\partial x} + \frac{\partial A_y}{\partial y} + \frac{\partial A_z}{\partial z}$$

* With notational changes, From Table II, H. H. Skilling, *Fundamentals of Electric Waves.* New York: John Wiley, 1942.

Curl $\nabla \times \mathbf{A} = \begin{vmatrix} \mathbf{i} & \mathbf{j} & \mathbf{k} \\ \dfrac{\partial}{\partial x} & \dfrac{\partial}{\partial y} & \dfrac{\partial}{\partial z} \\ A_x & A_y & A_y \end{vmatrix} = \mathbf{i}\left(\dfrac{\partial A_z}{\partial y} - \dfrac{\partial A_y}{\partial z}\right) + \mathbf{j}\left(\dfrac{\partial A_x}{\partial z} - \dfrac{\partial A_z}{\partial x}\right)$

$$+ \mathbf{k}\left(\dfrac{\partial A_y}{\partial x} - \dfrac{\partial A_x}{\partial y}\right)$$

Laplacian $\nabla^2 p = \nabla \cdot \nabla p = \dfrac{\partial^2 p}{\partial x^2} + \dfrac{\partial^2 p}{\partial y^2} + \dfrac{\partial^2 p}{\partial z^2}$

$$\nabla^2 \mathbf{A} = \mathbf{i}\nabla^2 A_x + \mathbf{j}\nabla^2 A_y + \mathbf{k}\nabla^2 A_z$$

Cylindrical coordinates (Mutually perpendicular unit vectors $\mathbf{1}_r$, $\mathbf{1}_\phi$, $\mathbf{k}$.)

$$x = r \cos \phi \qquad y = r \sin \phi \qquad z = z$$

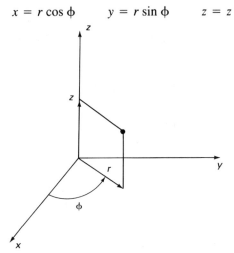

Gradient $\nabla p = \mathbf{1}_r \dfrac{\partial p}{\partial r} + \mathbf{1}_\phi \dfrac{1}{r}\dfrac{\partial p}{\partial \phi} + \mathbf{k}\dfrac{\partial p}{\partial z}$

Divergence $\nabla \cdot \mathbf{A} = \dfrac{\partial A_r}{\partial r} + \dfrac{A_r}{r} + \dfrac{1}{r}\dfrac{\partial A_\phi}{\partial \phi} + \dfrac{\partial A_z}{\partial z}$

Curl $\nabla \times \mathbf{A} = \begin{vmatrix} \dfrac{\mathbf{1}_r}{r} & \mathbf{1}_\phi & \dfrac{\mathbf{k}}{r} \\ \dfrac{\partial}{\partial r} & \dfrac{\partial}{\partial \phi} & \dfrac{\partial}{\partial z} \\ A_r & rA_\phi & A_z \end{vmatrix}$

(In 2 variables: $\nabla \times \mathbf{A} = \mathbf{k}\left(\dfrac{\partial A_\phi}{\partial r} - \dfrac{1}{r}\dfrac{\partial A_r}{\partial \phi} + \dfrac{A_\phi}{r}\right)$

Laplacian $\nabla^2 p = \dfrac{\partial^2 p}{\partial r^2} + \dfrac{1}{r}\dfrac{\partial p}{\partial r} + \dfrac{1}{r^2}\dfrac{\partial^2 p}{\partial \phi^2} + \dfrac{\partial^2 p}{\partial z^2}$

$$\nabla^2 \mathbf{A} = \mathbf{1}_r\left(\nabla^2 A_r - \dfrac{2}{r^2}\dfrac{\partial A_\phi}{\partial \phi} - \dfrac{A_r}{r^2}\right) + \mathbf{1}_\phi\left(\nabla^2 A_\phi + \dfrac{2}{r^2}\dfrac{\partial A_r}{\partial \phi} - \dfrac{A_\phi}{r^2}\right) + \mathbf{k}\left(\nabla^2 A_z\right)$$

Spherical coordinates (Mutually perpendicular unit vectors $\mathbf{1}_R$, $\mathbf{1}_\theta$, $\mathbf{1}_\phi$)

$$x = R \cos \phi \sin \theta \quad y = R \sin \phi \sin \theta \quad z = R \cos \theta$$

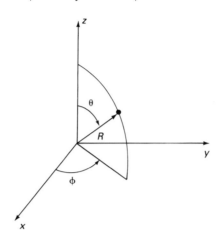

Gradient $\quad \nabla p = \mathbf{1}_R \dfrac{\partial p}{\partial R} + \mathbf{1}_\phi \dfrac{1}{R \sin \theta} \dfrac{\partial p}{\partial \phi} + \mathbf{1}_\theta \dfrac{1}{R} \dfrac{\partial p}{\partial \theta}$

Divergence $\nabla \cdot \mathbf{A} = \dfrac{1}{R^2} \dfrac{\partial}{\partial R} (R^2 A_R) + \dfrac{1}{R \sin \theta} \dfrac{\partial A_\phi}{\partial \phi} + \dfrac{1}{R \sin \theta} \dfrac{\partial}{\partial \theta} (A_\theta \sin \theta)$

Curl $\quad \nabla \times \mathbf{A} = \begin{vmatrix} \dfrac{\mathbf{1}_R}{R^2 \sin \theta} & \dfrac{\mathbf{1}_\theta}{R \sin \theta} & \dfrac{\mathbf{1}_\phi}{R} \\[2ex] \dfrac{\partial}{\partial R} & \dfrac{\partial}{\partial \theta} & \dfrac{\partial}{\partial \phi} \\[2ex] A_R & R A_\theta & R \sin \theta \, A_\phi \end{vmatrix}$

Laplacian $\quad \nabla^2 p = \dfrac{\partial^2 p}{\partial R^2} + \dfrac{1}{R^2} \dfrac{\partial^2 p}{\partial \theta^2} + \dfrac{1}{R^2 \sin^2 \theta} \dfrac{\partial^2 p}{\partial \phi^2} + \dfrac{2}{R} \dfrac{\partial p}{\partial R} + \dfrac{\cot \theta}{R^2} \dfrac{\partial p}{\partial \theta}$

Index